Alexander A. Petrov

Petroleum Hydrocarbons

With 90 Figures

Springer-Verlag
Berlin Heidelberg New York
London Paris Tokyo

Professor Dr. ALEXANDER A. PETROV
Institute of Geology and Exploitation of
Combustible Fuels
Fersman Str. 50
117312 Moscow, USSR

Translated from Russian by:
A. Y. SHOUMIKHIN

Title of the original Russian edition:
Uglevodorody nefti

ISBN 3-540-17329-3 Springer-Verlag Berlin Heidelberg New York
ISBN 0-387-17329-3 Springer-Verlag New York Berlin Heidelberg

Library of Congress Cataloging-in-Publication Data. Petrov, Aleksandr A. (Aleksandr Aleksandrovich) Petroleum hydrocarbons. Translation of: Uglevodorody nefti. Bibliography: p. 1. Hydrocarbons. 2. Petroleum products. I. Title. QD305.H5P34313 1987 547.8′3 86-33927

Printed in Germany

Typesetting: K+V Fotosatz GmbH, Beerfelden
Offsetprinting and Bookbinding: Konrad Triltsch, Graphischer Betrieb, Würzburg
2132/3130-543210

Preface

The monograph analyzes contemporary data on the composition, structure and ways of formation of various petroleum hydrocarbons: alkanes, cyclanes and arenes. Special attention is paid to biological markers, compounds that could preserve main structural features of original biogenic molecules. Reviewed are modern concepts on chemical classification of crude oils based on molecular mass distribution of the principle biological markers. Certain matters pertaining to the genesis and chemical evolution of petroleum hydrocarbons are also discussed.

The book may be found helpful to a wide range of specialists: geochemists, chemists involved in petroleum industry, petroleum chemists, as well as graduate and postgraduate students and professorial staff.

The reviewing of the book by T.P. Zhuze, P.I. Sanin and J. Rullkötter is gratefully acknowledged.

ALEXANDER A. PETROV

Contents

Introduction

This monograph is based on the research in the chemistry of petroleum hydrocarbons undertaken in the last 10–15 years by the Geochemical Laboratory of the Institute of Geology and Exploration of Combustible Fossiles (IG i GRI), although, naturally, full account was taken of the achievements of world science in this area.

Considerable success was registered in the last 15–20 years in elucidating the composition and structure of hydrocarbons in crude oils, coals and shales as well as dispersed organic matter. During this period, a new discipline evolved and obtained wide recognition, i.e. organic geochemistry, which entails the task of studying the composition and chemical evolution of organic molecules in the earth's crust. In effect, this field of science is a logical extension of the well-known chemistry of natural biological compounds.

Development of organic geochemistry proceeded on the basis of modern methodology, i.e. the application of molecular-level studies with the determination of both the structural and steric configuration of molecules, as well as the latest achievements of analytical and organic chemistry. Advancement of organic geochemistry is based on contemporary methods of analysis such as high resolution gas and liquid chromatography, gas chromatography-mass spectrometry with computerized data processing (including mass fragmentography) and ^{13}C NMR spectrometry.

However, the best research method in organic geochemistry is represented by the synthesis of standard samples. In effect, the majority of hydrocarbons to be discussed were first synthesized, then identified in crude oils. Naturally, initial evaluation of petroleum hydrocarbon. Structure can only be based on the analysis of crude oils.

As a result of investigations undertaken, over 700 individual C_1-C_{40} hydrocarbons of complex and unusual structure were identified in crudes. In the course of this research it was convincingly demonstrated that in their composition and structure petroleum hydrocarbons are complex and highly diversified organic molecules, which preserve information on the composition and construction of compounds comprising the lipid biosynthesis of ancient living organisms: algae, bacteria and higher plants. Of special significance was the discovery of the so-called relict hydrocarbons or biological markers (chemofossils), compounds which retained the basic structural features of the original biological molecules. Total concentrations of these compounds reach 35–40% in certain crudes, while their overall number exceeds 500. These biological markers are the focus of our attention.

Detailed information on crude oil composition is required not only in geochemical studies, but in petroleum chemistry as well. We could not completely agree with such a famous bio- and geochemist as Mr. Calvin who justly observed in his *Chemical Evolution* that "... petroleum composition should be established with great precision and in smallest details."[1]

In fact, methods of petroleum hydrocarbon separation are being constantly refined and improved, thus practical applications of a much wider variety of individual hydrocarbons may be foreseen in the future.

The main thesis of the present monograph in that petroleum (its hydrocarbons) is not only a raw material for fuels and oils, but also a most valuable mixture of complex organic compounds, which is after all not limitless. Apparently, one of the best ways leading to a more rational use of this natural wealth is the ability to demonstrate the value of this gift of nature in the chemical versatility of petroleum hydrocarbons, presently extracted at a rate of over 2 billion tons a year.

The monograph consists of six chapters. The first one reviews certain general properties of petroleum hydrocarbons and presents examples of new classification schemes (chemical classification according to type), based on a combination of molecular and group-type methods in analyzing petroleum hydrocarbons. The homology of petroleum hydrocarbons is also discussed.

Chapter 2 is devoted to C_5-C_{40} petroleum alkanes of various structures. It also gives examples of GC (gas chromatography) analysis of different petroleum fractions and presents retention indices for a large number of branched alkanes.

Chapter 3 contains information on the composition, structure and regularities in the distribution of alicyclic hydrocarbons (cyclanes, naphthenes), which is a most interesting and characteristic class of petroleum hydrocarbons. Analysis of cyclanes begings with the simplest monocyclic hydrocarbons and ends with complex polycyclic molecular fossils – steranes and hopanes.

Chapter 4 reviews aromatic hydrocarbons with a varying number of aromatic nuclei in a molecule, as well as hydrocarbons of mixed (naphtheno-aromatic) type, which are known to be widely represented in crude oils. Of special significance in this respect are as yet marginally investigated aromatic hydrocarbons with a biogenic carbon skeleton.

The study of petroleum hydrocarbon composition and structure in Chapters 2–4 is based on contemporary perceptions as to the sources and ways of their formation in nature. At the same time, these issues are comprehensively treated in Chapter 5, which is specially devoted to the chemistry of petroleum formation. The latter chapter presents information on thermodynamic and kinetic control in reactions of certain petroleum hydrocarbon formations. Also reproduced are some experimental data on laboratory simulation of petroleum formation reactions.

Chapter 6 is devoted to the processes of chemical evolution of crude oils in sediments. Factors determining the composition of petroleum hydrocarbons, such as thermal transformations and biodegradation, are discussed and the results of laboratory experiments are presented.

---

[1] Calvin M. Chemical Evolution. M.:Mir, 1971, p 240 (in Russian).

Within his means, the author took scientific publications appearing between 1975 and 1984 into account. Earlier papers are quoted less extensively, since appropriate material was already reflected in a number of monographs and reviews mentioned in the text.

A particular feature of this publication is the frequent use of gas chromatograms, which allows the presentation of ample information on the composition of complex mixtures of petroleum hydrocarbons.

In presenting factual material, where possible, the author presents data not only on the structure, but also on the stereochemistry of petroleum hydrocarbons (mostly on the steric structure of epimers). More detailed information on the stereochemistry of these hydrocarbons is presented in a recently published monograph.[2]

Less attention is paid to the geochemical problems of petroleum formation, such as the original biomass composition, structure and transformation of kerogen, geological factors, etc., since these issues were exhaustively treated by Tissot and Welte in their monograph.[3]

In conclusion, the author wishes to express his gratitude to the staff and postgraduate students working in the Laboratory of Petroleum Geochemistry, who participated directly both in the experimental analysis of the composition of petroleum hydrocarbons and in the synthesis of standard hydrocarbons.

The author is also thankful to I. A. Matveeva and N. N. Abryutina who helped to prepare the graphic material for the publication.

---

[2] Petrov A. A. Stereochemistry of Saturated Hydrocarbons. M.:Nauka, 1981, 254 pp (in Russian).

[3] Tissot B. P., Welte D. H. Petroleum Formation and Occurrence. Springer Verlag, Berlin, Heidelberg, New York, 2nd ed., 1984.

CHAPTER 1

General Characteristics of Petroleum Hydrocarbons Molecular and Group-Type Methods of Analysis and Classification

Hydrocarbons constitute the most important fraction in any crude oil. Although their proportion in different crudes varies significantly (e.g. from 30–40% to 100% in gas condensates), they comprise up to 70 mass % in all petroleums on the average. The history of petroleum chemistry, as a scientific discipline, is in effect the history of the chemistry of hydrocarbons. Research in petroleum chemistry was initiated in the 1860's by the well-known German chemist K. Shorlemmer, who discovered n-butane, n-pentane and n-hexane in crude oils from Pennsylvania (USA). Shorlemmer's success was due largely to his prior involvement in the synthesis of normal alkanes, conducted in the laboratory of his mentor, A. Würz. The Russian chemist, V. V. Markovnikov, while studying local crude oils from the Baku region 20–25 years later, concluded that it is not aliphatic, but alicyclic hydrocarbons, i.e. saturated hydrocarbons of the cyclopentane and cyclohexane series, which he called naphthenes, that prevail in crudes. And again this discovery was aided by Markovnikov's previous engagement in the synthesis and research of cycloalkane properties, undertaken in the laboratory of A. M. Butlerov. Thus, by the end of the 19th century, the methodological foundations for petroleum chemistry were already established, i.e. the synthesis of model hydrocarbons with their further identification in crudes. It was also then that original concepts on the chemical classification of oils emerged, suggesting their division into two main classes: paraffinic and naphthenic oils.

The "individual compound" (molecular) approach in the study of crude oil continued in the 20th century, though it was slowed down by the compositional complexity of petroleum hydrocarbons and inadequate analytical techniques. However, by the end of the 1950's investigations within Project 6 of the American Petroleum Institute, headed by F. D. Rossini, allowed the identification of over 150 hydrocarbons in a standard petroleum, belonging mostly to the light boiling fraction.

At the same time, the growing importance of oil in the world economy and the rapid development of oil processing necessitated at least some general knowledge of the chemical composition of a large number of crudes. Therefore, together with research on the molecular level, another avenue was explored, that of group-type analysis, based on the relative proportions of three different classes of petroleum hydrocarbons: alkanes, cyclanes and arenes.

Classification proceeded on the basis of information on the content of these three main classes of hydrocarbons, as well as the particulars of technological processes, such as the contents of resins, asphaltenes, sulfur, solid paraffins, etc.

In the 1940's, Nametkin (1950) and Dobryansky (1961) proposed a classification scheme of crude oils based on their group composition. Crudes were grouped in accordance with 14 indicators into 7 classes, varying in concentrations of aliphatic series hydrocarbons. Later, these schemes became more sophisticated. For example, the comprehensive formal-logical system of Kontorovitch (Kontorovitch and Stasova 1978) included 180 (!) classes of oils, varying in physical and chemical characteristics. Similar classification schemes were proposed by Vassoevitch (Vassoevitch and Berger 1968). An obvious methodological advantage of these classification schemes was that they were based on the analysis of all petroleum properties and not just those of light-boiling fractions, as often done before. However, only few of these classes actually exist in nature. Moreover, it is next to impossible to use these cumbersome schemes in their entirety. Therefore, Kontorovitch suggested only four basic types of crudes: (1) aliphatic, paraffinic; (2) aliphatic, low-paraffinic; (3) alicyclic-aliphatic; and (4) alicyclic. In a recently published comprehensive monograph, Tissot and Welte (1981) identified six different crude oil types: paraffinic, paraffinic-naphthenic, naphthenic, aromatic-intermediate, aromatic-naphthenic and aromatic-asphaltic. Various classification arrangements are reviewed in the monograph of Sokolov (1965). All the above classification schemes may apply to crudes as sources of fuels and lubricants, but may only partially be used in geochemistry, where the most important data can only be obtained by a classification based on the distribution of individual (primarily biological) hydrocarbons.

In the early 1960's new powerful analytical methods were elaborated (GC, gas chromatography-mass spectrometry) which completely revised our views on the composition and structure of petroleum hydrocarbons, hence, on the principles and methods of crude classification. The identification of a large number of biological markers (chemofossils) was an indisputable "discovery of the century". We include all hydrocarbons which retained specific structural features or the original biological molecules into this group, regardless of whether they were present in the initial biomass as hydrocarbons or were formed later from other compounds with functional groups (the chemistry of petroleum formation is analyzed in Chap. 5).

Biological Markers and Transformed Hydrocarbons

Hypothetically, all petroleum hydrocarbons can be subdivided into two main groups: (1) transformed hydrocarbons which have lost the structural features of the original biological molecules and (2) biological markers of chemofossils. Among the most important biomarkers we find: normal and isoprenoid alkanes and cyclic isoprenoids, e.g. steranes, triterpanes, etc.

Furthermore, all relict petroleum hydrocarbons can be subdivided into two main groups:

1. Biomarkers of isoprenoid type, both aliphatic and alicyclic, with one to five cycles per molecule.
2. Biomarkers of non-isoprenoid type represented mostly by aliphatic compounds with n-alkyl or slightly branched chains.

Biological markers of isoprenoid type are mostly represented by regular "head-to-tail" structures, although some hydrocarbons were found in crude oils with a "head-to-head" or "tail-to-tail" combination of isoprenoid units. Isoprenoid biomarkers are represented by a much greater variety of compounds than the non-isoprenoids.

Curiously, even in the early 1960's, a fairly pessimistic view was taken of the possibility of discovering biological markers in crudes. Thus, a well-known expert in petroleum chemistry, Dobryanskii (1961), wrote in his *Petroleum Chemistry:* " ... V. I. Vernadskii, speaking of petroleum properties conceived in organisms, was correct only as far as a limited quantity of highly stable compounds, such as porphyrins, is concerned. It its turn, the hydrocarbon material was reprocessed to such an extent, that it lost all inherited traces of the original matter."

However, already in 1962 aliphatic isoprenoids were discovered in petroleum, followed by steranes, hopanes, and other biomarkers.

Since then, biological markers of crudes, coals, shales and dispersed organic matter continue their virtually triumphant march on the pages of science magazines, monographs and congressional papers. As was mentioned, over 500 such hydrocarbons have been discovered, their number increasing annually.

Homology of Biological Markers

Homology is the most important property of biological markers. Homology depends on the fact that these hydrocarbons are usually present in series of homologs having a common structural moiety. It should be stressed that we regard homology from a wider angle, as compared to ordinary study courses in organic chemistry. Thus, apart from homologous series of normal alkanes, we can identify homologous series of 2-methylalkanes, 3-methylalkanes, 4-methylalkanes, etc. Homologous series of 1-methyl-2-alkylcyclohexanes, 1-methyl-3-alkylcyclohexanes, etc. can also be isolated. In all cases, a homologous series is formed by a common structural group, with an additional alkyl chain of varying length. The homology of petroleum biomarkers is explained by the peculiarities of their formation by the destruction of aliphatic chains at different positions in appropriate geopolymers (kerogen). Details of this reaction will be reviewed later (see Chap. 5). From this viewpoint is becomes clear why pseudo-homologous series of isoprenoid compounds are usually characterized by concentrational "dips" of each fifth homolog, since simultaneous rupture of two bonds is obviously impossible at the branching point (see Chaps. 2 and 3).

Another important property of biological markers is represented by their relatively high concentration in crudes, usually exceeding equilibrium concentrations of similarly structured isomers.

Biological markers are also important, since they serve as a source in the formation of a certain portion of severely transformed petroleum hydrocarbons, and hence, facilitate the analysis of this intricate part of petroleum.

As was mentioned, biological markers are closely connected to original biological molecules. This connection helps to elucidate the structures of petroleum hydrocarbons, because homologous series usually possess structural properties typical of the original material, reproduced later in petroleum derivatives. Earlier, this principle was applied in establishing the structures of isoprenoid compounds in accordance to different biogenic schemes (Waisberger 1967). Examples of such reconstructions are reviewed in Chapter 3.

The role of biological markers, especially in petroleum geochemistry, can hardly be overestimated. Firstly, their high concentration proves the biogenic nature of crude oils. Secondly, chemofossils are used as indicators of sedimentation conditions (Tissot and Welte 1981) in order to determine hydrocarbon source rocks, to reconstruct various oil-oil and oil-source rock relationships and to evaluate the thermal maturity of dispersed organic material. These compounds are also widely used in exploration in order to appraise the oil-bearing properties of various regions. A comprehensive review of the application of petroleum biomarkers for geochemical purposes was recently published by Mackenzie (1984).

Advances in organic and petroleum geochemistry created preconditions for the elaboration of new classification schemes of crude oils (according to chemical types), based on the molecular-level analysis of crude oils. Optimal results were achieved by combining molecular and group-type methods. One of these classification schemes was designed in the Laboratory of Petroleum Geochemistry (IG i RGI) and is based on a combination of GC data on the distribution of the most important alkane biomarkers, with MS (mass spectrometric) information on the quantitative distribution of saturated molecules, depending on the number of cycles per molecule (Zabrodina et al. 1978). The classification scheme proposed below is a refinement of our scheme suggested in the *Chemistry of Alkanes* (Petrov 1974).

Chemical Classification of Crude Oils

As was already mentioned, chemical classification is based on gas chromatography of whole oils using capillary columns with $25-30\times10^3$ theoretical plates and operating in the linear temperature programming mode. Experimental details have been presented (Zabrodina et al. 1978). Analysis of the bulk of a petroleum allows one to avoid quantitative errors often appearing in the separation of individual fractions, and helps to establish undistorted parameters of the relative concentrations of the most important biological markers: normal (C_{12}-C_{35} range) and isoprenoid alkanes (C_{14}-C_{25} range). Additionally, the group composition of the main fraction (the so-called petroleum body), i.e. a fraction boiling between 200° and 430°C (n-C_{11}-n-C_{27} can be determined.

The distribution of saturated molecules as a function of the number of cycles per molecule (0–5 cycles), determined by Polyakova (1973), gives a detailed characterization of saturated hydrocarbons in the crudes under investigation. If necessary, an appropriate MS analysis of aromatic hydrocarbons may also be conducted (see Chap. 4). Moreover, mass spectrometry helps to determine the total contents of both normal and branched alkanes separately. A similar scheme of analysis was applied by the French Institute of Petroleum in its studies of numerous crudes (Tissot and Welte 1981).

Table 1. The composition of the 200° – 430 °C boiling fraction in crudes of different chemical types

Oil Field	Depth (m)	Crude type	Fractional composition (%)			Alkanes (%)			Pristane/ phytane	C_i [a]	$\Sigma n\text{-}C_{13}$ $\text{-}n\text{-}C_{15}$ / $\Sigma n\text{-}C_{25}$ $\text{-}n\text{-}C_{27}$	Naphthenic fingerprint rel. %				
			Paraffins	Naphthenes	Aromatics	Normal	Isoprenoid	Isoalkane				Monocyclic	Bicyclic	Tricyclic	Tetracyclic	Pentacyclic
Timano-Pechora Oil and Gas Basin																
Solyukskoe	1603	A^1	32.1	28.9	39.0	29.9	11.8	58.3	1.0	0.8	1.8	27.3	32.2	20.4	13.5	6.6
Djyerskoe	800	A^2	9.9	22.1	68.0	15.2	22.2	62.6	1.1	3.3	0.8	39.6	28.6	15.9	8.7	7.2
Varandeiskoe	–	A^1	19.6	44.5	35.9	25.0	10.7	64.3	1.1	0.6	2.6	23.9	30.4	23.8	14.3	7.6
Yaregskoe	200	B^2	4.6	17.0	78.4	–	5.5	94.5	1.4	–	–	24.0	30.1	21.1	16.3	8.5
W. Tebuk	1943	A^1	33.0	32.2	34.8	28.5	9.4	62.1	1.1	0.7	1.1	31.5	30.9	17.4	13.2	7.0
	–	A^2	26.4	33.8	39.8	21.6	23.5	54.9	1.2	3.1	0.7	33.9	32.5	18.8	10.2	4.6
Vozeiskoe	1612	A^1	48.7	33.4	17.9	33.3	4.1	62.6	0.91	0.4	1.5	39.4	30.7	12.5	11.1	6.3
	1627	A^1	33.3	35.0	31.7	25.8	6.9	67.3	1.2	0.5	3.1	27.7	30.7	21.3	13.1	7.2
E. Grubeshor	3675	A^1	36.8	44.7	18.5	19.8	6.1	74.1	2.5	0.5	3.2	34.5	30.0	17.3	11.1	7.1
Shapkinskoe	1714	A^1	35.6	33.9	30.5	30.9	8.1	61.0	1.7	0.5	1.6	30.0	28.1	20.6	13.2	8.1
Pashninskoe	2742	A^1	32.9	28.9	38.2	24.0	5.8	70.2	1.0	0.8	1.0	31.0	32.1	20.3	12.9	3.7
Usinskoe	2300	A^1	42.2	28.9	28.9	25.2	5.9	68.9	0.8	0.6	2.2	34.6	29.8	16.3	13.4	5.9
Volga-Urals Oil and Gas Basin																
Zhirnovskoe	–	B^2	9.6	44.9	45.5	–	12.5	87.5	1.2	–	–	25.0	27.1	19.2	17.3	11.4
	1750	A^1	32.6	44.4	23.0	29.5	10.7	59.8	1.7	0.7	1.7	36.5	26.7	14.2	13.5	9.1
Tuyimazinskoe	–	A^1	32.5	23.5	44.0	25.9	6.4	67.7	0.8	0.4	3.1	31.8	29.0	18.7	13.0	7.5
Bavlinskoe	1880	A^1	35.0	30.2	34.8	32.6	6.6	60.8	0.6	0.4	3.1	34.3	31.0	16.9	11.2	6.6
Enaruskinskoe	1165	A^1	30.4	19.6	50.0	37.2	8.9	53.9	0.5	0.6	0.8	30.5	31.6	20.9	10.2	6.8
Stepnozerskoe	1356	A^1	30.9	21.9	47.2	23.6	5.8	70.6	0.5	0.5	1.3	27.9	42.8	15.5	9.4	4.4
Mordovo-Karmalskoe	80	B^2	20.1	34.5	45.4	–	24.9	75.1	0.6	–	–	29.9	31.5	21.3	11.9	5.4
Kasibskoe	–	A^1	27.8	35.0	37.2	43.2	8.6	48.2	1.4	0.4	1.5	30.6	28.4	17.9	15.0	8.1
Durinskoe	1870	A^2	7.2	20.8	72.0	13.9	11.1	75.0	1.1	1.5	1.4	28.5	32.1	20.9	13.1	5.4
Tikhovskoe	2012	A^1	41.7	24.6	33.7	28.8	5.0	66.2	1.4	0.3	1.3	31.0	30.4	18.3	12.3	8.0
	800	A^1	16.2	15.6	68.2	40.1	8.0	51.9	0.71	0.3	1.5	31.1	25.5	19.7	16.9	6.8

Sivinski	2806	B^2	9.2	53.3	37.5	–	19.6	80.4	1.0	–	–	13.5	30.3	28.1	18.9	9.2
Novo-Elkhov-skoe	1618	A^1	25.6	22.4	52.0	29.3	9.0	61.7	0.5	1.6	2.4	28.3	31.0	19.6	12.8	8.3
Tchapaevskoe	–	A^1	50.9	26.4	22.7	25.7	13.8	60.5	0.55	0.5	6.5	39.5	29.6	14.3	11.2	5.4
Near-Caspian Oil and Gas Basin																
Kursaiskoe	4410	B^1	7.6	67.7	24.7	–	–	–	–	–	–	16.3	29.1	25.3	18.5	10.8
Sagizskoe	800	B^1	7.9	66.4	25.7	–	–	–	–	–	–	17.8	28.2	25.5	17.8	10.7
Karajanbas	300	A^2	16.6	44.4	39.0	12.9	25.7	61.4	0.91	3.8	1.5	24.4	29.8	22.8	14.0	9.0
Koshkar	–	A^1	26.3	53.7	20.0	13.7	6.1	80.2	1.4	0.7	2.4	28.3	27.6	21.0	13.8	9.3
Kalamkas	847	A^2	25.2	44.1	30.7	13.0	8.3	78.7	1.0	1.6	0.5	30.4	29.8	20.8	13.3	5.7
N. Buzatchi	440	A^2	19.2	46.2	34.6	12.0	18.8	69.2	1.1	3.5	0.8	27.3	29.0	21.9	13.8	8.0
E. Prorva	3179	A^1	30.8	44.7	24.5	26.0	5.2	68.8	0.91	0.3	0.9	24.3	26.3	22.2	17.1	10.1
Biikjan	5712	A^1	27.0	40.6	32.4	18.1	5.2	76.7	1.6	0.6	3.0	31.9	28.4	18.6	13.7	7.4
Bekturly	2410	A^1	51.0	36.5	12.5	46.9	4.5	48.6	1.07	0.2	1.0	31.6	26.4	17.0	16.9	8.1
Kenkiyak	330	B^1	8.7	58.2	33.1	–	–	100	–	–	–	19.1	30.1	23.7	17.0	10.1
Dnepr-Donetsk-Pripyat Oil and Gas Basin																
Sagaidakskoe	4600	A^1	49.9	36.6	13.5	38.1	10.2	51.7	0.91	0.6	2.0	30.4	28.5	15.0	17.5	8.6
Prilukskoe	–	A^1	38.1	35.9	26.0	28.4	8.9	62.7	0.83	0.7	1.1	31.0	27.9	17.0	14.2	9.9
Retschitskoe	2193	A^1	31.1	23.1	45.8	28.6	5.5	65.9	1.0	0.4	2.0	24.7	34.0	20.4	14.6	6.3
Tishkovskoe	3688	A^1	41.0	28.9	30.1	29.0	7.8	63.2	0.83	0.6	1.8	26.8	29.0	16.9	21.5	5.8
Ostashkovitchi	2750	A^1	33.6	24.9	41.5	31.0	10.1	58.9	0.55	0.7	1.5	21.6	25.2	21.9	23.3	8.0
Ostashkovitchi	3663	A^1	35.3	25.6	39.1	26.6	8.2	65.2	0.77	0.6	1.2	27.4	30.2	17.6	19.3	5.5
Barsukovskoe	3504	A^1	38.8	21.6	39.6	32.5	4.6	62.9	1.25	0.3	1.1	27.0	27.9	19.8	17.3	8.0
Mozyrskoe	3350	A^2	31.2	30.8	38.0	18.6	10.3	71.1	1.0	1.4	5.4	24.8	23.3	24.9	16.5	10.5
Zavodya	3425	A^1	42.7	30.4	26.9	34.9	9.6	55.5	1.67	0.9	0.5	23.3	20.1	22.0	24.3	10.3
South Caspian Oil and Gas Basin																
a) Azerbaijan																
Balakhanskoe																
(Oleaginous)	–	B^2	17.4	59.6	23.0	–	16.4	83.6	1.1	–	–	20.6	26.5	21.6	21.2	10.1
(Heavy)	1302	B^2	10.2	69.8	20.0	–	20.4	79.6	1.25	–	–	22.9	26.8	18.6	21.1	10.6
Neftyanye Kamni	1400	A^2	20.0	48.0	32.0	15.4	65.2	19.4	1.1	14.6	1.0	24.9	29.9	20.6	16.3	8.3
Surakhanskoe	2574	B^2	14.3	44.7	41.0	–	15.0	85.0	1.0	–	–	24.4	26.6	19.5	19.9	9.6
	867	A^2	26.8	37.2	36.0	21.6	20.5	57.9	1.43	2.1	0.4	25.8	27.3	18.7	18.8	9.4
Sangatchaly	4782	A^1	35.4	40.2	24.4	31.6	8.8	59.6	1.67	0.7	1.4	22.6	26.6	20.7	20.3	9.8

Table 1 (continued)

Oil Field	Depth (m)	Crude type	Fractional composition (%)			Alkanes (%)			Pristane/ phytane	C_i[a]	$\Sigma n\text{-}C_{13}$ $\text{-}n\text{-}C_{15}$	Naphthenic fingerprint rel. %				
			Paraf- fins	Naph- thenes	Aroma- tics	Nor- mal	Iso- prenoid	Iso- alkane			$\Sigma n\text{-}C_{25}$ $\text{-}n\text{-}C_{27}$	Mono- cyclic	Bi- cyclic	Tri- cyclic	Tetra- cyclic	Penta- cyclic
Duvannyi- more	3440	A[1]	24.1	36.9	39.0	32.4	13.3	54.3	1.43	1.3	0.9	21.2	25.9	21.5	21.0	10.4
	3500	A[2]	17.2	50.8	32.0	23.8	61.6	14.6	1.43	6.6	1.5	19.4	24.6	21.4	23.5	11.1
	3904	A[2]	23.3	51.5	25.2	23.0	22.1	54.9	1.67	2.7	0.9	19.4	22.8	20.2	23.6	14.0
	4594	A[1]	28.8	37.7	33.5	38.9	12.5	48.6	1.43	0.8	1.0	20.2	25.8	20.6	22.1	11.3
	5203	A[1]	44.2	31.8	24.0	25.9	7.2	66.9	1.67	0.6	4.8	21.9	27.5	20.5	20.4	9.7
Isl. Garasu	–	A[1]	25.9	37.7	36.4	42.8	13.7	43.5	1.25	0.5	0.4	15.9	21.1	20.0	27.8	15.2
Gryazevaya Sopka	1006	B[1]	9.1	53.9	37.0	–	–	–	–	–	–	23.4	29.5	21.8	17.2	8.1
Surakhany	1674	B[1]	6.9	43.1	50.0	–	–	–	–	–	–	23.0	27.6	21.3	19.4	8.7
b) Turkmenistan																
Kotur-Tepe	1700	A[2]	22.7	52.3	25.0	18.1	13.7	68.2	1.25	2.1	0.9	22.9	26.6	19.7	20.6	10.2
	2500	A[1]	30.9	40.1	29.0	27.8	12.3	59.9	1.25	1.0	1.6	25.9	27.1	19.2	18.6	9.2
Dagadjikskoe	900	B[2]	23.7	50.3	26.0	–	14.8	85.2	1.25	–	–	24.5	28.0	19.7	19.6	8.2
	1000	A[2]	9.6	30.4	60.0	11.5	26.0	62.5	1.25	13.9	3.5	25.6	27.9	20.3	18.6	7.6
	1951	A[1]	41.9	33.1	25.0	42.2	8.6	49.2	1.43	0.5	1.2	24.3	28.3	19.6	19.1	8.7
Oval-Toval	1100	A[2]	28.5	52.2	19.3	12.8	18.3	68.9	1.43	3.4	1.6	24.9	25.8	17.2	22.6	9.5
Kamishldja	1000	B[2]	32.2	45.3	22.5	–	8.2	91.8	1.25	–	–	34.6	33.6	18.0	9.0	4.8
West Siberian Oil and Gas Basin																
Russkoye	837	B[1]	6.8	49.1	44.1	–	–	100	–	–	–	15.9	26.8	26.1	21.0	10.2
N. Kom- somolskoe	1000	B[1]	9.1	49.2	41.7	–	–	100	–	–	–	21.7	29.2	21.4	19.1	8.6
Novoportov- skoe	987	B[1]	8.0	47.0	45.0	–	–	100	–	–	–	18.4	24.1	25.1	21.8	10.6
	1828	A[2]	21.9	28.1	50.0	16.9	13.2	69.9	2.5	7.1	0.2	31.3	30.2	15.8	15.4	7.3
	1973	A[1]	31.1	28.0	40.9	36.3	10.0	53.7	2.0	0.5	1.4	31.5	27.1	19.1	15.7	6.6
	1870	A[1]	36.9	21.2	41.9	28.9	4.4	66.7	2.5	0.3	1.7	31.3	29.8	20.1	13.3	5.5
Taitymskoe	2600	A[1]	60.8	25.7	13.5	60.9	9.0	30.1	0.77	0.4	0.6	27.8	33.6	18.0	13.5	7.1

Maloitchskoe	2849	A^1	37.0	38.7	24.3	32.7	4.6	62.7	1.11	0.3	2.0	30.3	28.5	17.9	12.5	10.8
Samotlor	1777	A^2	19.0	30.5	50.5	12.6	16.3	71.1	0.91	3.9	1.0	30.0	25.6	16.0	19.8	8.6
	1724	A^2	20.7	30.3	49.0	9.2	18.8	72.0	0.91	7.7	1.0	36.0	28.3	16.9	14.2	4.6
	1748	A^1	27.5	24.9	47.6	20.0	6.2	73.8	0.91	0.7	2.4	34.6	27.1	15.4	15.9	7.0
	2208	A^1	29.4	27.6	43.0	24.1	8.5	67.4	1.0	0.8	0.9	34.8	26.1	15.5	16.7	6.9
Nijn.Taba-ganskoe	2995	A^1	33.9	29.3	36.8	32.7	4.5	62.8	1.0	0.2	2.3	25.9	33.3	19.9	13.5	7.4
Natalinskoe	2580	A^1	29.6	24.9	45.5	33.8	12.5	53.7	1.0	0.9	2.4	30.9	25.1	14.3	21.7	8.0
Rakitinskoe	2389	A^1	34.5	27.0	38.5	32.8	11.0	56.2	2.0	0.8	4.2	32.0	32.0	17.9	12.0	6.1
Salymskoe	2745	A^1	39.0	41.0	20.0	18.3	2.8	78.9	0.91	0.3	4.0	35.5	28.9	14.5	13.5	7.6
Soleninskoe	2408	A^2	25.9	34.1	40.0	20.1	12.2	67.7	2.5	4.0	0.2	31.9	29.1	18.1	14.5	6.4
Verkhne	2474	A^1	34.0	28.0	38.0	26.2	7.9	65.9	1.67	0.7	2.9	30.1	29.7	15.9	16.0	8.3
Tarskoe	2692	A^1	37.0	38.0	25.0	29.1	6.3	64.6	2.5	0.2	3.2	28.9	27.8	20.2	14.1	9.0
East Siberian Oil and Gas Basin																
Markovskoe	2112	A^1	35.5	35.5	29.0	13.0	6.5	80.5	0.67	0.9	4.4	27.6	32.0	19.4	13.4	7.6
Atovskoe	1943	A^1	46.8	34.9	18.3	16.9	2.1	81.0	1.43	0.1	3.9	33.7	23.0	12.9	16.5	13.9
Sukhotun-gusskoe	2472	A^1	45.7	36.1	18.2	21.2	7.7	71.1	1.0	0.5	6.4	29.0	31.5	19.1	12.5	7.9
Sr. Botu-binskoe	1470	A^1	34.0	29.0	37.0	31.5	12.6	55.9	0.67	0.8	4.8	31.0	27.2	20.4	13.9	7.5
Sr. Vilyuiskoe	2764	A^1	42.8	20.8	36.4	57.0	3.5	39.5	5.0	6.4	4.7	34.1	14.1	15.2	20.6	16.0
South Mangyshlak-Usturt Oil and Gas Region																
Uzenskoe	2858	A^1	35.7	34.3	30.0	35.0	16.5	48.5	1.67	0.9	2.4	38.6	29.0	14.6	11.0	6.8
	1330	A^1	44.3	30.7	25.0	50.8	5.4	43.8	1.67	0.2	1.4	33.0	28.9	18.8	12.9	6.4
Dunginskoe	2865	A^1	48.5	30.2	21.3	53.5	39.4	8.1	2.0	0.2	0.5	39.4	25.6	15.3	11.7	8.0
North Caucasus Oil and Gas Region																
Anastasiev-sko-Troitskoe, IV Horizon	–	B^2	12.7	47.3	40.0	–	26.5	73.5	0.67	–	–	18.8	30.2	22.5	18.0	10.5
Malgobek	–	B^2	9.8	55.2	35.0	–	30.7	69.3	0.62	–	–	17.6	28.5	24.3	19.2	10.4
Trans-Caucasus Oil and Gas Region																
Sartychala	2800	A^1	30.0	37.8	32.2	37.3	7.5	55.2	2.0	0.5	1.8	27.3	25.6	17.6	19.2	10.3
Norio	1400	B^2	8.3	45.9	45.8	–	7.8	92.2	2.0	–	–	22.1	27.4	19.7	19.9	10.9
Mirzaani	1000	B^2	13.8	48.7	37.5	–	9.1	90.9	0.31	–	–	25.7	28.2	18.5	18.8	8.8

[a] Isoprenoid coefficient (see explanation in text).

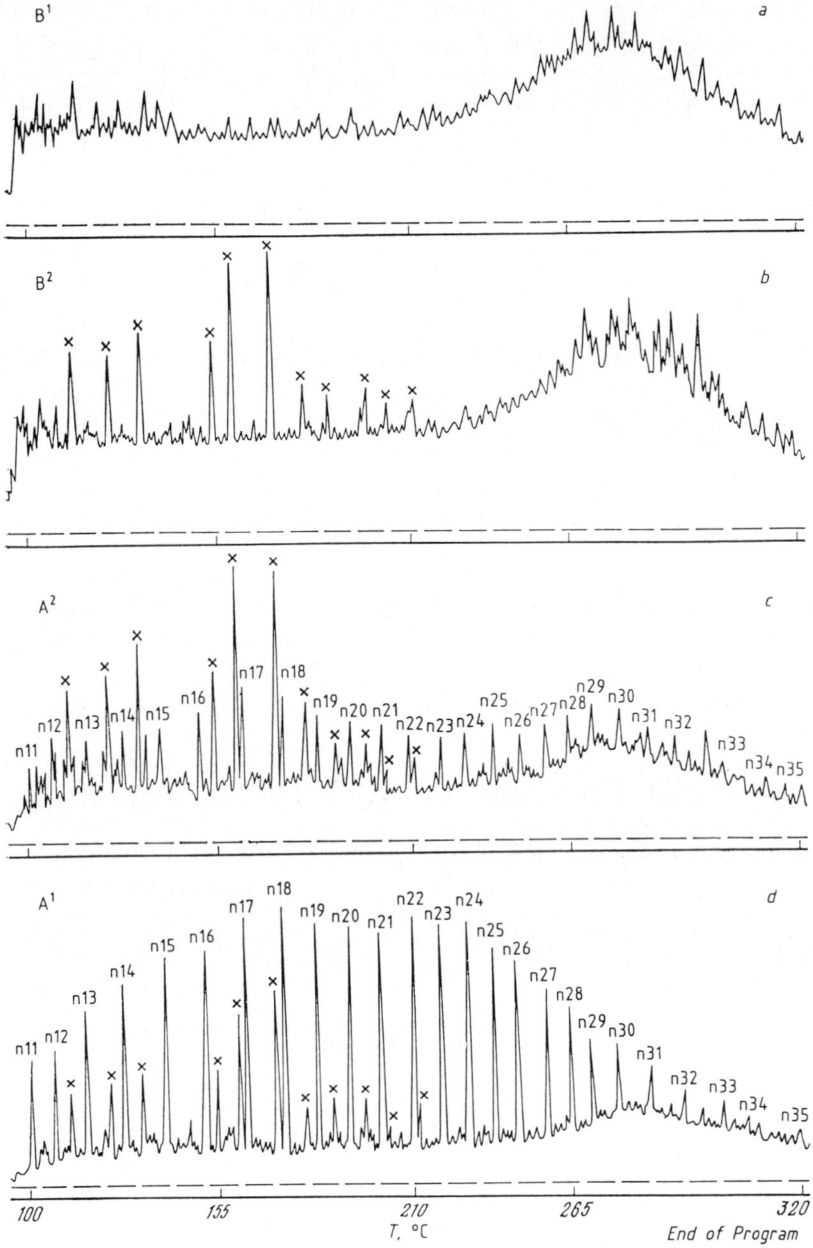

Fig. 1a–d. Chromatogram of different crude types. Deposits: **a** Surakhanskoe; **b** Balakhanskoe; **c** Kotur-Tepe; **d** Samotlor. The number of carbon atoms in the molecules are indicated. *X* denotes isoprenoid alkane peaks. Capillary column 30 m, Apiezon; linear temperature programming, 100°→3°/min

The above methodology was applied to the analysis of nearly 400 crude oils from various oil fields in the Soviet Union in actually every known oil and gas province: South Caspian, West Siberian, Timano-Petchora, Volga-Urals, Near Caspian, Dnepr-Donetsk-Pripyat, East Siberian, etc. Where possible, crudes from multilayer reservoirs were chosen in order to trace compositional alterations due to source-rock depth and age, as well as geothermal conditions.

Research data on 100 typical crude oils from various oil and gas basins in the Soviet Union are summarized in Table 1. Figure 1 presents chromatograms of different crude types to be described later.

Group composition data demonstrate that it is mostly saturated hydrocarbons (alkanes + cyclanes), which comprise 60–80% of the fraction that boil in the range of 200 °–430 °C. Ratios of alkane/cyclane concentrations may vary widely. Thus, alkane concentrations range within 6–60%, which led to the conclusion that this parameter is a decisive factor in the variation in the hydrocarbon compositions of crude oils. At the same time, there appears to be an approximate equality in the average contents of alkanes, cyclanes and arenes, which is compatible with the data presented by the French Institute of Petroleum, which confirmed equal correlations of these three types of hydrocarbons on the average (for 517 crude oils studied) in > 210 °C fractions (Tissot and Welte 1981). However, individual crudes vary considerably in their group composition. Table 2 presents variational limits and the most commonly occurring group composition parameters of different types of crudes, as listed in Table 1.

The variability of the group composition of crude oils, is depicted graphically in a triangular diagram (Fig. 2). The centre part is occupied by the most common group of crudes (category A), with equal average amounts of alkanes, cyclanes and arenes. Another group is characterized by low alkane contents (not exceeding 15%) and a wide spectrum of correlational variations between aromatic and naphthenic hydrocarbons. Crudes belonging to the second group (category B) are placed to the left of the triangle. Crudes of intermediate types are positioned between the two former groups.

Let us now examine the differences between these four crude types, identified in Tables 1 and 2 and Fig. 1 (i.e. crudes A^1, A^2, B^2, B^1), by GC analysis. As was

Table 2. Group composition of crudes of different chemical types in the 200° – 430° fraction (%)

Hydrocarbons	Crude type[a]			
	A^1	A^2	B^2	B^1
Total Alkanes	15 – 60 (25 – 50)	10 – 30 (15 – 25)	5 – 30 (10 – 25)	4 – 10 (6 – 10)
Normal alkanes	5 – 25 (8 – 12)	0.5 – 5 (1 – 3)	0.5	–
Isoprenoid alkanes	0.05 – 6.0 (0.5 – 3)	1 – 6 (1.5 – 3)	0.5 – 6 (0.2 – 3)	–
Cycloalkanes	15 – 45 (20 – 40)	20 – 60 (35 – 55)	20 – 70 (35 – 55)	20 – 70 (50 – 65)
Aromatic hydrocarbons	10 – 70 (20 – 40)	15 – 70 (20 – 40)	20 – 80 (20 – 45)	25 – 80 (25 – 50)

[a] Most commonly occurring values are given in parentheses.

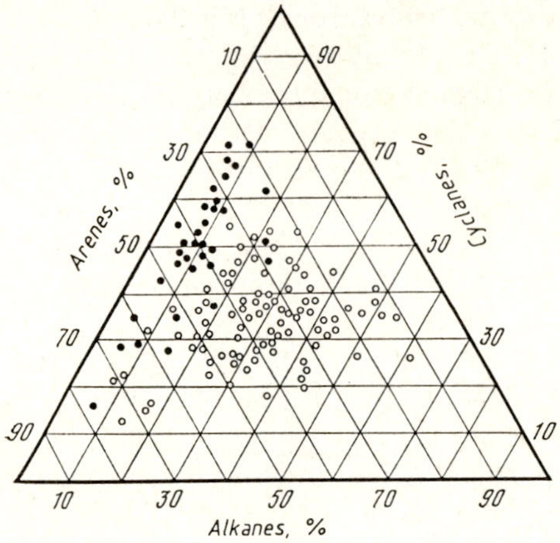

Fig. 2. Diagram of the distribution of hydrocarbons of various classes in crudes. ● type B crude; ○ type A crude

mentioned, this classifications is based on the distribution of normal and isoprenoid alkanes. Generally, all crude oils can be divided into two large groups: category A crudes, which show analytically determined quantities of normal and isoprenoid alkanes in their chromatograms, and category B crudes, which have no normal alkane peaks in their chromatograms. Depending on the relative concentrations of normal and isoprenoid alkanes in A crudes, and the presence or absence of isoprenoid alkanes in B crudes, each of these two categories are subdivided into two further types: A^1, A^2 and B^2, B^1.

The gaschromatograms of A^1 crudes show readily recognizable normal and isoprenoid alkane peaks, with the normal alkanes predominating over a moderate naphthenic background[1] consisting of hydrocarbons inseparable by GC. The background of the A^2 type becomes slightly increased in height, while normal alkane peaks are significantly diminished, with only isoprenoid alkane peaks standing out. B^2 chromatograms, with an even higher background, have only prominent isoprenoid alkane peaks. B^1 chromatograms completely lack normal or isoprenoid alkane peaks. Thus, these chromatograms represent a background consisting of inseparable hydrocarbons of all three classes. Later (Chap. 6), the genetic interrelationships between crudes of all four types, with their compositions reflecting various stages in petroleum evolution in the earth's crust will be demonstrated. A fairly consistent link exists between the chemical types (A^1, A^2, B^2, B^1) of crude oils and correlations of the alkanes and cyclanes in them (Fig. 3).

A concise description of all four crude types to be examined is given below.

Type A^1 Crude Oils correspond to paraffinic- and naphtheno-paraffinic-based crudes, according to the group composition of the 200°–430°C fraction. As a rule, these crudes have a high proportion of the gasoline fractions and a relatively low resin content. An important role among saturated high-boiling hydrocarbons is played by acyclic saturated hydrocarbons, which amount to 40–70% of the

[1] Background denotes the distance between the base line and the beginning of discrete peak.

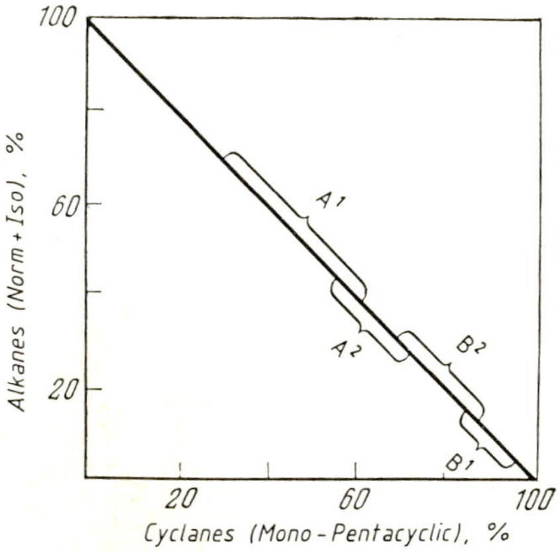

Fig. 3. Relative distribution of alkanes and cyclanes in crudes of different chemical types

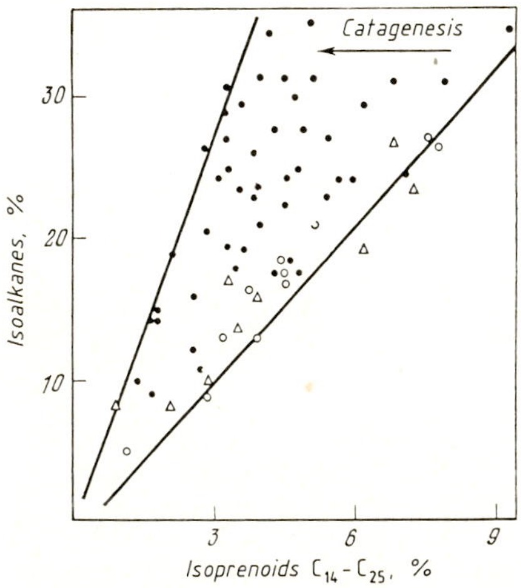

Fig. 4. Interconnection between total content of isoprenoid alkanes and other isoparaffins in crudes. ● type A^1 crudes; ○ type A^2 crudes; △ type B^2 crudes

saturated portion of the 200°–430°C fraction. These crudes are characterized by high concentrations of normal alkanes (5–25% of the fraction under examination). The content of isoprenoid alkanes ranges between 0.05–6%. The alkane content is always higher than that of isoprenoids and the isoprenoid coefficient $C_i = (i\text{-}C_{19} + i\text{-}C_{20})/(n\text{-}C_{17} + n\text{-}C_{18}) < 1$.

The total content of branched alkanes in such crudes is generally always higher than that of normal alkanes, while the isoalkane/normal alkane ratio ranges within 1–6. The presence of large quantities of resolved branched alkanes, which normally comprise a considerable part of the chromatographic "hump", i.e. hydrocarbons inseparable by GC, is a noteworthy phenomenon. There are reasons to believe that the branched alkanes are genetically related to isoprenoid alkanes, for which evidence may be derived from the graphic correlation shown in Fig. 4 (see also Table 1).[2]

[2] High concentrations of branched alkanes may probably result in errors in MS group-type analysis. Studies conducted in recent years have demonstrated that mass spectra of cyclanes with long isoprenoid chains are characterized by sets of fragment ions commonly observed for branched alkanes (i.e. m/z 113, 183, etc).

So far no convincing proof is available for the structural details of the majority of branched alkanes in fractions boiling above 200 °C. In any case, branched alkanes in the "hump" are represented by structures with numerous branching points, because all monomethyl-substituted alkanes elute in easily separable peaks (see Chap. 2, Fig. 18).

Bulk concentrations of cycloalkanes in type A^1 crudes are slightly inferior to those of alkanes. Cycloalkanes are mostly represented by mono- and bicyclic compounds, where the monocycloalkane content usually equals or exceeds that of the bicyclanes. Crudes of this type have wide occurrence in nature and are found in all oil and gas basins of the Soviet Union, in sediments of any geological age, but usually at depths over 1500 m. In accordance with A. A. Kartsev's classification, they belong to paleotype crude oils. Type A^1 oils are the principle high-value industrial crudes. Such oil fields as Romashkino or Somatlor are usually represented by these crudes. In view of their wide occurrence, A^1 crudes may certainly not be characterized by a single representative gaschromatogram, as the one repro-

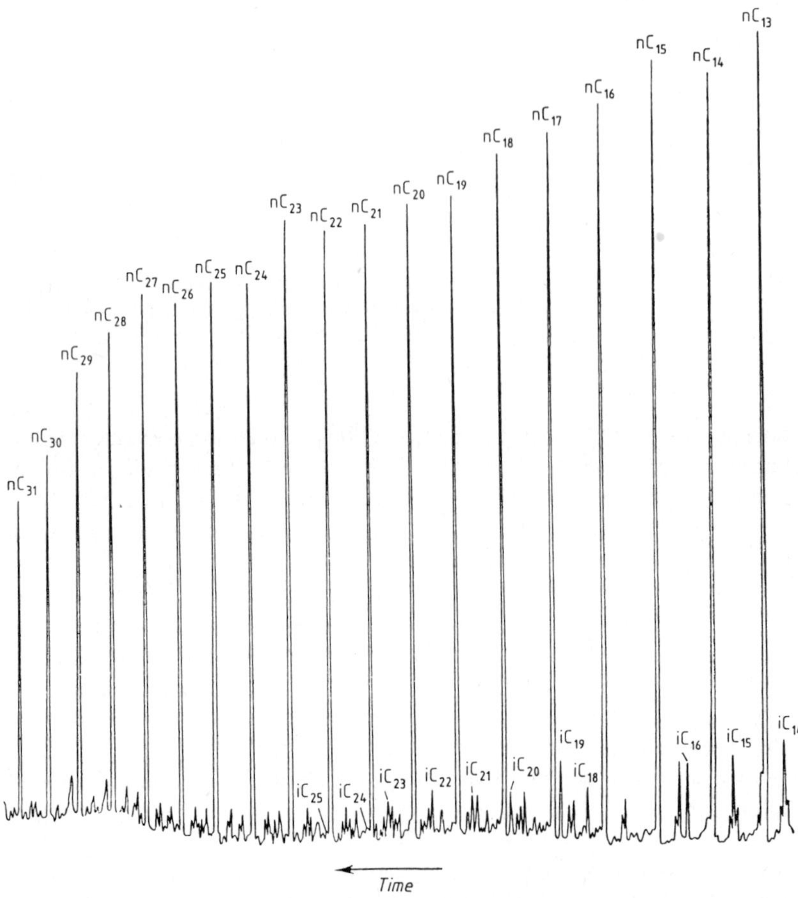

Fig. 5. Chromatogram of a highly paraffinic crude of type A^1. Analysis conditions are identical to those in Fig. 1. $i\text{-}C_{14}$, $i\text{-}C_{15}$, etc. indicate peaks of corresponding isoprenoid alkanes

duced in Fig. 1, which belongs to the most common A^1 subtype with a fairly even distribution of normal alkanes. However, this does not exclude other subtypes. One of these is characterized by a much lower concentration of $>C_{20}$ normal alkanes (which is also the case for the number of $>350\,°C$ fractions). This type may be called catagenetically modified and often occurs in primary gas condensates. In contrast another A^1 subtype is characterized by high concentrations of $>C_{20}$ normal and low concentrations of isoprenoid alkanes. This subtype may be called paraffinic. A typical chromatogram of a high paraffinic crude is presented in Fig. 5.

A^1 crudes often vary in concentrational distributions of normal alkanes. One subgroup is characterized by maximum in the n-C_{20} to n-C_{23} area; another has this maximum shifted to the n-C_{16} – n-C_{18} alkanes, while it is moved even further towards lighter hydrocarbons in a third group. Characteristic diagrams of normal alkane distributions in crudes of these subgroups are given in Fig. 6. In addition,

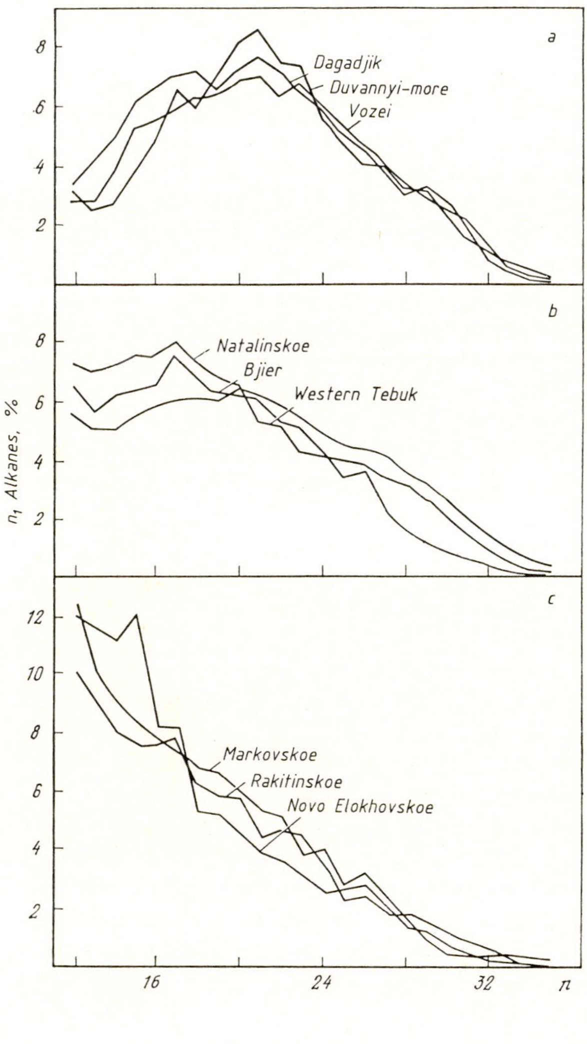

Fig. 6a–c. Diagram of normal alkane distributions in crudes belonging to different groups of type A^1. **a** group I; **b** group II; **c** group III; *n* number of carbon atoms in a molecule

extensive material on the regularities of normal alkane distribution in oils is presented in the comprehensive monograph of Safonova (1980). The ratio $\sum$n-C_{13}−n-C_{15}/$\sum$n-C_{25}−n-C_{27} may be used as a quantitative criterion in attribution A^1 crudes to a particular group. Typical values range from 0.5−1.2 for the first group, 1.2−3 for the second; and 3−8 for the third one (see Table 1). As is known, variations in high-molecular-weight normal alkane distributions are based on compositional peculiarities of the original organic matter (Tissot and Welte 1981; Martin et al. 1964; Guseva et al. 1976). This problem is extensively examined in Chapter 5. At this point, and in order to complete the study of normal alkane distributions, it should be noted that, as in the publication of Tissot and Welte (1981), we failed to discern any clear-cut regularities in the odd-even carbon number distribution of petroleum hydrocarbons. The oddness coefficient [C_{odd} = (n-C_{15}+n-C_{17})/2n-C_{16} = 1.1−1.15] has a range of 0.9−1.1, depending on the crude oil type and fractional composition, and may hardly be applied in any genetic or catagenetic models. As compared to other fossil fuels of dispersed organic matter, petroleum is a product of the extensive catagenetic transformations of the original material. Therefore, extensive destruction of the initial organic matter limits the informational potential of C_{odd}. However, increased concentrations of n-pentadecane and n-heptadecane were consistently observed in a number of crudes (mostly of marine origin). This indicator may be used in evaluating facial situations of types of initial material (Tissot and Welte 1981). However, the reasons for the increased concentrations of n-penta- and n-heptadecane remain as yet unclear. In any case, it may not be accounted for by simple decarboxylation of even-carbon number fatty acids, because thermal cracking of high-boiling hydrocarbons (C_{35} and higher) in those crudes inevitably produced similar distributions of n-C_{15}, n-C_{16} and n-C_{17}. Better results in establishing $C_{odd.}$ in the evaluation of the degree of catagenesis of continental organic matter are obtained by applying the ratio (n-C_{25}+n-C_{27})/2n-C_{26}. $C_{odd.}$ coefficients help to provide better results than CPI coefficients (carbon preference index) (Sokolov 1965), which take the correlation of numerous odd and even homologs (such as C_{21}−C_{34}) into account. However, it should be reemphasized that all that was said about $C_{odd.}$ relates exclusively to crude oils. In contrast, $C_{odd.}$ plays an exceptional role for alkanes in dispersed organic material or catagenetically immature coals and shales, since its absolute values increase to 3−5, and even higher levels, which are absolutely incompatible with the $C_{odd.}$ in crudes.

Type A^2 Crudes Oils, in their group composition, correspond to naphtheno-paraffinic or paraffino-naphthenic crudes. As compared to the type A^1, the alkane content is slightly inferior and reaches 25−40%. Normal alkane concentrations range from 0.5−5%, while those of isoprenoids vary from 1−6%. The majority of A^2 crudes are characterized by a significant predominance of isoprenoid over normal alkanes (C_i>1). The normal alkane content is smaller by one order than in A^1 crudes, although the character of the normal alkane relative distributions is preserved. The total concentration of isoparaffins is six to eight times higher than that of normal alkanes. The bulk concentration of cycloalkanes amounts to 60−75%. Mono- and bicyclic hydrocarbons still dominate among the naphthenes, although the amount of tricyclanes is slightly higher than in A^1

crudes. Type A^2 crudes have a more restricted occurrence. They are found in Cenozoic and occasionally Mesozonic sediments, at depths not exceeding 1500–2000 m. A^2 crudes have the common chromatographic feature of a moderate background ("hump")[3] and distributions identical with A^1 crude oils of normal, isoprenoid and monomethyl-substituted alkanes. Crudes from the Southern Caspian region (Surakhany, Neftyanye Kamni, Duvannyi-more), West Siberia (Samotlor, layer AB^2, Soleninskoe) and the Near-Caspian region (Koshkar, Kalamkas, Kara-Tube) may serve as typical representatives of A^2 crudes.

An N_b coefficient (reflecting the naphthenic background) may be introduced for category A crude oils; chromatographically, it represents a correlation of the sum of the $n\text{-}C_{17}+n\text{-}C_{18}$ peak heights and the height of the naphthenic background (b) at a given peak elution point (the distance between the base line and peak base of the normal alkane), calculated according to the formula $N_b = (n\text{-}C_{17}+n\text{-}C_{18})/b$. (The terms "naphthenic background" and "hump" are fairly figurative, since there is a certain amount of branched alkanes and aromatic hydrocarbons included in the naphthenic background).

N_b values range from 4–70 (usually from 6–15) for A^1 crudes, and 0.1–6 (mostly 0.5–4) for A^2 crudes.

Obviously, N_b values depend on the relative distributions of normal alkanes and may be used as an additional characteristic for crude oil classification.

Type B^2 Crude Oils correspond to crudes of a paraffino-naphthenic and especially naphthenic base. Cycloalkanes predominate among saturated hydrocarbons, with a content ranging from 60% to 75%. As a rule, mono-, bi- and tricyclic hydrocarbons dominate among cycloalkanes. Alkane hydrocarbons, with their contents varying from 5–30%, are exclusively represented by branched structures. A small quantity of normal alkanes was found only with the help of molecular sieves or thermal diffusion. Normal alkanes peaks fail to appear in chromatograms, since their total concentration does not exceed a few tenths of 1%. The concentration of isoprenoids amounts to 0.5–6% in the 200°–430°C fraction.

The distribution of $C_{14}-C_{25}$ isoprenoid alkanes in the majority of B^2 crudes is similar to that in A^1 and A^2 crudes, with certain atypical exceptions. For example, there are practically no pristane and phytane peaks in the crudes from the Norio and Anastasievo-Troitskoe reservoirs (horizon IV), although the concentrations of $C_{14}-C_{16}$ isoprenoids are considerable. B^2 crudes are characterized by the lack of monomethyl-substituted alkane peaks in their chromatograms.

Type B^2 crudes have a wider occurrence than type A^2 oils, mostly in cenozoic reservoirs at 1000–1500 m depths. Their typical representatives are crudes from Georgia (Norio, Mirzaani) and Northern Caucasus (Starogroznenskoe, Anastasievo-Troitskoe).

---

[3] A marked chromatographic background is a characteristic feature of crudes. This element is usually insignificant or nonexistent in gas chromatograms of hydrocarbons identified in catagenetically immature samples (such as brown coals, dispersed organic material).

Type B¹ Crude Oils belong to the naphthenic or naphtheno-aromatic group, according to their compound class composition. As a rule, they contain few light fractions. They are characterized by the total absence of normal and isoprenoid alkanes and only insignificant quantities of other branched alkanes (4–10%). Bicyclic hydrocarbons dominate over monocyclic hydrocarbons among the cycloalkanes. In general, B^1 crudes occur in cenozoic sediments of many oil and gas basins of the USSR at 500–1000 m depth. Typical are crudes of the Southern Caspian region and the northern part of West Siberia (Gryazevaya Sopka, Surakhany, Balakhany, Russkoe and other fields. According to Kartsev's classification (1978), they belong to cenozoic crudes.

Figure 7 shows the mean statistical distribution of different classes of crudes according to the depth and geological age of their source rocks (the latter indicator is less correlative). A clear zoning can be easily observed: A^2, B^2 and B^1 crudes are mainly concentrated in the upper zone at 500–1500 m depth, which usually corresponds to the hypergene zone. A^1-type crudes begin to predominate at 2000 m depth. Such a distribution according to zones is sometimes observed not only regionally, but within a single field. It is important that crudes of different chemical types occur in different temperature regimes which clearly proves the significant role played by the temperature conditions in the geochemical alterations of crudes oils (see Chap. 6). A^2, B^2 and B^1 crudes usually occur at moderate temperatures (40°–70 °C), while the majority of A^1-type crudes are characterized by reservoir temperatures >90 °C. Naturally, there are exceptions to these general regularities in alteration of crude oil types as a function of depth, however, they are rare and limited to certain regions. Some unique crudes with low alkane contents were found at 1500 m depth, while others had high alkane contents at very shallow depth. For example, there is a B^1 crude, belonging to the Kursai field in the Near-Caspian region, at a depth of 4410 m; furthermore, a B^2-type crude was found at 2806 m depth in the Sivinskoe field (the Volga-Urals). An A^1 crude was located at 800 m in the Tikhovskoe deposit, and an A^2 oil at 3900 m in the Duvannyi-more (southern Caspian region). As mentioned earlier, the geological age of source rocks may have a more indirect influence on

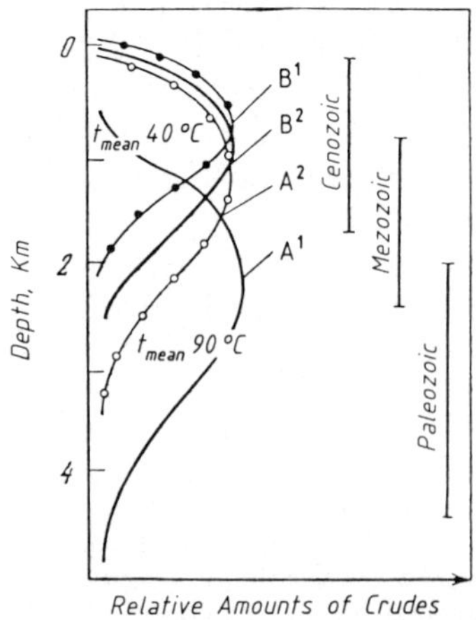

Fig. 7. Mean statistical distribution of different crude types depending on depth of occurrence and source-rock age

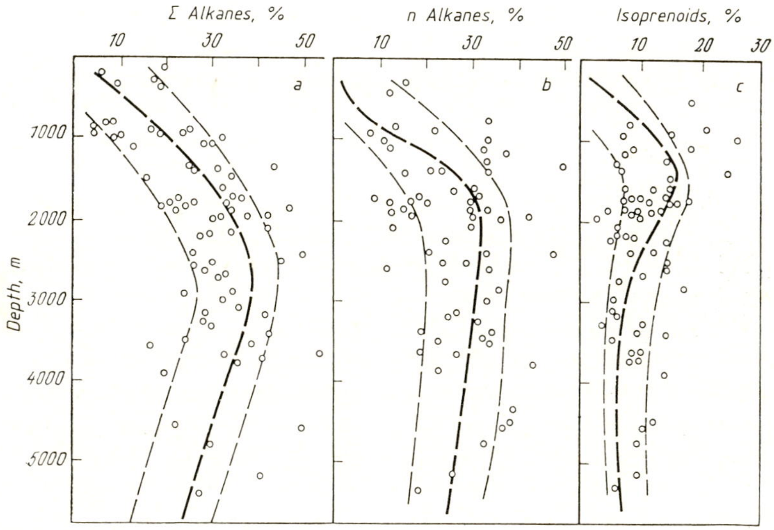

Fig. 8a–c. Dependence of total contents of alkanes (**a**), with normal (**b**) and isoprenoid (**c**) alkanes given separately, on the depth of petroleum occurrence

the hydrocarbon content of crude oil which has been often demonstrated in the literature (Tissot and Welte 1981; Andreev et al. 1958; Proskuryakov and Drabkina 1981).

On the whole, regardless of the geological age, of the source rocks, a gradual alteration from type B^1 to A^1 usually takes place with an increase in the depth of petroleum deposit. Alteration regularities in the total contents of alkanes and, separately, normal and isoprenoid alkanes are represented in Fig. 8. The latter regularity is, however, less obvious. Increased depth and, consequently, severity of actual temperature and pressure conditions are accompanied by an increase of alkane content in the majority of the crudes examined. This increase is especially significant at 2000-m depths, to subside at greater depths (Fig. 8a), accompanied by logical modifications in normal and isoprenoid alkane composition. As can be seen in Fig. 8b, normal alkanes are generally absent from curdes at 500–600 m, but starting from the 800-m level, their proportion grows rapidly, down to approximately 2000 m. Further increase in depth does not affect this parameter. Isoprenoid alkane concentrations begin to increase from barely traceable to maximum quantities, with their maximum (in terms of depth) preceding that of normal alkanes. Then, within the 1500–2000 m range, a decrease of almost twofold occurs in their concentration, while any further increase in depth affects the isoprenoid concentrations less markedly. Such a regularity was also observed in the publication of Solodkov et al. (1977).

The reasons for all these transformations are fairly complex. New formation of these hydrocarbons, due to thermal cracking proceeds concurrently with their elimination through biodegradation. All these factors are extensively examined in Chaps. 5 and 6. Classifications of crudes (according to chemical type) which will be presented schematically, as well as the research methods on which they are based, find wide application in the scientific literature. Thus, for example, characteristics of various crudes in the monograph of Tissot and Welte (1981) are usually illustrated either by gaschromatograms or by diagrams of the relative of normal and isoprenoid alkane distributions as a function of their molecular mass (the number of carbon atoms per molecule). Naturally, the proposed classification is far from being perfect. In particular, no special type is foreseen for highly by aromatic crudcs containing ≥50% aromatic hydrocarbons in the 200°–430°C fraction. It is true, however, that we could identify only a small number of such crudes (approximately 10–12 of 400 examined), while their chromatograms were usually commensurate with those of types A^1, A^2, B^2, and B^1.

It should be emphasized in conclusion that the method of chemical classification devised by us is highly appropriate for geochemical research. However, this scheme may also be found useful in chemical processing of petroleum, since it provides a rather comprehensive idea on the contents of hydrocarbons belonging to various groups, as well as on the possible future processing and utilization of specific crudes. For example, B crudes may be used in the production of high-quality lubrication oils, etc.

Group-Type Distribution of Naphthenic Hydrocarbons in Crudes (Naphthenic Fingerprint)

Naphthenic hydrocarbons, which comprise a significant part of the high-boiling fractions in any crude, have been less adequately studied due to their complex composition. It appears impossible to fully identify a large number of structural hydrocarbons or stereoisomers at the molecular level, although appreciable results were obtained in this area (see Chap. 3). Mass spectrometric determination of concentrations of cyclanes, which may contain one to five cycles per molecule, is a most important group-type method of naphthene analysis. The relative concentrational distribution of naphthenes, as a function of the number of cycles per molecule, will in the following be called a naphthenic fingerprint. Naphthenic fingerprint data were already presented in Table 1; however, their interpretation is a challenging task in view of the amount of figures. Therefore, additional naphthenic fingerprints of oils from various regions are diagrammatically reproduced in Figs. 9 and 10. Shaded sections correspond to changes in the naphthenic fingerprints of crudes from the corresponding regions.

Naphthenic fingerprints of the different crudes examined are quite distinctive in each region. Thus, for example, contrary to other regions, bi- and tetracyclic naphthenic hydrocarbons dominate in crudes from the southern Caspian region. Comparison of naphthenic fingerprints of oils from different regions allows one to establish their similarities and distinctions. Mention should also be made of the extensive research devoted to naphthenic fingerprints, as reported in the publication of Kartsev (1978). A general regularity of mono- and bicyclane predominance over other naphthenes is found in most cases. On the average, these hydrocarbons amount to 50–60% of all naphthenes in most crudes, while there is not more than 10% pentacyclic structures. Petroleum naphthenic fingerprints differ mainly in the relative proportions of mono- and bicyclic naphthenes.

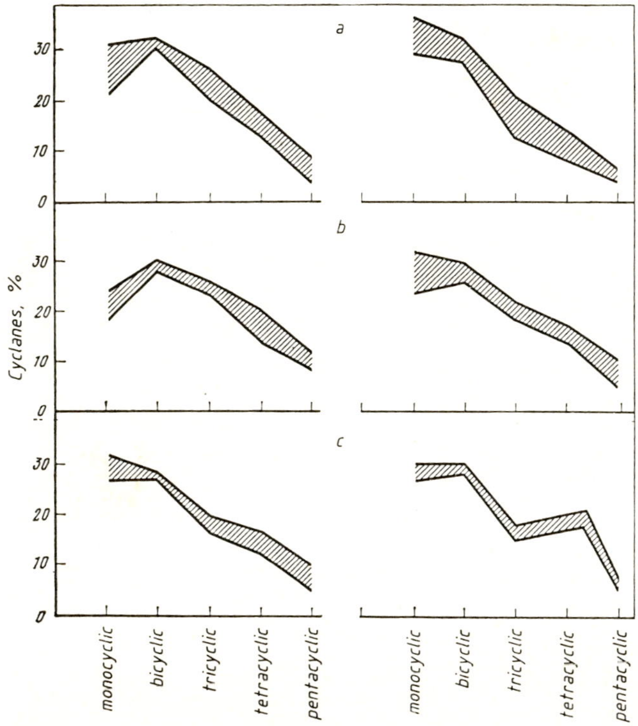

Fig. 9a–c. Naphthenic fingerprints of the 200°–430 °C fractions in crudes of various provinces. **a** Timano-Petchorskaya oil province; **b** Near-Caspian depression; **c** Dneprovo-Donetskaya depression. Here and in Fig. 10 cyclane contents are cited for the naphthenic part of the 200°–430 °C fraction

Crudes with equal concentrations of these cyclane groups may be found, or with a predominance of monocyclic naphthenes, while the reverse, i.e. predominance of bicyclic structures, occurs in crudes with a low content of normal alkanes. In almost all of the crude oils the concentration of polycyclic hydrocarbons tends to decrease with the increase in the number of cycles per molecule.

Characteristically, crudes of different chemical types have similar concentrations of mono-, bi-, tri-, tetra- and pentacyclanes if they belong to the same region or area within this region. Moreover, it was discovered that crudes coming from closely spaced reservoirs with similar pristane/phytane ratios have identical naphthenic fingerprints. Hence, it follows that the relative distribution of cyclanes with a different number of cycles in a molecule (for genetically related crude oils) is independent of the total content of naphthenes. This peculiarity is important in establishing the reasons and regularities in the formation of differnt crude oil types. Compared to alkanes, naphthenic hydrocarbons are characterized by their structural stability and preservation of genetic features, which are inherited from peculiarities of the original organic matter and the conditions of its transformation into petroleum hydrocarbons. As with alkanes, the main shifts in the concentrations of naphthenes occur at depths not exceeding 2000 m. The petroleum

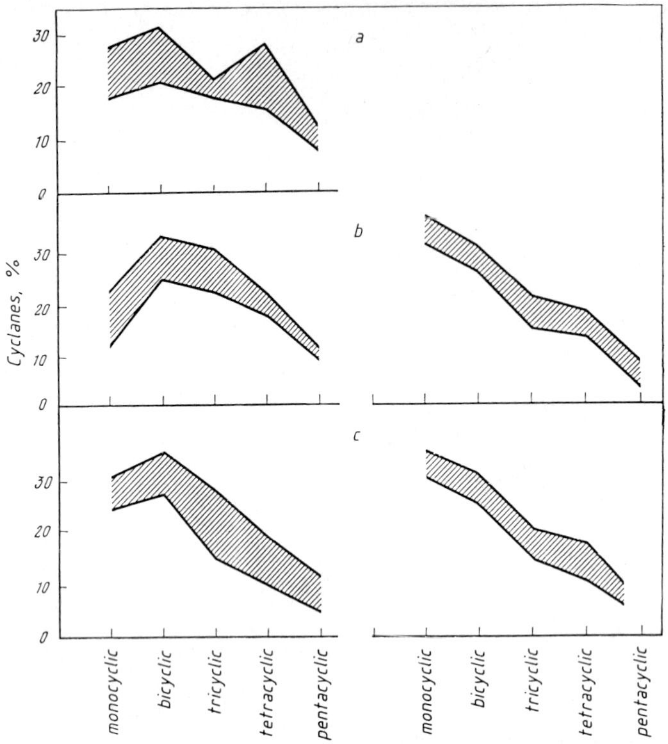

Fig. 10 a–c. Naphthenic fingerprints of the 200°–430 °C fractions in crudes of various provinces. **a** South Caspian depression; **b** Western Siberia; **c** Volga-Urals

composition changes much more moderately at greater depths. Of particular interest are the changes in the naphthenic fingerprint as a function of depth. It was found that crudes different in chemical composition and reservoir depth, but related genetically (according to the pristane/phytane ratio), have identical naphthenic fingerprints in virtually all of multilayer reservoirs in regions with different oil-bearing levels.

Table 3 serves as an illustration of the naphthenic fingerprints of crudes belonging to different chemical types in certain Soviet oil fields.

Thus, it may be concluded that oils inherit their naphthenic fingerprints from the original organic matter, these fingerprints also depend on lithologic-facial conditions of their transformations. Both pristane/phytane ratios and naphthenic fingerprints may be used as additional criteria in determining the genitic ties of crudes. The chemical type of oils and their naphthenic fingerprints have no obvious genetic connection.

Therfore, we should underline the uniformity in hydrocarbon compositional changes which occur in multilayer reservoirs in different regions. Moreover, we may note that contrary to naphthenic hydrocarbons, the content of normal and isoprenoid alkanes responds more directly to the geochemical conditions of petroleum occurrence.

Table 3. Relative distribution of saturated hydrocarbons in 200° – 430 °C fractions of crudes belonging to multilayer deposits

Reservoir	Depth (m)	Crude oil type	Pristane/Phytane	C_i	N_b	Naphthenic passport					Fractional composition (%)		Isoprenoids (%)			Alkanes (%)	
						Mono-cyclic	Bi-cyclic	Tri-cyclic	Tetra-cyclic	Penta-cyclic	Alkanes	Naph-thenes	Light $C_{14}-C_{18}$	Inter-mediate $C_{19}-C_{20}$	Heavy $C_{21}-C_{25}$	Norm.	Iso.
Kotur-Type	2500	A^1	1.25	1.0	12.0	25.9	27.1	19.2	18.8	9.0	43.6	56.4	38.9	48.6	12.5	12.1	31.5
	1700	A^2	1.25	2.1	3.5	24.1	26.6	19.7	19.6	10.0	30.3	69.7	39.2	47.7	13.1	5.5	24.8
Dagadjikskoe	1951	A^1	1.33	0.5	21.9	24.3	28.3	19.6	19.1	8.7	55.8	44.2	37.4	49.7	12.9	23.6	32.2
	1000	A^2	1.25	13.9	0.8	26.6	27.9	20.3	17.6	7.6	24.0	76.0	39.9	43.2	16.9	2.8	21.2
	900	B^2	1.25	–	–	24.5	28.0	19.7	19.6	8.2	32.0	68.0	34.9	51.8	13.3	–	32.0
Samotlor	2212	A^1	1.0	0.8	6.1	34.8	26.1	15.5	16.7	6.9	48.4	51.6	46.4	40.2	13.4	8.5	39.9
	1756	A^1	0.91	0.7	6.5	34.6	27.1	15.4	15.9	7.0	52.4	47.6	44.3	40.5	13.2	10.5	41.9
	1728	A^2	0.93	7.7	0.7	36.0	28.3	16.9	14.2	4.6	40.6	59.4	44.2	43.8	12.0	3.7	36.9
Novoportov-skoe	1975	A^1	2.1	0.5	11.4	31.5	28.2	18.1	15.7	6.5	36.2	63.8	49.1	37.2	13.7	19.1	17.1
	1830	A^2	–	7.1	0.9	31.3	30.2	15.8	15.4	7.3	43.9	56.1	48.6	38.7	12.7	7.5	36.4
	985	B^1	2.2	–	–	18.4	24.1	25.1	21.8	10.6	14.5	85.5	–	–	–	–	14.5
Djyerskoe	1586	A^1	1.04	0.8	8.5	44.7	27.3	15.1	8.1	4.8	49.6	50.4	35.1	51.3	13.6	14.5	35.1
	800	A^2	1.15	3.3	1.2	39.6	28.6	15.9	8.7	7.2	30.9	69.1	45.5	43.5	11.0	4.6	26.1
W. Tebuk	1945	A^1	1.09	0.75	8.8	31.5	30.9	17.4	13.2	7.0	50.6	49.4	48.0	39.5	12.5	14.4	36.2
	1377	A^2	1.08	3.1	2.6	32.6	31.5	18.8	11.6	5.5	43.9	56.1	47.7	41.4	10.9	9.5	34.4
Enaruskinskoe	1165	A^1	0.54	0.6	6.2	30.5	31.6	20.9	10.2	6.8	60.7	39.3	44.7	42.1	13.2	22.5	38.2
Mordovo-Karmalskoe	80	B^2	0.62	–	–	28.9	31.5	21.3	11.9	5.4	36.8	63.2	36.1	46.5	17.4	–	36.8
Duvannuj-more	5203	A^1	1.67	0.6	12.3	21.9	27.5	20.5	20.4	9.7	58.2	41.8	50.8	39.1	10.1	15.1	43.1
	3904	A^2	1.56	2.7	3.1	19.4	22.8	20.2	23.6	14.0	31.3	68.7	38.2	47.7	14.1	7.1	24.2
	3500	A^2	1.43	6.6	1.1	19.4	24.6	21.4	23.5	11.1	25.3	74.7	36.5	51.1	12.4	6.1	19.2
	3440	A^1	1.43	1.3	7.2	21.2	25.9	21.5	21.0	10.4	39.5	60.5	37.0	49.6	13.4	12.8	26.7

In making a general conclusion we should reemphasize the existence of two zones where petroleum transformations occur with varying intensity. The first, upper zone ranges from 0 to 1800–2000 m. Crudes, occurring at similar depths in this zone, vary considerably in their composition. The second, lower zone lies at depths at or over 2000 m and is characterized by crudes of greater compositional homogeneity; modifications of crudes depend les on variations of depth in this zone.

From the view point of the transformations of alkane biomarkers, the above zones coincide with the well-known zones of dia- and catagenesis in the sediments, as comprehensively discussed in the works of N.B. Vassaevitch, V.A. Uspenski, A.F. Dobryanskii, A.I. Bogomolov, A.A. Kartsev, A.G. Gabrielyan, B. Tissot, D. Welte, and other geochemists.

Detailed experimental data on the character and causes of compositional alterations of crude oil in these zones are analyzed in Chapter 6.

It remains to be added that extensive material on the regional characteristics of the chemical and molecular composition of crude oils in various oil fields of the Soviet Union is contained in the book of Maksimov and Botneva (1981).

CHAPTER 2

$C_5 - C_{40}$ Alkanes[1]

The previous chapter presented a review of certain group characteristics of petroleums. The present chapter, as well as the two that follow, are devoted to individual petroleum hydrocarbons, i.e. the results of research undertaken at the molecular level are presented. All data were obtained with the help of the most modern research methods, such as GC using capillary columns and temperature programming, as well as gas chromatography-mass spectrometry with computerized analysis and chromatogram reconstruction on the basis of individual characteristic fragment ions (mass fragmentography or mass chromatography). Wide use was also made of ^{13}C NMR spectroscopy. The majority of the petroleum hydrocarbons reviewed were also synthesized as reference standards in the laboratory. Standard methods of synthesis were applied, as well as catalytic equilibration, which provides easily separable mixtures of structurally related hydrocarbons whose composition was determined by ^{13}C NMR spectroscopy. Identification of any hydrocarbon was considered as proven, if the peaks in the gas chromatograms (two phases were generally used), coincided with the standard, while at the same time, the mass spectra of the crude oil compound and the standard hydrocarbon were identical.

The present chapter primarily deals with alkanes of relict structures, particularly with hydrocarbons of isoprenoid structures. It should be recollected that the stereochemistry of alkanes, synthesis and thermodynamic stability of these compounds as well as the analysis of alkanes in crudes were extensively treated in our monograph (Petrov 1974).

First, let us review certain methods pertaining to the analysis of alkane hydrocarbons in oils. The most effective and informative method of alkane analysis is gas chromatography.

Gas Chromatography in the Analysis of Alkanes

Contemporary methods of analyzing petroleum hydrocarbons (at the molecular level) involve the preliminary separation of petroleums into two or three fractions of different boiling temperatures, afterwards each fraction is subjected to separation (liquid chromatography on silica gel) into paraffin-naphthenic (saturated)

[1] Gaseous $C_1 - C_4$ alkanes are analyzed in detail in a number of publications (Kartsev 1978; Proskuryakov and Drabkina 1981; Sokolov et al. 1972).

and aromatic hydrocarbons. It is recommended that the latter, especially in the case of high-boiling fractions, are separated, depending on the task assigned (and again by liquid chromatography, using alumina), into mono-, bi- and polyaromatic hydrocarbons.

Fraction I (30°–200 °C) – C_5-C_{11} Hydrocarbons. Analysis proceeds in a capillary column coated with squalane and with a capacity of $80-100\times10^3$ theoretical plates. Linear temperature programming is applied (heating rate 1 °C/min). The initial temperature is 50 °C. The end of the analysis (after elution of n-un-

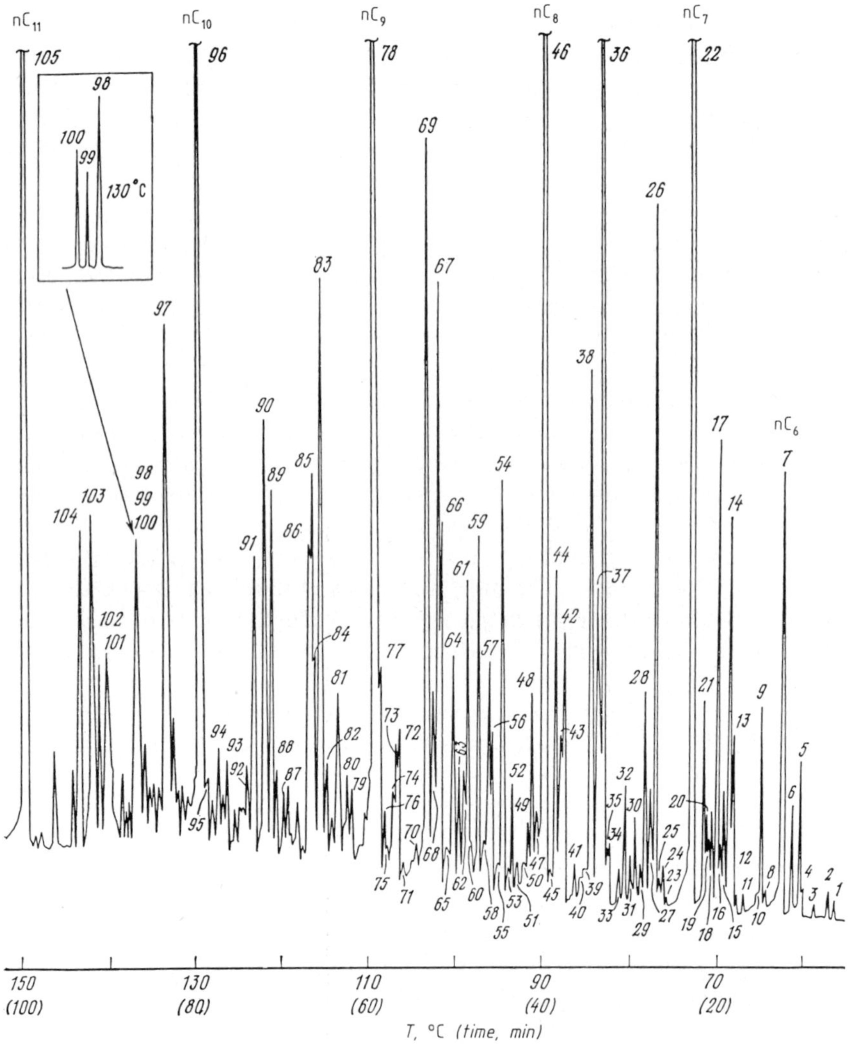

Fig. 11. Chromatogram of a C_5-C_{11} saturated hydrocarbon mixture of the Surgut crude. Peaks are identified in the text. Capillary column 100 m, squalane; linear temperature programming, 50°→1°/min

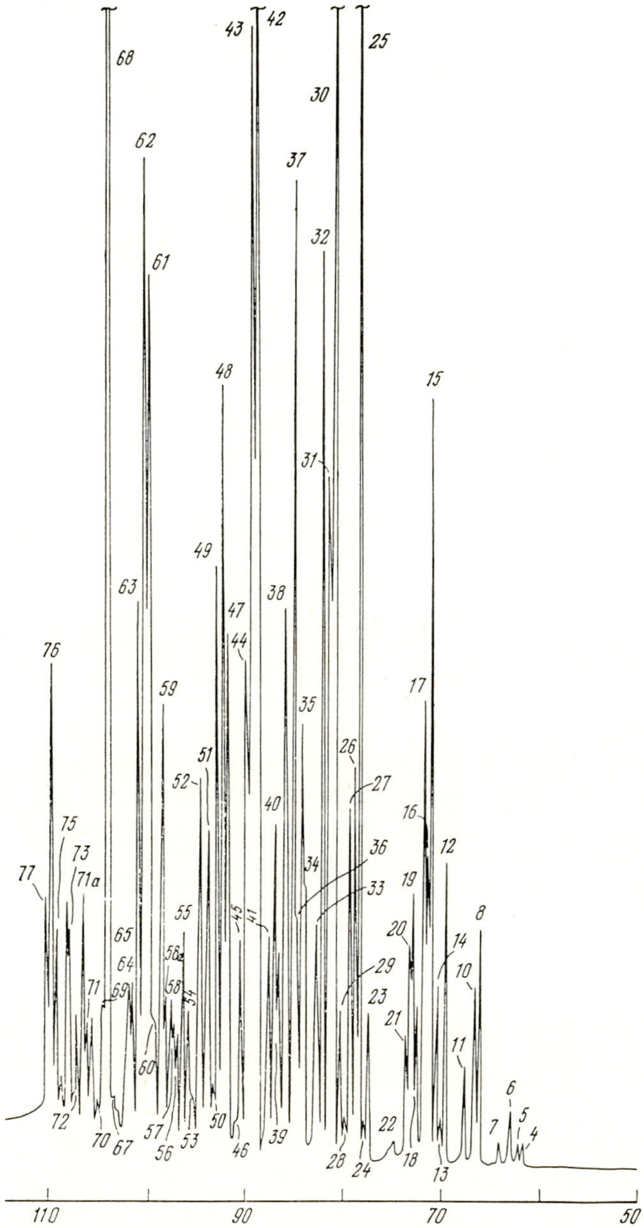

Fig. 12. Chromatogram of a saturated hydrocarbon mixture from the Anastasiyevsko-Troyitskoe deposit. Conditions of analysis and peak identification are the same as in Fig. 11

decane) is usually at 150 °C (n-decane elutes at 130 °C, n-nonane at 110 °C, etc.). The column inlet pressure is set depending on column type.

Figure 11 shows a typical chromatogram of a gasoline range (saturated hydrocarbons) of a paraffinic oil (type A[1]). Similar analysis of gasoline hydrocarbons of naphthenic crude oils is more complicated and feasible only in the case of fractions distilling at a temperature not higher than 150 °C (C_5-C_9).

Figure 12, as an example, provides a chromatogram of saturated C_5-C_9 hydrocarbons isolated from a type B[2] crude. Notice should be taken of the unusual distribution of alkanes according to structural type, but mostly of the high concentrations of compounds with geminal and vicinal substituents.

Peak identifications from Figs. 11 and 12 are as follows:

1. Isopentane
2. n-pentane
3. 2,2-Dimethylbutane
4. Cyclopentane
5. 2-Methylpentane + 2,3-dimethylbutane
6. 3-Methylpentane
7. n-Hexane
8. 2,2-Dimethylpentane
9. Methylcyclopentane
10. 2,4-Dimethylpentane
11. 2,2,3-Trimethylbutane
12. 3,3-Dimethylpentane
13. Cyclohexane
14. 2-Methylhexane
15. 2,3-Dimethylpentane
16. 1,1-Dimethylcyclopentane
17. 3-Methylhexane
18. *cis*-1,3-Dimethylcyclopentane
19. 3-Ethylpentane
20. *trans*-1,3-Dimethylcyclopentane
21. *trans*-1,2-Dimethylcyclopentane
22. n-Heptane
23. 2,2-Dimethylhexane
24. *cis*-1,2-Dimethylcyclopentane
25. 1,1,3-Trimethylcyclopentane
26. Methylcyclohexane + 2,5-dimethylhexane
27. 2,4-Dimethylhexane
28. Ethylcyclopentane
29. 2,2,3-Trimethylpentane
30. *trans, trans*-1,2,4-Trimethylcyclopentane
31. 3,3-Dimethylhexane
32. *trans, cis*-1,2,3-Trimethylcyclopentane
33. 2,3,4-Trimethylpentane
34. 2,3-Dimethylhexane
35. 2,3,3-Trimethylpentane + 2-methyl – 3-ethylpentane
36. 2-Methylheptane
37. 4-Methylheptane + 1,1,2-trimethylcyclopentane
38. 3-Methylheptane + 3-ethylhexane
39. *cis, trans*-1,2,4-Trimethylcyclopentane
40. *cis, cis*-1,2,4-Trimethylcyclopentane + 2,2,5-trimethylhexane
41. *cis, trans*-1,2,3-Trimethylcyclopentane
42. *cis*-1,3-Dimethylcyclohexane + *trans*-1,4-dimethylcyclohexane
43. 1,1-Dimethylcyclohexane + *cis*-1-methyl-3-ethylcyclopentane
44. *trans*-1-Methyl-3-ethylcyclopentane + *trans*-1-methyl-2-ethylcyclopentane
45. 1-Methyl-1-ethylcyclopentane
46. n-Octane
47. *trans, trans, trans*-1,2,3,4-Tetramethylcyclopentane
48. *trans*-1,2-Dimethylcyclohexane
49. *trans*-1,3-Dimethylcyclohexane + *cis*-1,4-dimethylcyclohexane
50. 2,3,5-Trimethylhexane
51. 2,2-Dimethylheptane + 1,1,2,4-tetramethylcyclopentane – *trans* + *cis*

52. 2,4-Dimethylheptane
53. 2-Methyl-4-ethylhexane
54. 2,6-Dimethylheptane
55. 1,1-Dimethyl-3-ethylcyclopentane
56. 2,5-Dimethylheptane
57. 3,5-Dimethylheptane + n-propylcyclopentane
58. *cis*-1,2-Dimethylcyclohexane
59. Ethylcyclohexane
60. *trans, cis*-1,4-Dimethyl-2-ethylcyclopentane
61. 1,1,3-Trimethylcyclohexane + 1,4-dimethyl-2-ethylcyclopentane *trans, trans* + 1,3-dimethyl-1-ethylcyclopentane
62. 1,1,4-Trimethylcyclohexane + 1,3,5-trimethylcyclohexane – *cis, cis* + 1,3-dimethyl-2-ethylcyclopentane – *trans, trans*
63. *trans, trans*-1,2-Dimethyl-3-ethylcyclopentane
64. 2,3-Dimethylheptane
65. 4-Ethylheptane
66. 4-Methyloctane
67. 2-Methyloctane
68. 1,2,4-Trimethylcyclohexane – *trans, cis* + 3-ethylheptane
69. 3-Methyloctane
70. *trans, cis*-1,2-Dimethyl-3-ethylcyclopentane
71. *cis, trans*-1,2-Dimethyl-3-ethylcyclopentane
72. *trans*-1-Methyl-2-propylcyclopentane
73. 1,2,3-Trimethylcyclohexane – *trans, trans* + 1-methyl-3-propylcyclopentane – *cis*
74. *trans*-1-Methyl-3-propylcyclopentane
75. 1,1,2-Trimethylcyclohexane
76. 1,2-Diethylcyclopentane – *trans* + 1,1,3,5-tetramethylcyclohexane – *cis*
77. *cis*-1-Methyl-3-ethylcyclohexane
78. n-Nonane + *trans*-1-methyl-4-ethylcyclohexane
79. *trans*-1-Methyl-2-ethylcyclohexane
80. 2,4-Dimethyloctane
81. 2,5-Dimethyloctane + 1,1,3,4-tetramethylcyclohexane – *trans*
82. Isopropylcyclohexane + 2,7-dimethyloctane
83. 2,6-Dimethyloctane
84. n-Butylcyclopentane + 3,6-dimethyloctane
85. n-Propylcyclohexane
86. 2-Methyl-3-ethylheptane
87. 4-Ethyloctane + 2,3-dimethyloctane + 1,1-dimethyl-3-ethylcyclohexane
88. 5-Methylnonane
89. 4-Methylnonane
90. 2-Methylnonane + 3-ethyloctane
91. 3-Methylnonane
92. *trans, cis*-1,4-Dimethyl-2-ethylcyclohexane
93. *trans*-1-Methyl-2-butylcyclopentane
94. *cis*-1-Methyl-3-isopropylcyclohexane
95. *trans*-1-Methyl-4-isopropylcyclohexane
96. n-Decane
97. 2,6-Dimethylnonane

98.–100. 3,7-Dimethylnonane + amylcyclopentane + n-butylcyclohexane (triplet peak 98–100 is easily separable on the same column in an isothermal mode at 130 °C)

101. 5-Methyldecane
102. 4-Methyldecane
103. 2-Methyldecane
104. 3-Methyldecane
105. n-Undecane

Reliable identification in the chromatogram was accomplished by introducing reference hydrocarbons and by gas chromatography-mass spectrometry. In reproducing this analysis one may use the retention indices of the branched alkanes presented in Table 20 at the end of this chapter. Application of retention indice to the analysis of petroleum alkanes is always useful, because normal alkanes are found regularly in most oils and may serve as references. Experience shows that retention index values for branched alkanes are fairly well reproducible and depend insignificantly on the chromatographic conditions, which is unfortunately not the case for the retention indices of cyclanes and aromatic hydrocarbons.

The presented identification of compounds in the 30°–200 °C fraction shows that some peaks are composite in nature and include two, sometimes three components. In order to separate a composite peak, the analytical conditions should be varied. Accordingly, it should be kept in mind that an increased column inlet pressure would result in alkanes being eluted later relative to cyclanes. Reduction of pressure would lead to a reverse effect. Different isothermal procedures (*Methods of Analysing* ... 1969) are also recommended. The first hydrocarbons indicate in all of the doublet peaks of paraffinic (Surget) crude the prevailing components.

Fraction II (200°–430 °C) – $C_{12}-C_{27}$ Hydrocarbons. Analysis proceeds in a capillary column coated with Apiezon and with a capacity of $40-60\times10^3$ theoretical plates. Carrier gas is hydrogen (application of hydrogen in high-temperature GC is always preferable, because it protects the stationary phase from oxidation). The initial temperature is 100 °C, the final temperature 300°–310 °C. The heating rate is 2 °C/min. Furthermore, normal alkanes, monomethyl-substituted alkanes are identified within this range as well as isoprenoid alkanes.

A typical chromatogram of a mixture of saturated $C_{12}-C_{27}$ hydrocarbons (fraction 200°–430 °C) is presented in Fig. 13 (retention indices are also listed in Table 20). Notice should be taken of the "homologous" sequence of elution and retention index values of monomethyl-substituted alkane isomers. For all practical purposes, retention indice in the sequence 6-methylalkanes, 5-methylalkanes, 4-methylalkanes, 2-methylalkanes, 3-methylalkanes, are practically identical for hydrocarbons of any molecular mass. Retention indices are applicable only for the linear temperature programming mode; they are calculated according to a slightly simplified equation on the basis of the publication of Habgood and Harris (1964):

$$I_{\text{retention}} = n\times100+L/M\times100,$$

where n is the number of hydrocarbon atoms of the closest, lower boiling, normal alkane; L is the distance between the peaks of a given hydrocarbon and a lower boiling, normal alkane; and M is the distance between the closest, higher and lower boiling, normal alkanes.

The main regularity in the eluting sequence of methyl-substituted alkanes is the following: within the interval between the preceding and the following n-alkane all mono- and di-substituted isomers of the preceding n-alkane are eluted. The group of di-substituted alkanes is the first to be eluted, followed by the group of mono-substituted alkanes. Trimethyl- and tetramethyl-substituted alkanes (i.e. isoprenoid hydrocarbons) are eluted much earlier. Thus, for example, pristane and

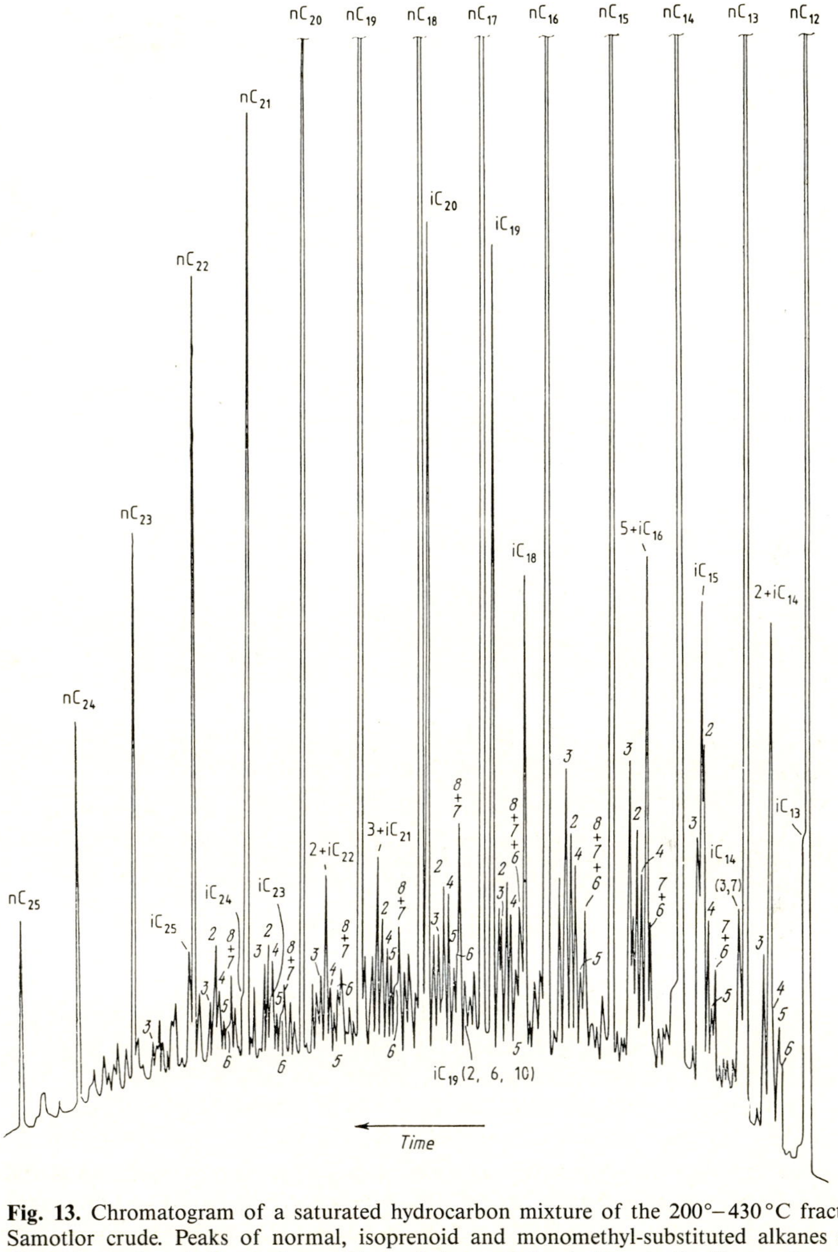

Fig. 13. Chromatogram of a saturated hydrocarbon mixture of the 200°–430 °C fraction of Samotlor crude. Peaks of normal, isoprenoid and monomethyl-substituted alkanes are indicated; *numbers* indicate the position of the methyl substituent. Capillary column 80 m, Apiezon; linear temperature programming, 100°→2°/min

Table 4. Boiling temperatures for normal alkanes (recalculated for 760 mmHg)

Hydrocarbon	C_{11}	C_{12}	C_{13}	C_{14}	C_{15}	C_{16}	C_{17}	C_{18}
Boiling temp., °C	196	216	235	254	271	287	303	317
Hydrocarbon	C_{19}	C_{20}	C_{21}	C_{22}	C_{23}	C_{24}	C_{25}	C_{26}
Boiling temp., °C	327	345	357	371	383	395	406	417
Hydrocarbon	C_{27}	C_{28}	C_{29}	C_{30}	C_{31}	C_{32}	C_{33}	C_{34}
Boiling temp., °C	427	437	447	457	466	475	484	493
Hydrocarbon	C_{35}	C_{36}	C_{37}	C_{38}	C_{39}	C_{40}		
Boiling temp., °C	501	509	517	525	532	539		

phytane are eluted earlier than normal alkanes having two carbon atoms less, while lycopane (C_{40}) elutes prior to the n-C_{35} alkane.

Fraction III (>430 °C). $C_{20}-C_{40}$ Hydrocarbons. Analysis of this fraction is conducted mainly in order to identify steranes and hopanes. A capillary column coated with Apiezon is used. Depending on the problem to be solved, programmed or isothermal modes are chosen. Details of such experiments are reviewed in Chapter 3. Fractions II and III are also often used in research in the linear temperature programming mode.

To facilitate the separation of individual fractions the boiling temperatures of normal alkanes are presented in Table 4.

Let us now turn directly to the review of the qualitative and quantitative distributions of alkanes of different structures in petroleums.

Quantitative Distribution of Alkanes in Petroleums. Possible Ways of Alkane Formation

Alkanes belong to the better studied hydrocarbons of any crude. A characteristic feature of petroleum hydrocarbons of this class is their relatively high concentration, especially of certain key structures. These structures include, for example, normal alkanes, monomethyl-substituted alkanes with different positions of substitution, as well as isoprenoid hydrocarbons or isoprenanes (Sanin 1976). This allows us to speak of different homologous series of alkanes in petroleums. This homology spreads over a rather broad distillation range of petroleum hydrocarbons. However, regardless of certain achievements in the study of alkanes at the molecular level, it should be noted, as proven by mass spectrometric data, that a certain portion of branched alkanes elutes in the chromatographically unresolved "hump". The composition and structure of these hydrocarbons is as yet little investigated. It may only be speculated that they consist of extensively branched structures (having at least two substituents in the chain). There is reason to believe that these structures represent slightly modified (isomerized) isoprenoid hydrocarbons.

Several methods of quantitative analysis of individual hydrocarbons in oils exist. The simplest of these is the measurement of the absolute concentration of a given compound in a crude. However, in this case we ignore important regularities connected with the concentrational distribution of isomers. In addition, figures which represent concentrations of certain hydrocarbons and are small in absolute terms (i.e. 0.02%, 0.005%, etc.), are difficult to understand and memorize. Evaluation of the relative distribution of isomers and homologs is more attractive. This method, elaborated in the publication of Smith and Rall (1953), is helpful in making various theoretical generalizations and visualizing connections between alkane structures and their concentrations in different crudes.

Together with the relative distribution of isomers, an important role is played by the evaluation of the relative distribution of alkanes in different homologous series, i.e. the determination of the content of hydrocarbons of particular structures as a function of their molecular mass. Both methods are applied for finding quantitative regularities in the occurrence of alkanes of different structures in petroleum. Presently, reliable data on the quantitative content of various C_5-C_{40} alkanes in crudes are available, which were obtained with the help of gas chromatography and are sufficiently accurate.

Normal Alkanes

For the first time ever quantitative data on the distribution of normal alkanes in petroleum were reported by Martin et al. (1964). Extensive information on the distribution of these alkanes in various Soviet crudes is presented in the monograph of Safonova (1980).

The analysis of the most typical regularities of the distributions of normal alkanes in different crudes was discussed in Chapter 1. Undoubtedly, the distribution of these alkanes in oils reflects the composition of the source material. Thus, transformation of the original biomass of continental origin, i.e. lipids of higher plants, leads to the formation of crudes with a relatively high component of normal $C_{25}-C_{33}$ alkanes (Tissot and Welte 1981) (for details, see Chap. 5). Additional information on the relative concentration on normal alkanes (among isomers) is presented below.

Branched Alkanes. Relative Concentration of Isomers (C_6-C_{10} Hydrocarbons)

Nowadays the compositional determination of light petroleum fractions is relatively easy. Available data provide the distribution of isomeric alkanes in scores, if not hundreds, of crudes. In summarizing the results of this research, we can identify three main types of gasolines which, naturally, correspond to the chemical types of petroleum as described in Chapter 1.

1. *Gasolines of paraffinic crudes (crudes of type A^1).* The composition of the light fractions is fairly uniform in this case. Slight variations exist only in the relative concentrations of normal and branched isomers.

2. *Gasolines of type A^2 (crudes much rarer occurrence).* These gasolines differ in their composition from gasolines of A^1 crudes in that they have much lower con-

Table 5. Relative distribution of hexanes in crudes (%)

Hydrocarbons	Type A[1] crude			Type A[2] crudes	Type B crudes		Gas condensates	
	Surgut	Romashkinskoe	Groznenskoe paraffinic	Starogroznenskoe	"Gryazevaya Sopka"	Anastasiyevsko-Troyitskoe	"Shur-Tepe"	"Oval-Toval"
n-Hexane	58.8	71.0	52.0	12.0	19.0	4.0	67.8	1.0
2-Methylpentane	24.3	14.5	26.6	38.4	8.0	13.0	14.7	1.5
3-Methylpentane	15.0	12.5	18.4	37.0	49.0	34.0	14.3	1.0
2,3-Dimethylbutane	1.6	1.0	2.7	12.3	12.2	31.0	2.7	64.5
2,2-Dimethylbutane	0.3	1.0	0.3	0.3	11.8	18.0	0.5	32.0
Σ Monosubstituted / Σ Disubstituted	20.7	13.5	15.0	3.9	2.37	0.96	9.0	0.026

centrations of normal alkanes. However, the distribution of branched alkanes is approximately the same as in A[1] crudes.

3. *Gasolines of type B crudes.* These samples are often characterized by a peculiar distribution of isomers. Relatively high concentrations of *gem*-substituted as well as *vic*-substituted structures are observed quite frequently. As was pointed out, these specific features are more prominent in the composition of C_5-C_8 hydrocarbons. Tables 5–8 present data on the distribution of isomers in a number of characteristic crudes types of A[1], A[2] and B. The following are used as typical crudes: (1) the Surgut crude (Cretaceous), Western Siberia; (2) the Romashkinskoe crude, Tatariya (Devonian); (3) the Groznenskoe paraffinic crude (Cretaceous); (4) the Starogroznenskoe crude derived from Tertiary sediments (Chokrak); (5) the crude from the Gryazevaya Sopka reservoir, derived from Tertiary sediments of the Apsheron; (6) Anastasiyevsko-Troyitskoe (horizon IV), Western Near-Caucasus (Miocene) and gas condensates from Shur-Tepe and Oval-Toval-Western Turkmenistan (from Tertiary sediments).

The information in Tables 5–8 underlines the compositional differences between the gasolines of category A and B crudes. The former group of crudes, especially those of type A[1], is characterized by relatively high concentrations of normal alkanes. Methyl-substituted structures dominate among branched isomers. The concentrations of *gem*-substituted hydrocarbons are insignificant.

Table 6. Relative distribution of heptanes in crudes (%)

Hydrocarbon	Type A[1] crude			Type A[2] crudes	Type B crudes		Gas condensates
	Surgut	Romashkinskoe	Grozneнskoe paraffinic	Starogrozneнskoe	"Gryazevaya Sopka"	Anastasiyevsko-Troyitskoe	"Shur-Tepe"
n-Heptane	55.9	61.0	56.0	9.5	6.0	2.0	53.8
2-Methylhexane	15.9	11.4	11.0	27.0	6.4	7.0	18.3
3-Methylhexane	18.1	16.1	22.0	33.8	21.6	17.0	19.0
3-Ethylpentane	2.2	2.6	2.5	9.6	12.0	7.0	0.7
2,3-Dimethylpentane	4.0	6.0	6.0	15.5	40.0	31.0	5.1
2,4-Dimethylpentane	1.5	1.0	2.0	0.6	3.0	8.5	1.6
2,2-Dimethylpentane	2.1	1.0	0.5	1.2	2.5	12.0	0.7
3,3-Dimethylpentane	0.3	0.6	0.3	1.1	6.8	11.5	0.5
2,2,3-Trimethylbutane	traces	0.3	0.2	1.2	1.7	4.0	0.3
$\frac{\sum \text{Monosubstituted}}{\sum \text{Disubstituted}}$	4.6	4.0	3.9	3.8	0.74	0.5	4.8

In contrast, dimethyl-substituted structures, both geminal and vicinal, are present in high concentrations in group B crudes. It is noteworthy that among monosubstituted structures in these crudes 3-methylalkanes have higher concentrations than 2-methyl-alkanes.

At first glance it may appear that isomeric hydrocarbons with the same type of substitution are present in the category A crudes in near-equilibrium concentrations. Doubtless, a certain tendency towards equilibration in the distribution of isomeric petroleum hydrocarbons exists. However, a true equilibrium among isomers is not found. Crudes are only characterized by a predominance of thermodynamically stable structures.

Certain methylalkanes and dimethylalkanes (Petrov 1974) approach equilibrium more closely. Nevertheless, even in these cases the apparent equilibrium among monomethylalkanes is largely influenced by the thermodynamic control during the formation of certain isomers and does not signify that a real equilibrium state is reached, since between mono- and dimethylalkanes there is no equilibrium (for more details, see Chap. 5).

Table 7. Relative distribution of octanes in crudes (%)

Hydrocarbon	Type A[1] crudes			Type A[2] crudes		Type B crudes	Gas condensates
	Surgut	Romashkinskoe	Grozne-nskoe paraffinic	(Staro-grozne-nskoe)	"Gryazevaya Sopka"	Anastasi-yevsko-Troyitskoe	("Shur-Tepe")
n-Octane	54.8	41.0	43.3	4.3	7.0	3.2	41.3
2-Methylheptane	18.6	22.2	23.8	31.3	9.3	6.4	21.5
3-Methylheptane	10.7	17.1	14.0	22.6	8.8	12.8	15.7
4-Methylheptane	5.6	7.5	5.8	13.7	19.6	7.2	5.9
2,3-Dimethylhexane	0.8	1.6	2.0	3.1	7.0	4.8	2.5
2,4-Dimethylhexane	3.2	2.6	4.0	8.7	10.4	11.2	4.4
2,5-Dimethylhexane	2.8	2.9	4.0	8.6	3.1	4.0	4.2
3,4-Dimethylhexane	1.7	0.8	1.1	–	7.5	8.0	1.0
2,2-Dimethylhexane	0.4	0.3	0.5	0.8	5.6	4.8	1.0
3,3-Dimethylhexane	0.6	0.7	0.5	0.9	1.4	16.8	0.9
2,3,4-Trimethylpentane	Trace	0.5	0.9	0.5	5.7	6.4	0.5
2,2,3-Trimethylpentane	0.13	0.5	0.1	1.2	1.2	4.8	0.5
2,3,3-Trimethylpentane	0.5	2.3	Trace	4.3	13.4	9.6	0.6
Σ Monosubstituted / Σ Disubstituted	3.7	5.9	3.7	3.2	1.06	0.53	3.1

Tables 9 and 10 provide published data on equilibrium concentrations of heptane and nonane isomers.[2]

[2] High concentrations of biological markers in crudes excludes the possibility of equilibrium conditions for most isomers. Therefore data presented here and in further chapters on equilibrium conditions of structural isomers should be regarded rather as substantiating the lack of equilibrium in petroleum hydrocarbon isomers; although data on equilibrium concentrations of isomers deserve special scientific attention.

Table 8. Relative distribution of nonanes in crudes (%)

Hydrocarbons	Type A[1] crudes			Type A[2] crudes	Type B crudes		Gas condensates
	Surgut	Romashkinskoe	Groznenskoe paraffinic	(Starogroznenskoe)	"Gryazevaya Sopka"	Anastasiyevsko-Troyitskoe	("Shur-Tepe")
n-Nonane	38.4	30.7	34.4	5.2	Trace	Trace	28.7
2-Methyloctane	12.9	9.2	10.3	11.3	4.5	1.2	16.4
3-Methyloctane	14.2	15.6	9.7	15.3	17.5	1.4	15.9
4-Methyloctane	7.8	6.3	6.7	8.3	8.6	1.3	9.1
2,3-Dimethylheptane	5.3	5.5	12.0	20.3	17.7	4.5	7.0
2,4-Dimethylheptane	2.9	3.2	3.0	6.1	7.1	23.9	3.1
2,5-Dimethylheptane	3.2	6.3	3.7	6.7	7.0	7.3	5.9
2,6-Dimethylheptane	9.1	16.2	10.0	14.6	11.2	8.9	9.0
3,4-Dimethylheptane	3.1	2.1	2.5	–	14.4	5.0	1.4
3,5-Dimethylheptane	1.5	3.5	1.7	5.0	3.2	7.0	1.7
2,2-Dimethylheptane	0.7	0.7	1.6	3.4	4.6	23.3	0.9
3,3-Dimethylheptane	0.6	1.0	2.4	1.6	3.0	9.7	1.1
2,2,5-Trimethylhexane	–	–	–	1.8	2.6	6.5	–
$\frac{\sum \text{Monosubstituted}}{\sum \text{Disubstituted}}$	1.07	0.82	0.68	0.6	–	0.05	1.04

From the geochemical point of view, i.e. for the understanding of their sources and the mechanisms of their formation, we should concentrate on those hydrocarbons which appear in petroleums in amounts exceeding equilibrium concentration. These hydrocarbons, representing biological markers, provide the most valuable information on petroleum genesis and source-rock composition.

Normal alkanes deserve obvious attention from this point of view. However, even among branched alkanes of type A crudes, beginning with nonanes, groups of hydrocarbons emerge, which are significant from the standpoint of oil formation. Data presented in Table 8 give a convincing illustration of the considerable relative increase of dimethyl-substituted structures among the C_9 alkanes.

Table 9. Equilibrium composition of heptanes mixture at 300 K (%)

n-Heptane	2-Methyl-hexane	3-Methyl-hexane	2,3-Dimethyl-pentane	2,4-Dimethyl-pentane	3-Ethyl-pentane
5.5 (55.5)	30.7 (13.8)	20.8 (19.2)	10.9 (6.1)	30.7 (1.7)	1.4 (2.6)

Note: Summary data on 18 crudes are tabulated in parentheses (Martin et al. 1964). 2,2- and 3,3-Dimethylpentanes were not identified in the experiments.

Table 10. Equilibrium composition of nonanes (%)[a] (experiment)

Isomer	300 K	600 K	Isomer	300 K	600 K
n-Nonane	3.7	16.5	2,2-Dimethyl-heptane	65.0	51.8
2-Methyl-octane	42.8	30.1	3,3-Dimethyl-heptane	33.4	39.4
3-Methyl-octane	36.4	35.9	4,4-Dimethyl-heptane	1.6	8.8
4-Methyl-octane	16.5	24.1			
3-Ethyl-heptane	2.4	5.7			
4-Ethyl-heptane	1.9	4.2			
Σ Mono-substituted	100 (21.2)	100 (54.4)	Σ-*gem*-di-substituted	100 (12.6)	100 (5.6)
2,3-Dimethyl-heptane	5.6	13.6	Σ Di-substituted other than *gem*	(47.8)	(27.7)
2-Methyl-3-ethyl-hexane	0.6	1.4	2,2,4-Tri-methylhexane	2.8	7.1
2,4-Dimethyl-heptane	13.7	15.8	2,2,5-Tri-methylhexane	79.3	35.7
2-Methyl-4-ethylhexane	1.1	0.5	2,3,4-Tri-methylhexane	8.8	–
2,5-Dimethyl-heptane	39.5	30.0	2,3,5-Tri-methylhexane	7.7	43.0
2,6-Dimethyl-heptane	26.7	14.0	2,4,4-Tri-methylhexane	0.7	7.1
3,4-Dimethyl-heptane	3.0	10.7	2,4-Dimethyl-3-ethyl-pentane	0.7	7.1
3,5-Dimethyl-heptane	9.8	14.0			
Σ Di-substituted (other than *gem*)	100 (35.2)	100 (22.1)	Σ Tri-substituted	100 (27.3)	100 (1.4)

Note: Relative concentrations of hydrocarbons of a particular substitution type are given in parentheses.

[a] For a group of similar type isomers; for instance monomethyl Σ = 100%, dimethyl Σ = 100%, etc.

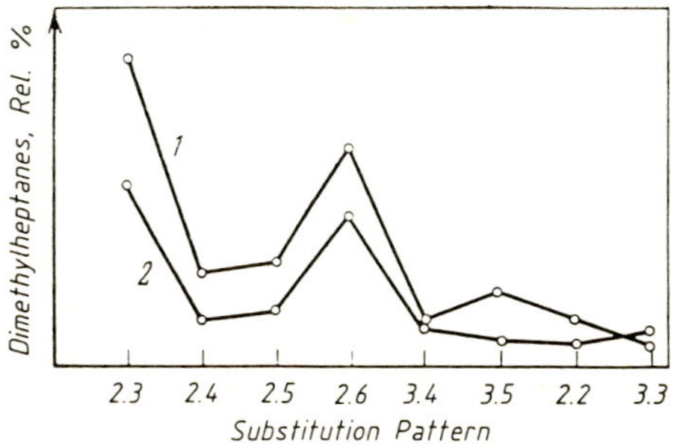

Fig. 14. Relative distribution of dimethylheptanes in various crudes. *1* Starogroznenskoe; *2* Groznenskoe paraffinic. *Numbers* on the abscissa indicate the positions of the methyl substituents

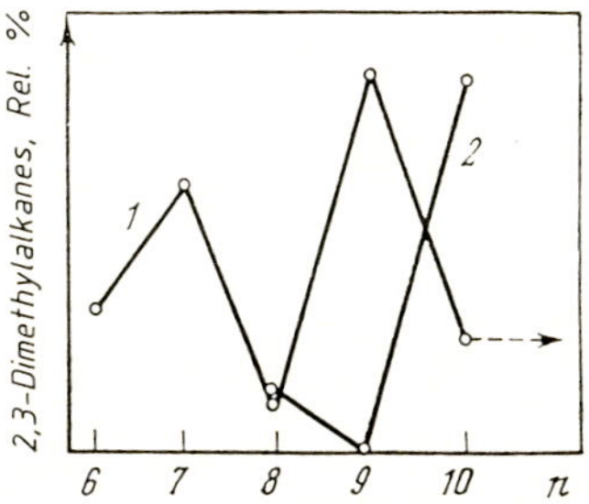

Fig. 15. Relative distribution of 2,3-dialkylalkanes in Surgut crude. *1* 2,3-Dimethylalkanes; *2* 2-methyl-3-ethylalkanes; *n* denotes the number of carbon atoms in a molecule

Table 11. Relative distribution of decanes in crudes (%)

Hydrocarbons	Type A^1 crudes		Type A^2 crudes (Starogroznenskoe)
	Surgut	Groznenskoe paraffinic	
n-Decane	39.4	39.9	2.0
2-Methylnonane	8.9	7.7	9.0 (38.8)
3-Methylnonane	7.7	10.7	6.0 (25.9)
4-Methylnonane	7.5	8.9	6.9 (29.7)
5-Methylnonane	1.6	0.9	1.3 (5.6)
2,3-Dimethyloctane	1.1	1.8	6.3 (8.3)
2,4-Dimethyloctane	1.6	1.8	11.1 (14.7)
2,5-Dimethyloctane	3.4	4.1	15.7 (20.8)
2,6-Dimethyloctane	12.9	12.8	20.6 (27.3)
2-Methyl-3-ethylheptane	15.8	13.5	21.8 (28.9)

See note to Table 10.

Table 12. Equilibrium concentrations of decanes (%) (experiment)

Isomer	300 K	700 K
2-Methylnonane	36	29
3-Methylnonane	29	32.5
4-Methylnonane	21	19
5-Methylnonane	11	8
4-Ethyloctane[a]	3	11.5
Σ Mono-substituted	100	100
2,4-Dimethyloctane	11	12
2,5 + 3,5-Dimethyloctanes	23	23
2-Methyl-5-ethylheptane[b]	5	8
2,7-Dimethyloctane	19	9
2,6-Dimethyloctane	26	25.5
3,6-Dimethyloctane	9	13.5
2,3 + 3,4-Dimethyloctane	7	9
Σ Di-substituted	100	100
2,4,6-Trimethylheptane	1	[c]
2,2,6-Trimethylheptane	45	[c]
2,2,5-Trimethylheptane	32	[c]
2,5,5-Trimethylheptane	22	[c]
Σ Tri-substituted	100	

[a] The most stable ethyloctane.
[b] The most stable methylethylheptane.
[c] Formation of hydrocarbons not registered at 700 K.

This may be due to the relatively high concentrations of two hydrocarbons: 2,3- and 2,6-dimethylheptane. High relative concentrations of these alkanes are fairly obvious in Fig. 14. Structurally, both hydrocarbons belong to geochemical fossils preserving fragments of source-rock structures, which explains their relatively high concentrations. Characteristically, 2,3-dimethylalkanes are not found in high concentrations in all instances. Thus, besides 2,3-dimethylheptane, there is a fair amount of 2,3-dimethylpentane in petroleum (Fig. 15).

In the C_{10} alkane series 2,3-dimethyloctane coelutes with the structurally similar 2-methyl-3-ethylheptane, discovered in crudes by Mair et al. (1966). As a rule, this hydrocarbon is commonly present in high concentrations.

Table 11 gives the distribution of certain isomers of C_{10} alkanes characteristic of category A crudes.

The most pronounced regularity is the strong dominance of 2,6-dimethyloctane and 2-methyl-3-ethylheptane, which accounts for almost 75% of all disubstituted C_{10} alkanes in the crudes investigated. These figures in Table 11 clearly exceed equilibrium concentrations, which (for structures of similar substitution pattern) are shown in Table 12. (Note that 2-methyl-3-ethylheptane possesses a particularly low thermodynamic stability.)

Scheme 1

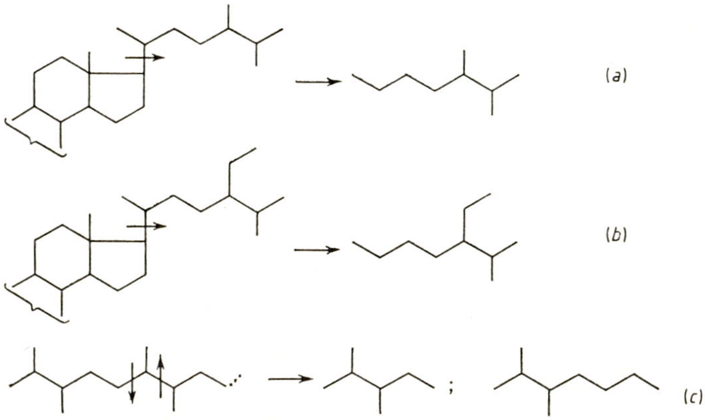

Undoubtedly, 2,6-dimethyloctane as well as 2,6-dimethylheptane are the first representatives of petroleum isoprenoid hydrocarbons, which will have to be thoroughly analyzed further. At the same time, high concentrations of 2-methyl-3-ethylheptane, as well as those of C_7 and C_9 2,3-dimethylalkanes, deserve special attention. In our opinion, these hydrocarbons may be generated in two ways (Scheme 1). The first possibility (a) or (b) is the loss of aliphatic chains from biological polycyclic hydrocarbons, i.e. steranes, or from their precursors, phytosteriols. It is known that steranes are found in large quantities in various crudes. The other possible way of 2,3-dimethylalkane formation (c) is the rupture of the aliphatic chain in botryococcane, a hydrocarbon also identified in crudes.

Mair, who found considerable amounts of 2-methyl-3-ethylheptane, suggested that limonene participates in the formation of this hydrocarbon (Mair 1966). High concentrations of these di-substituted alkanes again indicate the lack of equilibrium among isomers, even more so because kinetically these structures are rather reactive and could easily be transformed into more stable isomers with a similar degree of substitution (Petrov 1974).

Mono- and Dimethyl-Substituted $C_{11}-C_{15}$ Alkanes

The relatively simple composition of the methyl-substituted alkanes in group A^1 crudes allowed the qualitative and quantitative determination of this type of hydrocarbon in the higher boiling fractions as well. Krasavtchenko et al. (1971) reported on the determination of these hydrocarbons by gas chromatography on high resolution capillary columns. Methyl-substituted alkanes of higher molecular weight were determined by mass spectrometry (Hoeven et al. 1966). Typical chromatograms of the saturated hydrocarbons of a medium fraction are reproduced in Figs. 13 and 16. Data on the content and quantitative distribution of monomethylalkanes (Krasavtchenko et al. 1971) in crudes are presented in Table 13.

Despite certain relative and possible qualitative errors in determining individual methylalkane concentrations, the general order of figures is similar for

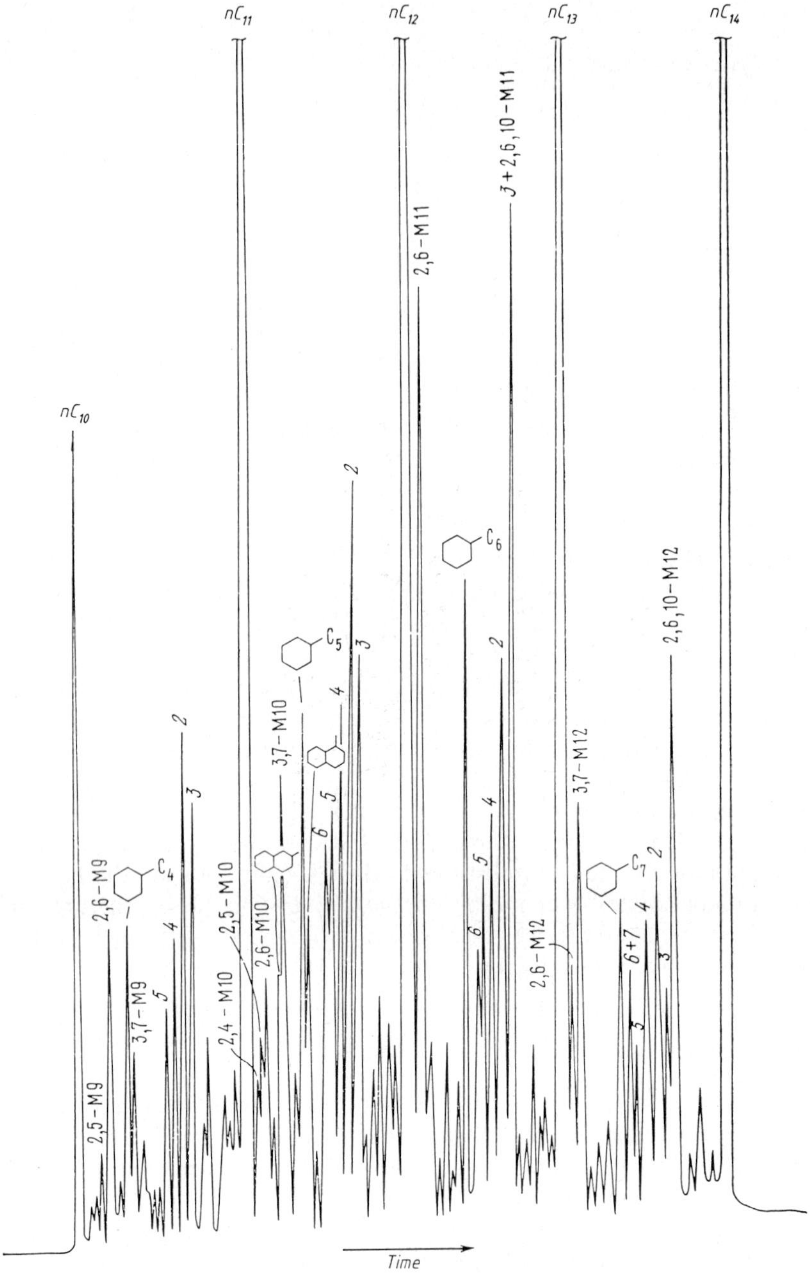

Fig. 16. Chromatogram of a saturated $C_{10}-C_{14}$ hydrocarbon mixture in Surgut crude. Peaks of the most important alkanes and certain cycloalkanes are reproduced. *Numbers* indicate the positions of the methyl substituents: 2,5-M9 : 2,5-dimethylnonane; 2,6,10-M12 : 2,6,10-trimethyldodecane, etc. Capillary column 50 m, squalane; linear temperature programming, 50°→1 °/min

Table 13. Content and relative distribution of monomethyl-substituted alkanes in crudes (%)

Isomer	C_{11}			C_{12}			C_{13}			C_{14}			C_{15}		
	I	II	III	I	II	III	I	II	III	I	II	III	I	II	III
2-Methyl-alkane	32	26	35	30	34	37	29	38	36	32	26	42	26	27	29
3-Methyl-alkane	20	30	21	22	17	16	22	29	25	30	22	17	27	38	23
4-Methyl-alkane	21	27	17	19	20	13	19	14	20	13	16	20	17	10	13
5-Methyl-alkane	27	17	27	17	16	20	17	10	10	13	10	9	20	11	27
6-Methyl-alkane	–	–	–	12	13	14	12	9	9	–	–	–	–	–	–
6- and 7-Methyl-alkanes	–	–	–	–	–	–	–	–	12	26	12	12	10	14	8
Σ Methyl-alkanes[a]	–	–	–	0.52	0.43	0.70	0.54	0.42	0.63	0.45	0.47	0.43	0.60	0.44	0.55
Methyl-alkanes (n-Alkanes)	0.60	0.58	–	0.72	0.60	0.66	0.59	0.51	0.78	0.47	0.59	0.53	0.64	0.63	0.51

Note: Crudes: *I* Surgut, *II* Romashkinskoe, *III* Groznenskoe (paraffinic).

[a] % petroleum.

hydrocarbons of different crudes. Methyl-substituted alkanes with symmetrical structures (5-methylnonane, 6-methylundecane) are present in considerably lower concentrations than other isomers. A similar situation was observed earlier in the case of 4-methylheptane and 3-methylpentane.

There also appears to be a general tendency to a decrease in the concentration of isomers with the shift of the methyl group towards the centre of the molecule. Thus, the average concentration ratio for 2-methylalkanes : 3-methylalkanes : 4-methylalkanes : 5-methylalkanes : 6-methylalkanes is 1 : 0.75 : 0.52 : 0.34 : 0.28 (symmetrical structures were not analyzed). The ratio between the overall content of monomethyl-substituted alkanes and the content of normal alkanes of different molecular weights is preserved throughout, although a certain tendency to decreasing values with decreasing molecular weight exists. The reported relationships among monomethylalkanes bear a certain analogy to the composition of an equilibrium mixture of these isomers. However, this situation is most likely due to the thermodynamic control of their formation from alk-1-enes and this will be thoroughly discussed in Chapter 5. Table 14 and Fig. 17 present data on the composition on equilibrium mixtures of $C_{13}-C_{16}$ monomethylalkanes. Chromotograms of the equilibrium mixtures of these hydrocarbons are given in Fig. 18.

It is much more complicated to establish the composition of dimethyl-substituted $C_{11}-C_{13}$ alkanes even in relatively uncomplicated paraffinic crudes. Basic

Table 14. Equilibrium concentrations of methyl-substituted $C_{13}-C_{16}$ hydrocarbons (%).

Isomer	300 K	700 K	Isomer	300 K	700 K
Tridecanes			*Pentadecanes*		
2-Methyldodecane	25	22	2-Methyltetradecane	27	18
3-Methyldodecane	22	23	3-Methyltetradecane	21	21
4-Methyldodecane	19	19	4-Methyltetradecane	13	17
5-Methyldodecane	20	19	5-Methyltetradecane	18	17
6-Methyldodecane	14	17	6-Methyltetradecane	11[b]	14
			7-Methyltetradecane	10[b]	13
2-Methyl / 3-Methyl	1.20	0.90	2-Methyl / 3-Methyl	1.27	0.82
Tetradecanes			*Hexadecanes*		
2-Methyltridecane	26	19	2-Methylpentadecane	23	15
3-Methyltridecane	19	22	3-Methylpentadecane	19	20
4-Methyltridecane	15	17	4-Methylpentadecane	14	14
5-Methyltridecane	20	19	5-Methylpentadecane	16	15
6-Methyltridecane	13[b]	15	6-Methylpentadecane	11[b]	14
7-Methyltridecane[a]	7[b]	8	7-Methylpentadecane	11[b]	14
			8-Methylpentadecane[a]	6[b]	8
2-Methyl / 3-Methyl	1.37	0.87	2-Methyl / 3-Methyl	1.28	0.75

[a] Structures with a symmetry plane. [b] Determined summarily. The concentration of individual isomers are estimated in analogy to C_{12} and C_{13} alkane isomers with similar structures.

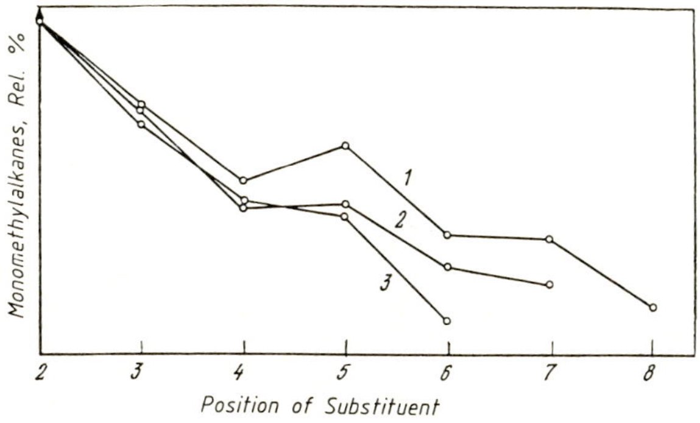

Fig. 17. Relative distribution of monomethylalkanes with different positions of the methyl substituent in equilibrium (300 K) mixtures of C_{12}, C_{15} and C_{16} hydrocarbons. Substituent position is marked on the abscissa. *1* C_{16}, *2* C_{15}, *3* C_{12}

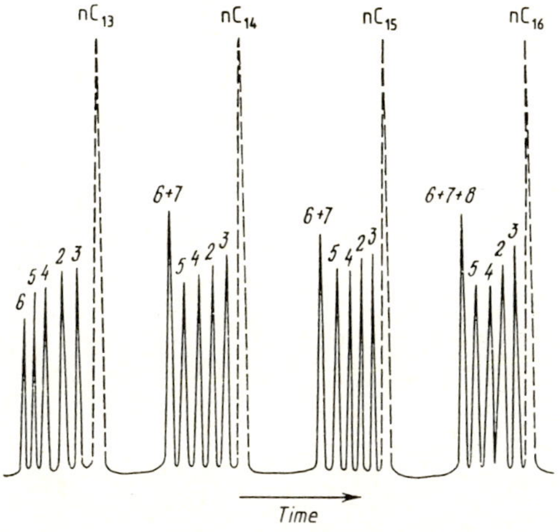

Fig. 18. Chromatogram of equilibrium (300 K) mixtures of $C_{13}-C_{16}$ monomethylalkanes. *Numbers* indicate the position of the methyl substituent. Capillary column 80 m, Apiezon; linear temperature programming, 100°→2°/min

Table 15. Equilibrium concentrations of methylundecanes (%)

Isomer	300 K	450 K	700 K
2-Methylundecane	32	28	25
3-Methylundecane	24	27	28
4-Methylundecane	16	17	19
5-Methylundecane	19	18	20
6-Methylundecane[a]	9	10	8
2-Methyl / 3-Methyl	1.33	1.04	0.72

[a] Structure with a symmetry plane.

compounds in this case are represented by isoprenoid-type hydrocarbons which complicate the investigation of other isomers due to their high concentrations.

In the case of dimethyl dodecanes, due to the low concentration of the isoprenoid 2,6-dimethyldecane, it was possible to establish the following relative isomer distribution for the Surgut crude; 2,4-dimethyl- (12%); 2,5-dimethyl-

Table 16. Equilibrium concentrations of dimethyldecanes

Isomer	Equilibrium concentration (%)	
	300 K	700 K
2,3-Dimethyldecane	4	4
2,4-Dimethyldecane	7	7
3,5-Dimethyldecane	3	4
2,5-Dimethyldecane	10	9
4,7-Dimethyldecane	3	4
2,6-Dimethyldecane	12	11
2,7-Dimethyldecane	11	10
2,8-Dimethyldecane	14	11
2,9-Dimethyldecane[a]	11	9
3,6-Dimethyldecane	6	8
3,7-Dimethyldecane	8	8
3,8-Dimethyldecane[b]	6	9
4,6-Dimethyldecane	5	6

[a] Structure with a symmetry plane.
[b] *Trans*-isomers have symmetry centres in 3,8-dimethyldecane as well as in 5,6 and 4,7-dimethyldecanes; *cis*-isomers are chiral, but have a C_2 symmetry axis.

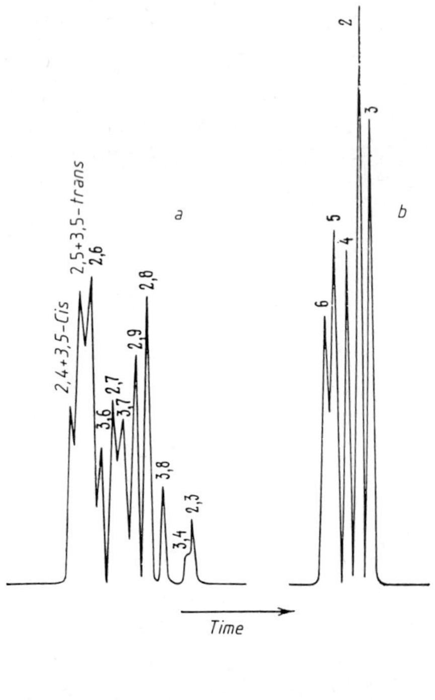

Fig. 19a, b. Chromatogram of equilibrium (300 K) mixtures of dodecanes. **a** Dimethyldecanes; **b** methylundecanes. *Numbers* indicate the position of the methyl substituents. Capillary column 50 m, squalane, 90 °C

(17%); 2,6-dimethyl- (20.6%); 3,6-dimethyl- (8.5%); 3,7-dimethyl- (24.2%, isoprenoid); 2,9-dimethyl- (9.2%); 2,8-dimethyl- (8.7%). The overall concentration of these dimethyldecanes amounted to 0.25% to the crude, i.e. almost one-half of the summed concentration of isomeric methylundecanes. Equilibrium concentrations of methylundecanes and dimethyldecanes are presented in Tables 15 and 16. Chromatograms of equilibrium mixtures are given in Fig. 19.

$C_{16}-C_{23}$ Methylalkanes (Iso- and Anteisoalkanes)

The determination of all C_{16} methylalkanes and higher is fairly complicated. Still a group of hydrocarbons of this structural type exists that is important for the chemistry of petroleum hydrocarbons, namely 2- and 3-methylalkanes. The material presented above may have already led to the conclusion that 2- and 3-methylalkanes usually differ from other monomethyl isomers by slightly higher concentrations. This fact provided a possibility for their determination in fractions with boiling temperatures exceeding 300 °C.

Thus, 2- and 3-methylalkanes play a certain role in proving the genetic connections between petroleum hydrocarbons and a number of natural compounds, in particular aliphatic iso- and anteiso-carboxylic acids. Organic geochemists denote these hydrocarbons as iso- or anteisoalkanes (Calvin 1971), in accordance to the nomenclature of similarly structured fatty acids. It should be noted that the carboxyl group of these branched acids occupies the opposite (unbranched) and of the aliphatic chain, which predetermines the naming of the hydrocarbons corresponding to them. For example, 8-methylnonanoic acid (isoacid) is transformed into 2-methyloctane (isoalkane) in the process of decarboxylation, while 7-methylnonanoic (anteisoacid) becomes 3-methyloctane (anteisoalkane) after decarboxylation. [The prefix ante ("before") means that the methyl group in anteisoacids is positioned one carbon atom closer to the carboxyl group than in isoacids].

Figure 13 presents a typical chromatogram of hydrocarbons in the 200°–430 °C fraction with peak identifications for to 2- and 3-methylalkanes.

Table 17 deals with the ratio between the overall content of these hydrocarbons and the content of isomeric normal alkanes for hydrocarbons of various molecu-

Table 17. Correlation of iso- and anteisoalkanes and normal alkanes in the Romashkinskoe crude

Hydrocarbon	2- and 3-Methylalkanes / n-Alkane	Hydrocarbon	2- and 3-Methylalkanes / n-Alkane
C_7	0.60	C_{15}	0.24
C_8	0.53	C_{16}	0.21
C_9	0.43	C_{17}	0.19
C_{10}	0.44	C_{18}	0.18
C_{11}	0.33	C_{19}	0.20
C_{12}	0.25	C_{20}	0.24
C_{13}	0.21	C_{21}	0.23
C_{14}	0.22	C_{22}	0.23

lar weight. It is obvious that this ratio is higher for low-molecular-weight alkanes, but then it gradually decreases with increasing number of carbon atoms in the molecule. Starting with $C_{12}-C_{13}$, it changes only insignificantly.

Still another hydrocarbon of this series, namely 7-methylheptadecane, deserves special consideration. The relative retention time of this alkane is depicted in Fig. 13. The absolute concentration of this compound in petroleums is relatively small, although it exceeds the amount of its 2-methylheptadecane isomer almost by a factor of two. It should be emphasized that these relations are reversed in the case of the nearest homologs. The reason for the relatively high concentration of 7-methylheptadecane is found in its relict character; this hydrocarbon occurs in appreciable concentrations in blue-green algae. It is assumed that 7-methylheptadecane exists as a mixture with its isomer 8-methylheptadecane, inseparable by GC methods. Details of the determination of these hydrocarbons in sedimentary organic material may be found in the monograph of Calvin (1971).

$C_{24}-C_{30}$ 12- and 13-Methylalkanes

Recently, a group of monomethyl-substituted alkanes was found in a number of East Siberian crudes of Pre-Cambrian deposits (Vend): $C_{24}-C_{30}$ 12- and 13-methylalkanes (Makushina et al. 1978). These relict compounds are of interest since they represent hydrocarbons of relatively high molecular weight in crudes. It is known that the higher the molecular mass of the relict, the more unique and characteristic is its structure, and the more valuable its inherited geochemical information.

The hydrocarbons under review form two homologous series: $C_{24}-C_{30}$ 12-methylalkanes and $C_{26}-C_{30}$ 13-methylalkanes. It is noteworthy that the hydrocarbon concentrations are identical in both series. Besides "true" 12- and 13-methylalkanes, obvious products of their destruction ($C_{19}-C_{24}$ hydrocarbons) were also found in crudes. Accordingly, the alkane series under examination may be represented by the following general formulae:

$$\underset{\text{I}}{C_{11}-\underset{\substack{|\\ C}}{C}-R' \; (R' = C_6-C_{17})} \qquad \underset{\text{II}}{C_{12}-\underset{\substack{|\\ C}}{C}-R'' \; (R'' = C_5-C_{16})} \; .$$

In gas chromatograms the hydrocarbons of series I and II elute at a retention time common for monomethylalkanes, having methyl substituents in the centre of the molecule. The composition and structures of these hydrocarbons were established by gas chromatography-mass spectrometry. Their amounts in crude oils range from 10% to 90% of the content of normal alkanes eluting within the respective intervals. A chromatogram of the $>200°$ saturated fraction, characteristic of the analyzed crudes, is presented in Fig. 20. This series of methylalkanes found in crudes is listed in Table 18.

The high content and the details of the concentrational distribution as well as the homology of methylalkanes identified in crudes leave no doubt as to their

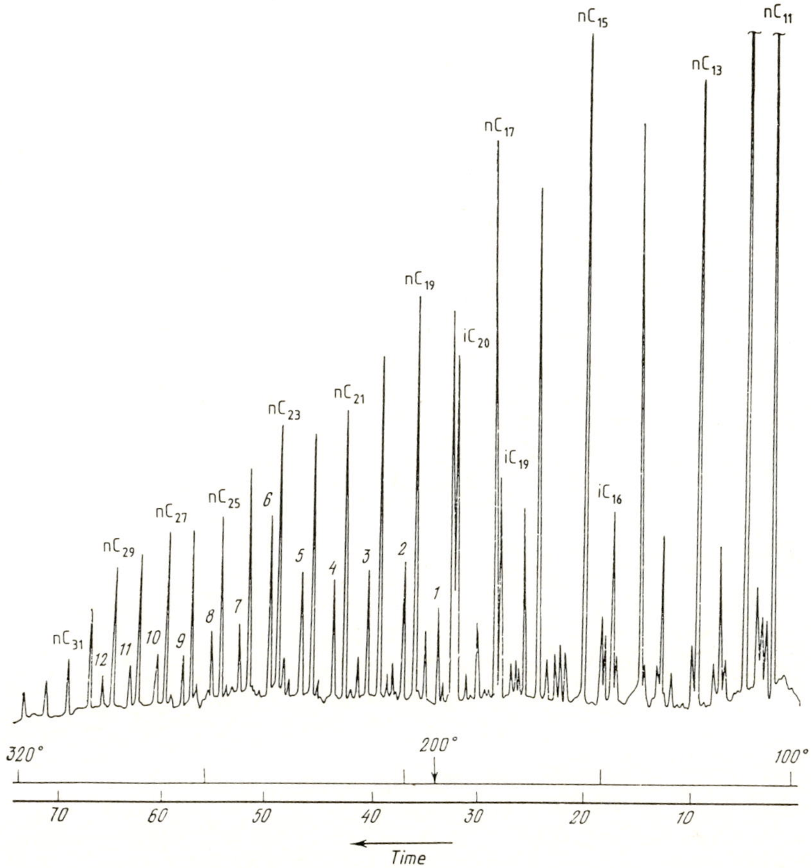

Fig. 20. Chromatogram of saturated hydrocarbons of Yaraktin crude. Normal and isoprenoid alkane peaks are identified, as well as 12- and 13-methylalkane peaks and their degradation products (*1–12*). For peak identifications, see Table 18. Capillary column 40 m, Apiezon; linear programming, 100°→2°/min

relict nature. Hydrocarbons similar to those identified were found, for example, in beeswax, where small quantities of C_{26}, C_{28}, and C_{30} monomethylalkanes were detected, and where the C_{28} alkanes are represented by a mixture of 9-methyl-, 11-methyl- and 13-methylheptacosanes (Stransky et al. 1966). However, high concentrations of 12- and 13-methylalkanes in crudes necessitate the search for other possible biological precursors.

The genesis of these hydrocarbons is most satisfactorily explained by the presence of a C_{25} cyclopropanecarboxylic acid, namely 12,13-methylenetetracosanoic acid, in source rocks. For possible primary transformations of the acid in sediments leading to the formation of the alkanes under examination see Scheme 2.

Scheme 2 is based on two reactions easily performed in the presence of aluminosilicates:

Table 18. Relict methyl-substituted $C_{19}-C_{30}$ alkanes in crudes

Carbon number	Peak number in Fig. 20	Series I	Series II
C_{19}	1	7-Methyloctadecane	6-Methyloctadecane
C_{20}	2	8-Methylnonadecane	7-Methylnonadecane
C_{21}	3	9-Methyleicosane	8-Methyleicosane
C_{22}	4	10-Methylheneicosane	9-Methylheneicosane
C_{23}	5	11-Methyldocosane	10-Methyldocosane
C_{24}	6	12-Methyltricosane[a]	11-Methyltricosane
C_{25}	7	12-Methyltetracosane	12-Methyltetracosane
C_{26}	8	12-Methylpentacosane	13-Methylpentacosane[a]
C_{27}	9	12-Methylhexacosane	13-Methylhexacosane
C_{28}	10	12-Methylheptacosane	13-Methylheptacosane
C_{29}	11	12-Methyloctacosane	13-Methyloctacosane
C_{30}	12	12-Methylnonacosane	13-Methylnonacosane

[a] Precursors of the corresponding homologous series of 12- and 13-methylalkanes. Hydrocarbons of lower molecular weight are regarded as destruction products of higher alkanes of the respective series.

Scheme 2

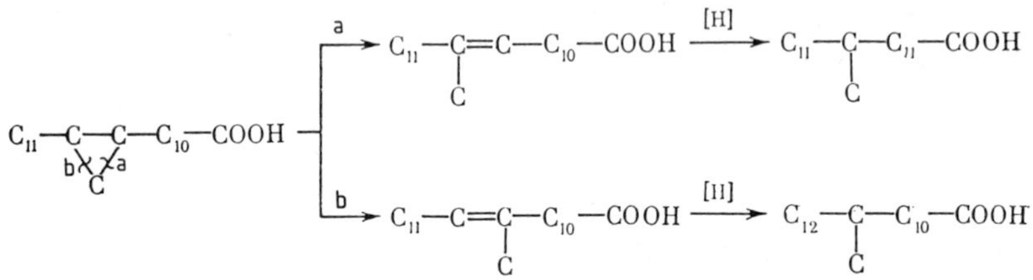

1. Equistatistical cleavage of the three-membered ring in directions a and b.
2. Saturation of the double bond through hydrogen redistribution (Petrov 1960).

Experimentally, using 12-methyltricosene-11 it was shown that radicals migration in the process of 12-methyltricosane formation does not occur as opposed to alk-1-enes (see below).

Further transformation of saturated acids into hydrocarbons of higher or lower carbon numbers proceeds according to common mechanisms of decarboxylation, ketonization and cleavage (see Chap. 5). Scheme 2 gives a plausible explanation for the equal concentrations of the homologous series I and II in crudes.

Unfortunately, the available literature so far fails to present information on the presence of the above mentioned acid in nature. At the same time, we possess numerous records of the presence of analogous acids with lower molecular weight in lipids of living microorganisms. Thus, 11,12-methyleneoctadecanoic acid, 9,10-methylenehexadecanoic and 9,10-methyleneoctadecanoic acids (Kreps 1981; Tocanne 1972) were found in lipids of different bacteria. The latter two acids were found in contemporary sediments in Britain (Granwell 1973). Apparently, these acids are methylenated homologs of unsaturated fatty acids widespread in nature.

For example, we should point out that 11,12-methyleneoctadecanoic acid (lactobacillum) is a possible source of the 7- and 8-methylalkanes found in practically all paraffinic crudes.

East Siberian crudes are unique in the quantitative content of 12- and 13-methylalkanes. Gas chromatograms of crudes from other Soviet petroleum provinces (see Table 1) show a peak corresponding to the C_{24} number of this hydrocarbon series in concentrations not exceeding 3 – 10% of the n-tetracosane concentration. This means that the actual concentration of these biological markers in other crudes is one order of magnitude lower than in East Siberian crudes, which complicates the determination of the other hydrocarbons in this series. It should be added, as was already pointed out, that crudes of Eastern Siberia (southern part of the Siberian platform) are also unique from the viewpoint of their geological age. Thus, reservoir ages vary here from Vend to lower Cambrian. Apparently, there is a connection between the age of the crudes, the specific source rock composition and the ensuing high content of relict methylalkanes. These hydrocarbons may serve as a characteristic qualitative indicator of the syngenetic character of crudes and bitumens, as well as the genetic homogeneity of crudes from a given petroleum area.

However, it should not be assumed that the described hydrocarbons are characteristic of all Pre-Cambrian oils. Thus, Pre-Cambrian crudes of the Volga-Urals area (the Perm district) and some other Pre-Cambrian petroleums do not contain high concentrations of the above hydrocarbons, although in certain instances, interesting and specific structures belonging to hydrocarbons of a different structure type were detected.

Isoprenoid Alkanes (Isoprenanes)

Beyond any doubt, the most important discovery in petroleum chemistry and organic geochemistry in the last 2 decades was the detection of a large number of aliphatic isoprenoid hydrocarbons in crudes, coals, shales and dispersed organic material. It was determined that the entire suite of sentiments is literally permeated with compounds of isoprenoid type, although previously only a considerable amount of aliphatic compounds with unbranched chains had been determined. These two basic structural types – the unbranched aliphatic chain and the isoprenoid unit – comprise the bulk of the original biological lipid material and hydrocarbon fossil fuels. It is difficult to estimate which of these structural types played the greater role in the formation of petroleum hydrocarbons. One fact is, however, clear, i.e. that the "variety" of isoprenoid compounds is incomparably larger, and the number of isoprenoid-type compounds discovered in crudes increases each year. The structures of these compounds are rather complicated and unusual. Therefore, isoprenoid hydrocarbons will be the focus of attention in the following chapters of our monograph.

Initial publications dealing with the presence of the isoprenoids pristane and phytane (2,6,10,14-tetramethylpentadecane and 2,6,10,14-tetramethylhexadecane) in petroleum go back to the early 1960's. Next, isoprenoid hydrocarbons were discovered in numerous crudes and other natural samples. The number of publications devoted to the determination of these compounds increases each year. There hardly seems to be any place in the sedimentary record where these compounds were not found. A compendium of earlier publications on isoprenoids may be found in the papers of Petrov (1974); Calvin (1971); Eglinton and Murphy (1966).

Due to their special structure, which is characteristic of a saturated polyisoprenoid chain, these compounds were named "biological markers". In effect, their distinctive structure and high concentration in different crudes convincingly support their biogenic origin.

Isoprenoid alkanes were first identified in Soviet crudes in 1969. At present, extensive analyses to determine these hydrocarbons in crudes have been undertaken, their results thus contributing to a better understanding of the geochemical conditions of the formation of petroleum deposits (Safonova 1980). If, in the beginning, only $C_{10}-C_{20}$ isoprenoid alkanes were identified, shortly afterwards higher, regular isoprenoids with 21 to 25 carbon numbers (Petrov et al. 1973; Han and Calvin 1969) were found, followed by isoprenoids ranging up to C_{40} (Albaiges 1979; Albaiges et al. 1978). Finally, in recent years, isoprenoid alkanes with irregular and pseudoregular structure types were identified (Chappe et al. 1979; Vorobyova et al. 1982).

Which branched alkanes may be included in the isoprenoid hydrocarbon series? Strictly speaking, the terminology is rather arbitrary in this case, since isoprenoid petroleum alkanes do not have to consist of complete polyisoprenoid units. These hydrocarbons, as typical relicts [biomarkers], are characterized by homology and obvious "inequilibrium". A criterion used to attribute alkanes to the isoprenoid series consists in the regular alternation of methyl groups. As always, homology is the result of destruction of higher molecular weight compounds. However, contrary to unbranched alkanes, there are always concentration "gaps" of certain homologs in isoprenoid series. These gaps (absence or small relative concentrations) of certain homologs occur when the chain is broken next to places where methyl substituents are positioned. Such a peculiarity is of cardinal importance in determining the sources of the formation of individual isoprenoid alkanes. It is the lack of homologs that sometimes offers information of utmost value.

Regular and Irregular Isoprenoid Alkanes. Currently, there are two main sources of isoprenoid alkane regular and irregular structures. It is obvious that destruction of molecular chains of regular isoprenoids, as in the case of phytol (C_{20}) and solanesane (C_{45}), which have branches at every fourth carbon atom in the chain, may only result in the formation of regular isoprenoid alkanes (Scheme 3). Here, and in following sections we examine only the possibility of breaking just one C-C bond.

Irregular structures, such as crocetane, squalane and lycopane may produce two new types of isoprenoid alkanes. If the cleavage occurs in the irregular link, then so-called pseudoregular structures may emerge, i.e. structures, which have a regular arrangement of methyl substituents, but are impossible to be derived from ordinary regular sources (for example, C_{17} and C_{19} 2,6,10-trimethylalkanes). However, when an irregular isoprenoid chain is broken away from the irregular

Scheme 3

Regular structure:

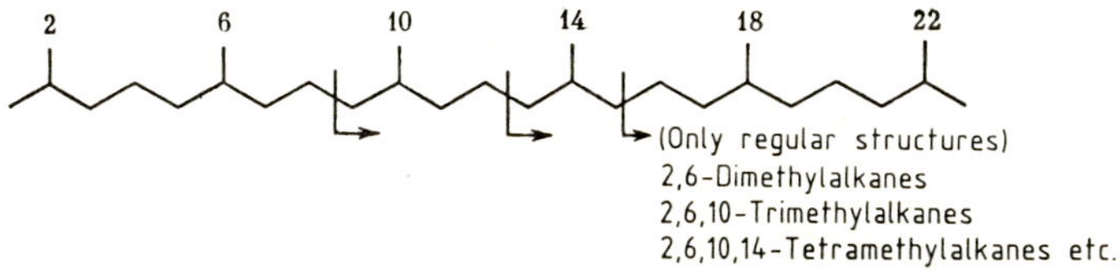

Scheme 4

Squalane (2,6,10,15,19,23-Hexamethyltetracosane)

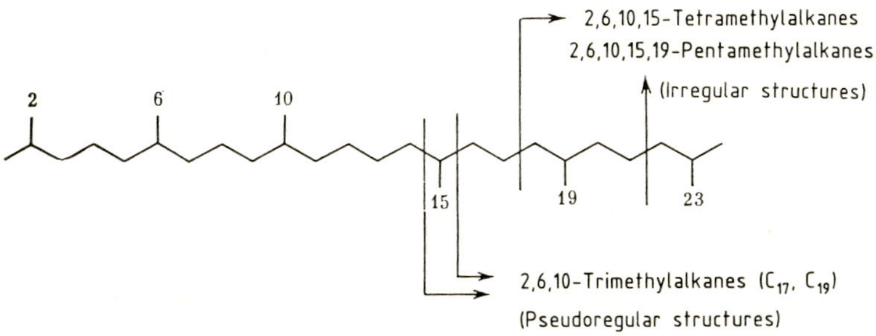

Scheme 5

Lycopane (2,6,10,14,19,23,27,31-octamethyldotriacontane)

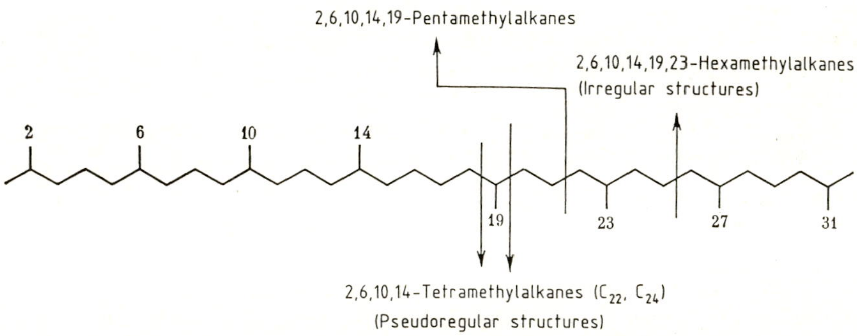

link, the latter is naturally preserved, and the emerging isoprenoid alkane will typically be of the irregular type. Regular isoprenoids are composed exclusively by "head-to-tail" links of isoprene units, whereas irregular isoprenoids have a "tail-to-tail" link. (Later, we will provide examples of irregular isoprenoids found in crudes, having a head-to-head link.) Schemes 4 and 5 deal with the formation of various pseudoregular and irregular isoprenoids. Certain regular, mostly low-molecular-weight structures, may be produced from irregular precursors. These are identical to structures emerging from regular sources.

Let us begin with the analysis of the distribution of classic, regular C_9-C_{25} isoprenoids in petroleums supported by a wealth of factual material. The following compounds are included: 2,6-dimethylalkanes, C_9-C_{14}; 2,6,10-trimethylal-

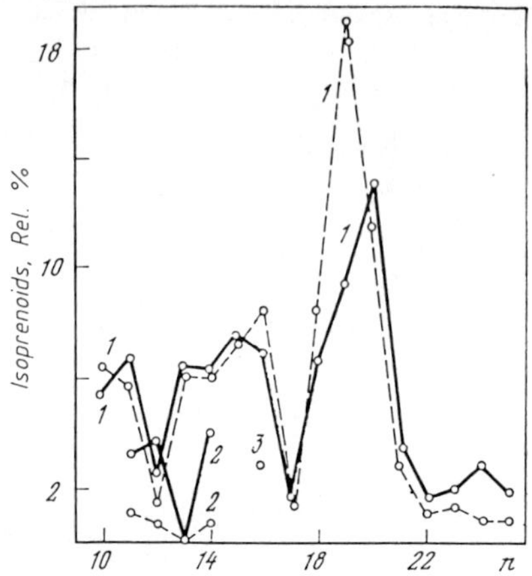

Fig. 21. Relative distribution of C_9-C_{25} isoprenoid alkanes in type A[1] crudes; ——— Surgut crude; – – – Groznenskoe paraffinic crude. *1* 2,6-Dimethyl-2,6,10-trimethyl-2,6,10,14-tetramethylalkanes; *2* 3,7-dimethylalkanes; *3* 3,7,11-trimethyltridecane; *n* denotes the number of carbon atoms per molecule

kanes, $C_{14}-C_{19}$; 2,6,10,14-tetramethylalkanes, $C_{19}-C_{24}$ and 2,6,10,14,18-pentamethyleicosane. Concentrations of all the above hydrocarbons in crudes are far from being identical. (As can be easily observed, pseudoregular structures with usually low concentrations are included.)

The best method of determining isoprenoid hydrocarbons is either GC with linear temperature programming using high resolution capillary columns or gas chromatography-mass spectrometry. Satisfactory results may also be obtained with a prior concentration of isoprenoid alkanes through clathrate formation with thiourea. Isoprenoid petroleum alkanes vary widely in molecular weight and, hence, belong to fractions of different distillation temperatures. The lowest molecular weight petroleum isoprenoid (2,6-dimethylheptane) has a boiling point of 135 °C, while that of the highest boiling isoprenoid (lycopane) is 496 °C.

Category A and B crudes differ to a certain extent in content and relative distribution of isoprenoid alkanes. Pristane and phytane usually prevail in type A[1] crudes (the correlation of these hydrocarbons is an important genetic indicator and depends on several factors, which will be analyzed later); other C_9-C_{25} isoprenoids are also present in noticeable concentrations. Figure 21 presents typical isoprenoid distribution curves in type A[1] crudes (pristane dominates in one of them; phytane, in the other). Also clearly discernable are the concentrations "gaps" of the C_{12} and C_{17} pseudoregular compounds. This is the case in the so-called classical type of isoprenoid alkane distributions in crudes. Data on the relative concentrations of isoprenoid alkanes in category A petroleums are also presented in Table 19.

It is characteristic that with the exception of phytane, a predominance of isoprenoid alkanes with carbon numbers being a multiple of five is not observed. The same applies to the majority of other isoprenoid hydrocarbons.

As a rule, a high count of normal alkanes in medium A[1] type is connected with a high isoprenoid alkane content. The concentration of the latter reaches 3–4% by volume, calculated for petroleum, occasionally even more. For example, the crude oil from the Azevo-Salaushskoe field (Tataria) contains 9% isoprenoids, including 2.4% pristane and 3.3% phytane (Safonova 1980). Highly paraffinic

Table 19. Relative distribution of isoprenoid hydrocarbons in crudes (%)

Hydrocarbons	Number of carbon atoms	Type A[1] crude			Type A[2] crude (Starogroznenskoe)
		Surgut	Romashkinskoe	Groznenskoe Paraffinic	
2,6-Dimethylheptane	9	3.80	3.66	5.74	–
2,6-Dimethyloctane	10	4.26	4.14	6.44	–
2,6-Dimethylnonane	11	6.78	5.75	5.74	–
3,7-Dimethylnonane	11	3.26	1.30	1.19	–
2,6-Dimethyldecane	12[a]	1.44	1.62	1.43	–
3,7-Dimethyldecane	12	3.80	1.04	0.71	–
2,6-Dimethylundecane	13	6.49	6.80	6.20	–
2,6,10-Trimethylundecane	14	6.25	6.55	6.20	10.3
2,6-Dimethyldodecane	14[b]	Trace	Trace	Trace	–
3,7-Dimethyldodecane	14	4.06	2.62	0.71	–
2,6,10-Trimethyldodecane (Farnesane)	15	7.58	10.40	7.40	11.2
2,6,10-Trimethyltridecane	16	7.05	11.70	8.60	13.8
3,7,11-Trimethyltridecane	16	1.90	1.83	1.43	–
2,6,10-Trimethyltetradecane	17[a]	1.63	2.35	1.43	–
2,6,10-Trimethylpentadecane	18	6.78	8.10	8.60	12.2
2,6,10,14-Tetramethylpentadecane (Pristane)	19	9.48	9.40	19.10	24.2
2,6,10-Trimethylhexadecane	19[b]	Trace	Trace	Trace	–
2,6,10,14-Tetramethylhexadecane (Phytane)	20	13.25	12.00	11.70	15.7
2,6,10,14-Tetramethylheptadecane	21	3.52	4.45	2.86	2.2
2,6,10,14-Tetramethyloctadecane	22[b]	1.63	1.04	1.19	2.2
2,6,10,14-Tetramethylnonadecane	23	2.03	1.30	1.43	2.9
2,6,10,14,18-Pentamethylnonadecane	24	2.98	1.57	0.95	2.2
2,6,10,14-Tetramethyleicosane	24[b]	Not found			–
2,6,10,14,18-Pentamethyleicosane	25	2.03	2.35	0.95	2.2
Σ, % calculated for petroleum	–	3.69	3.83	4.19	3.72

[a] Formation of these structures from phytane and phytanic acid is unlikely.
[b] Compounds formed only from isoprenoids with irregular structures.

crudes, i.e. crudes with high contents of C_{20} and higher normal alkanes are an exception, because the abundance of isoprenoids in their case is usually insignificant.

A different situation occurs in the case of isoprenoids in category B crudes. Some of these crudes contain no isoprenoids (type B[1] crudes); some contain isoprenoids in ratios differing from those in category A crudes. Category B crudes also exist with isoprenoid distributions resembling those of category A crudes.

A peculiar and unusual distribution of isoprenoid alkanes was registered in the crudes from the Anastisievsko-Troyitskoe field (horizon IV), produced from Miocene reservoirs. Of the total isoprenoid content of 2.5%, calculated for petroleum, 24.6% belongs to 2,6,10-trimethylundecane; 28.5% to 2,6,10-trimethyldodecane; and 40.5% to 2,6,10-trimethyltridecane. At the same time, the concentration of pristane and phytane was insignificant and amounted to only 5.6% of the total isoprenoid alkanes.

Let us now analyze the regularities in isoprenoid distributions in type A^1 crudes (see Fig. 21). Already in early papers devoted to the determination of isoprenoid compounds in fossil fuels, it was suggested that their main source of formation was phytol, an unsaturated aliphatic alcohol, which is known to be part of the plant chlorophyll. In fact, the isoprenoid hydrocarbon distribution presented in Fig. 21 convincingly support this assumption.

Thermal or thermocatalytic transformation of phytol (for experimental investigation of this reaction, see Chap. 5) may lead to the production of the entire range of C_9-C_{20} isoprenoids (Scheme 6), with the exception of the C_{12} and C_{17} isoprenoids, the formation of which is theoretically possible only through a much less probable rupture of two C-C bonds.

2-Methylheptane may also be partially included into the isoprenoid hydrocarbone series, although the bulk of this hydrocarbon was probably produced differently. The commonly observed predominance of 2-methylheptane (relative to 3-methylheptane) in paraffinic crudes and close correlations of this compound to the nearest homology may, however, be explained by the reproduced scheme of phytol degradation (see Scheme 6).

A significantly smaller amount of isoprenoid hydrocarbons is formed by the cleavage of phytol C-C bonds away from the functional group. This is no surprise, because it is more difficult to obtain saturated aliphatic hydrocarbons without structural rearrangement from that part of phytol containing the double bond and the hydroxyl group (Scheme 7).

Scheme 6

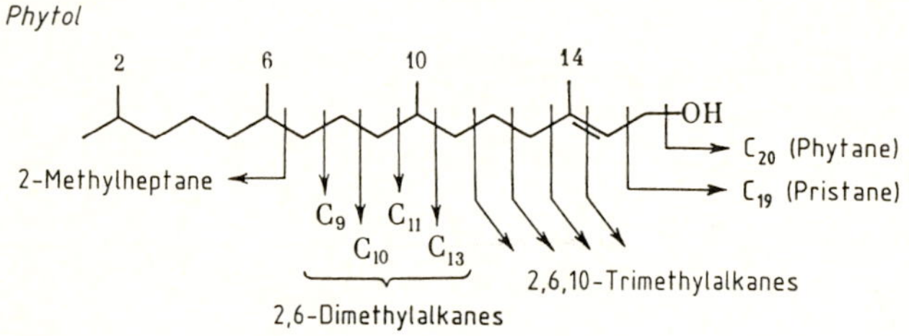

Scheme 7

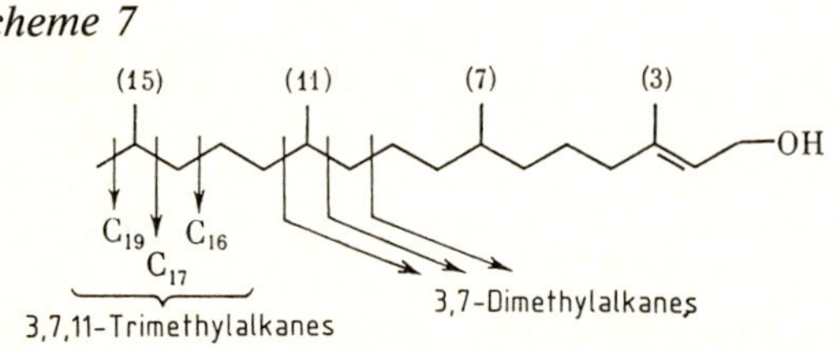

However, 3,7-dimethyl- and 3,7,11-trimethylalkanes may be formed as a result of the degradation of phytane dimers with a head-to-head link (see below).

It should not be assumed that initially a phytane molecule is formed, which undergoes further degradation, although this procedure may be partially involved; this process is actually more complex.

Stereochemistry of Isoprenoid Alkanes. An important indication of the formation of isoprenoid hydrocarbons from phytol could be found in the stereochemical relationship between phytol and the corresponding $C_{15}-C_{20}$ alkanes with chiral centres at C-6 and C-10. It is known that biosynthesized phytol possesses a rigidly defined R-configuration at the chiral centres C-7 and C-11 (which correspond to centres C-10 and C-6 in isoprenoid alkanes).

With respect to the spatial arrangement of the methyl substituents at C-7 and C-11, a phytol with the biogenic configuration (Petrov 1981) is a *cis*-isomer. If farnesane (2,6,10-trimethyldodecane), pristane and phytane are formed from phytol without configurational changes at C-7 and C-11, then C-6 and C-11 atoms in the alkanes should have the same absolute configuration, i.e. the same relative arrangement of methyl radicals as the biogenic configuration. Scheme 8 gives a graphic presentation of the anticipated transformation products (only one enantiomer is reproduced for *dl*-pairs).

Scheme 8

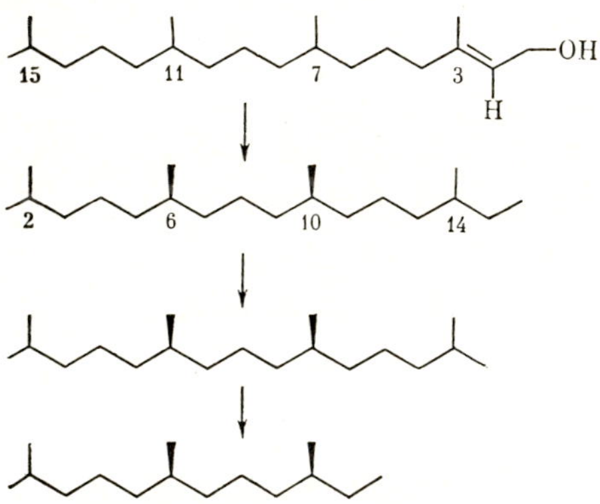

Phytol: 3,7*R*,11*R*,15-Tetramethylhexadec-2-enol

6*R*, 10*S*-Phytane

Meso-pristane (6*R*,10*S*)

6*R*, 10*S*-Farnesane

Extensive experimental research reveals (Maxwell et al. 1971; Patience et al. 1978; Cox et al. 1972; Brooks et al. 1975; Patience et al. 1979) that, unfortunately, epimerization of chiral centres in petroleums happens fairly fast, and pristane is an equilibrium mixture of *dl*- (I) and *meso*-forms (II) of approximately equal concentrations:

I *dl*-(*trans*-) II *meso*-(*cis*-)

Farnesane (two *dl*-pairs) has an analogous stereochemical composition. And it is only in petroleum phytane[3] (four *dl*-pairs) that a slight predominance of stereoisomers having a 6R configuration was registered. Let us note that C-6 is the farthes removed from the molecule's labile part.

At the same time, up to 80% pristane is found in young shales (Maxwell et al. 1971; Patience et al. 1978), and pristane of immature organic material consists of the *meso*-form inherited from biological synthesis. However, phytol, although being most important, is not the only source of petroleum isoprenoid alkanes. A certain amount of these compounds may originate from isoprenoid acids (farnesanic, phytanic acid and others). A considerable number of these acids was found in different crudes (Eglinton and Murphy 1969; Blumer and Snyder 1965; Seifert 1975). The formation of isoprenoid alkanes from acids is obviously identical to the process of fatty acid transformation into normal alkanes. These reactions are analyzed in Chap. 5.

The discovery of isoprenoid hydrocarbons $>C_{20}$ created considerable interest. The relative concentrations of the C_{21} isoprenoid (see Table 19) was especially high. According to earlier publications, C_{21} and higher isoprenoid alkanes identified in crudes usually belonged to the regular isoprenoid type. The problem of the sources of these hydrocarbons are profoundly treated in the monograph of Calvin (1971). We should add that sometimes $C_{21}-C_{25}$ isoprenoids are called sesterterpanes (Albaiger et al. 1978).

Interesting examples of possible precursors of high-molecular-weight isoprenoids of regular structures (III–VI) are also presented in the publication of Ten Fu Yen and Chilingarian (1976):

Tocopherol

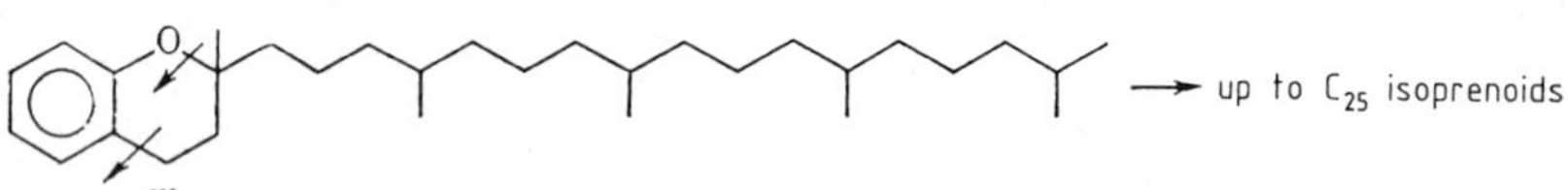

III

Vitamin K_2

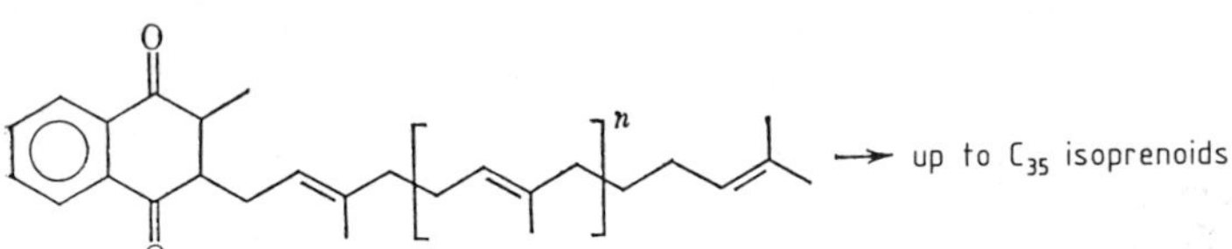

IV $n = 5$

Ubiquinone

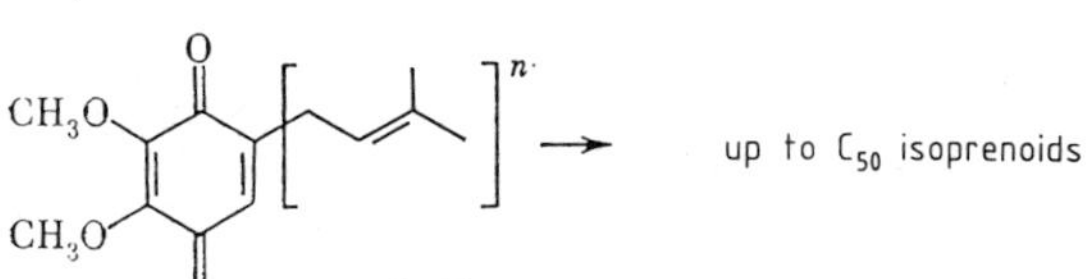

V $n = 6-10$

---

[3] The configuration of the C–7 (C–10) atom remains unchanged in the transformation from phytol to this phytane stereoisomer. Letters R and S in this instance reflect the Cahn-Ingold-Prelog nomenclature convention.

Plastoquinone

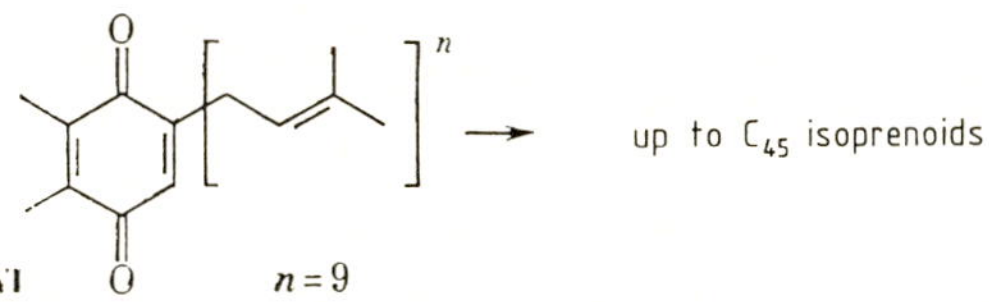

VI

However, in recent years crudes were discovered which had relatively high concentrations of higher isoprenoids as well as isoprenoids of irregular and pseudoregular structure (Vorobyova 1982). These works widened significantly the sphere of possible sources of isoprenoid alkanes. Apparently, squalane and lycopane should be included with these sources.

Figure 22a presents a gas chromatogram of an isoprenoid alkane concentrate (200°–500 °C fraction) isolated by thiourea adduction from a petroleum deposit of Karajanbas (Buzachi peninsular, northeastern Near-Caspian Lowlands). A type B^2 crude occurs in Jurassic sediments at 800-m depth. The crude is characterized by a high concentration of branched alkanes which, according to mass spectrometric analysis, reaches 32%. The chromatogram shown is analogous to a reconstructed mass fragmentogram of ion m/z 71. Such a recording allows one to obtain a simpler chromatogram of a petroleum fraction, i.e. a chromatogram represented in this case entirely by branched alkane peaks and, in effect, exclusively by isoprenoid hydrocarbons.

As can be observed from the figure, regular (R) $C_{11}-C_{40}$ isoprenoids as well as C_{17}, C_{19}, C_{22}, C_{24} pseudoregular (P) and $C_{21}-C_{40}$ irregular (I) structures, including squalane and lycopane, were determined in the fraction analyzed. Of special significance are the relatively high concentrations of squalane and lycopane, as well as C_{17}, C_{19}, C_{21} and C_{24} pseudoregular isoprenoids. Schemes of the possible formation of these hydrocarbon from squalane and lycopane were presented above. At the same time, small concentrations of C_{12} and C_{14} 2,6-dimethylalkanes suggest that an irregular isoprenoid like crocetane (2,6,11,15-tetramethylhexadecane) does not participate in the formation of petroleum isoprenoids in any significant manner. Irregular structures found in the Karajanbas crude have 20 and more carbon atoms. The formation of these hydrocarbons is possible from squalane ($C_{20}-C_{24}$ 2,6,10,15-tetramethylalkanes and $C_{26}-C_{29}$ 2,6,10,15,19-pentamethylalkanes), as well lycopane ($C_{26}-C_{29}$ 2,6,10,14,19-pentamethylalkanes and $C_{30}-C_{39}$ 2,6,10,14,19,23-hexamethylalkanes). Many of these components were discovered in the Karajanbos crude. At the same time, low-boiling, irregular isoprenoids, such as 2,6,11-trimethylalkanes, were not registered, which again excludes crocetane as a possible precursor of petroleum isoprenoid alkanes. Let us emphasize again that throughout the analysis we only considered variations in the cleavage of only a single C-C bond in precursor molecules.

In order to determine sources of any irregular isoprenoid alkanes, we present gas chromatograms of the products of the thermal decomposition of squalane (Fig. 22c) and lycopane (Fig. 22b), which visually underline the peaks of evolving hydrocarbons. Incidentally, this method can be applied for producing standards to be used in GC analysis of petroleum mixtures.

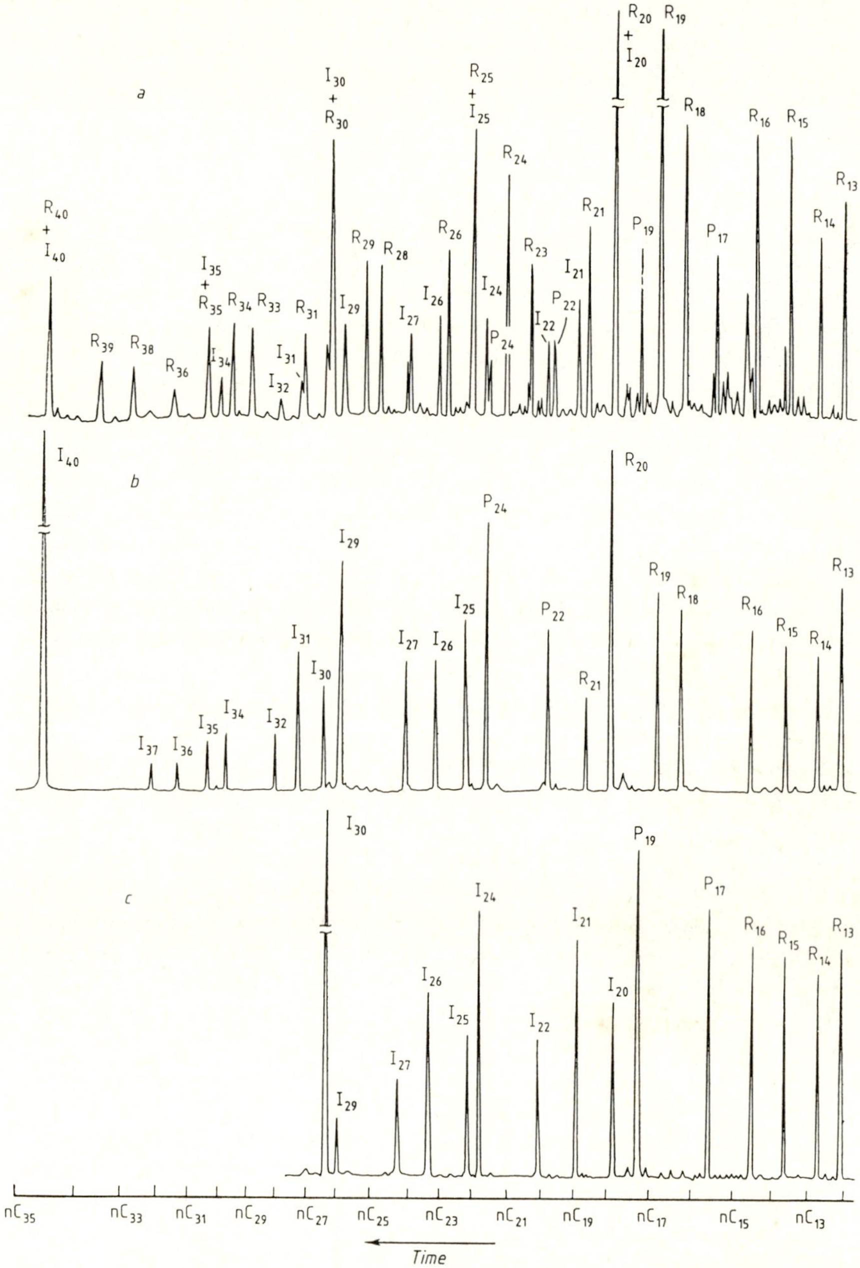

Fig. 22a–c. Chromatogram of isoprenoid alkane mixtures. **a** Isolated from Karajanbas crude; **b** produced by thermal decomposition of lycopane; **c** produced by thermal decomposition of squalane; *R* regular isoprenoids; *P* pseudo regular isoprenoids; *I* irregular isoprenoids. *Digits* indicate the number of carbon atoms per molecule (hydrocarbon structures are discussed in the text). Points of elution of normal $C_{13}-C_{35}$ alkanes are identified on the abscissa. Capillary column 80 m, Apiezon; linear temperature programming, 100°→2°/min

Thermal decomposition is performed by heating squalane or lycopane in closed vessels at 400 °C for 2 h. The extent of the transformation of the original hydrocarbons does not exceed 5 – 10%, however, in the case of a moderate extent of transformation, the reaction products are usually free of unsaturated hydrocarbons or secondary products of transformation, which interfere with the analysis.

Peculiar isoprenoids (as well as other branched alkanes) were reviewed in the publications of Chappe et al. (1979); Michaelis and Albrecht (1979); Moldowan and Seifert (1979). These papers describe isoprenoid structures having head-to-head linkages. Of special interest is diphytyl (or bisphytyl), a C_{40} hydrocarbon (VII):

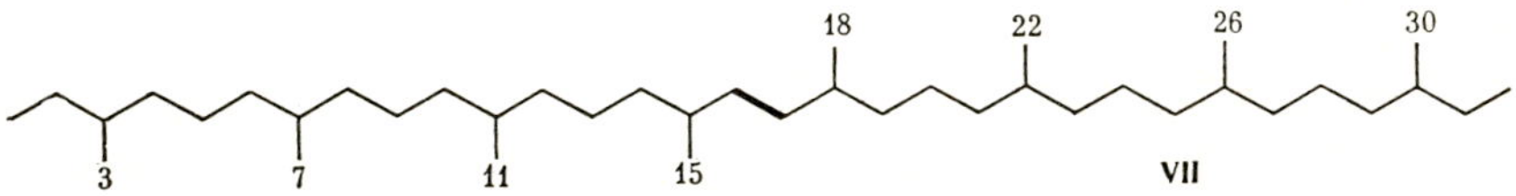

VII

This hydrocarbon may also be a precursor of 3,7-dimethylalkanes and 3,7,11-trimethylalkanes discovered earlier in crudes. Besides diphytyl, 13,16 dimethyloctacosane was found, most probably a *trans (meso)*-isomer, which is the "dimerization" product of 2-methyltetradecane. It is assumed that both hydrocarbons were produced in the course of bacterial elimination of hydrogen from methyl groups positioned at the front end of the phytane chain or of 2-methyltetradecane, with consequent formation of a C-C bond (accentuated by the thick line in the formula). The hydrocarbons under review are part of membrane lipids of Archaebacteria, comprising a considerable portion of kerogen that serves as a possible source for petroleum formation.

Several representatives of $C_{35} - C_{40}$ isoprenoid alkanes having head-to-head links are also described in the paper of Moldowan und Seifert (1979). Besides diphytyl, a series of isoprenoids formed by the combination of different isoprenoid alkanes of lower molecular weight was found. Thus, $C_{20} - C_{20}$, $C_{19} - C_{20}$, $C_{19} - C_{19}$, $C_{18} - C_{20}$ and $C_{18} - C_{19}$, i.e. C_{40}, C_{39}, etc. isoprenoids were identified in Californian crude oil (Miocene). The overall concentration of the above compounds in petroleum was ≈ 0.1%. In all cases head-to-head polymerization is believed to have occurred by elimination of hydrogen atoms in methyl groups, located at the first carbon atom in the original "monomers" (Scheme 9).

Scheme 9

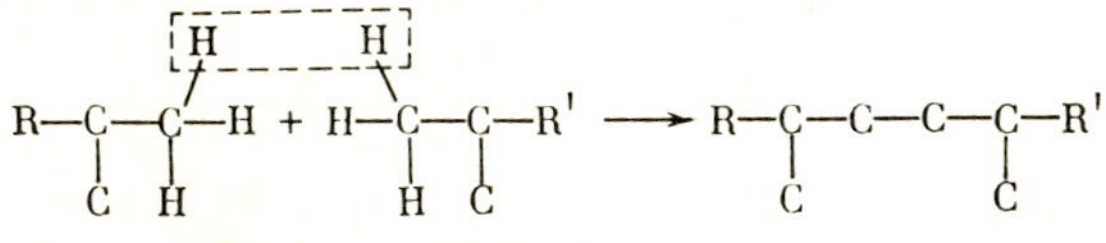

There are reasons to believe that this process is microbially mediated in nature, because, as was already stated, the resultant hydrocarbons are included in the lipid part of Archae bacterial membranes (although the process is much more complicated in real terms).

Let us also analyze an interesting hydrocarbon (botryococcane; 2,3,6,7,10,13,16,17,20,21-decamethyl-13-ethyldocosane, VIII), comprising 1.4%

in one Indonesian crude (Moldovan and Seifert 1980). This hydrocarbon originates from the unsaturated botryococcene (IX), which is a constituent of certain algae:

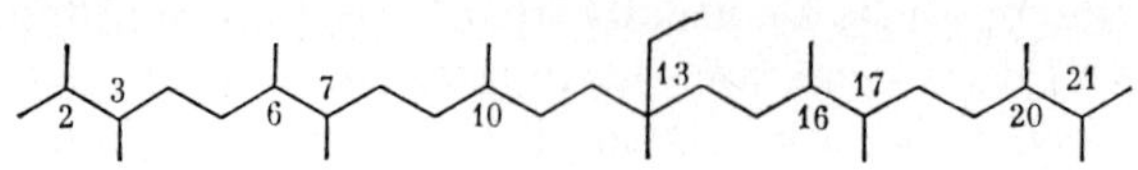

VIII

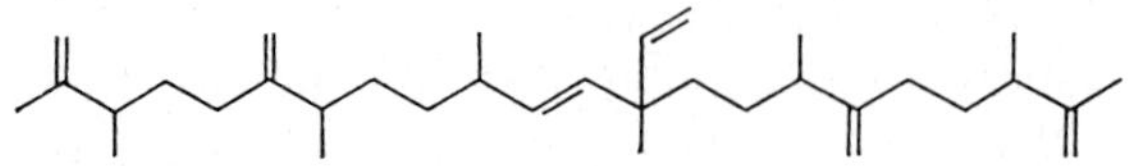

IX

Botryococcane may be regarded as an additional source of C_7 and C_9 2,3-dimethylalkanes.

Botryococcane is not the only representative of isoprenoid alkanes having a geminal methylethyl grouping. Isolycopane, (X) with an analogous structural element, was identified in a number of crudes (Vorobyovra et al. 1985):

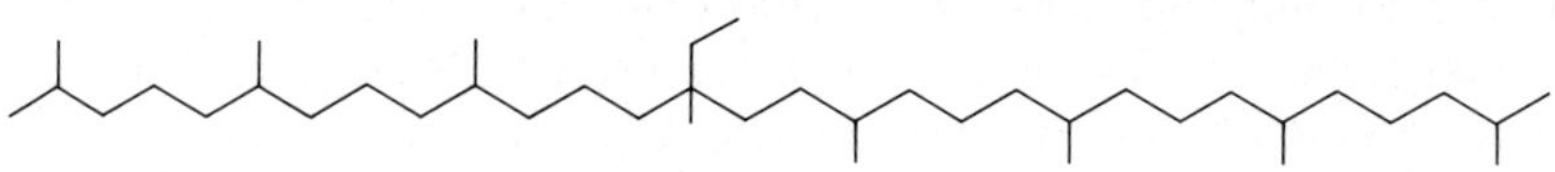

X

Isolycopane is most probably produced by the dimerization of phytene according to a carbonium-ionic mechanism:

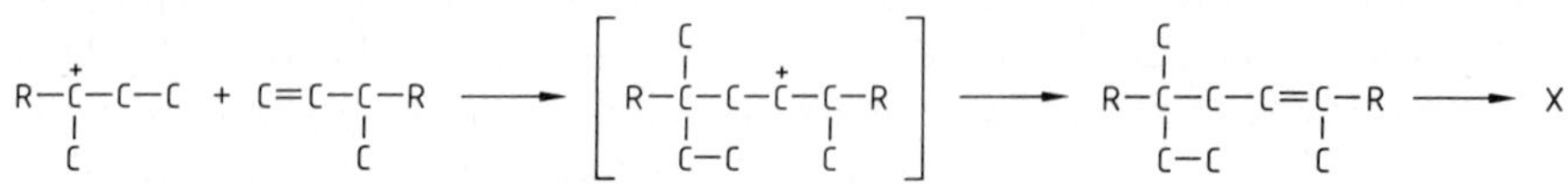

In the paper of Yon et al. (1982), a peculiar C_{20} isoprenoid alkane (XI) with a T-shape configuration, was described as occurring on petroleum:

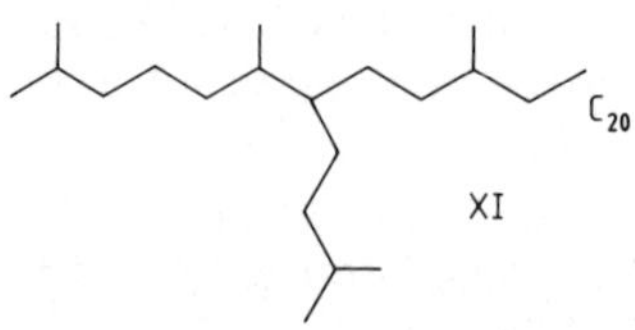

XI

Its C_{25} homolog (XII) was also identified in petroleum and has the following tentative structure (Rowland et al. 1985):

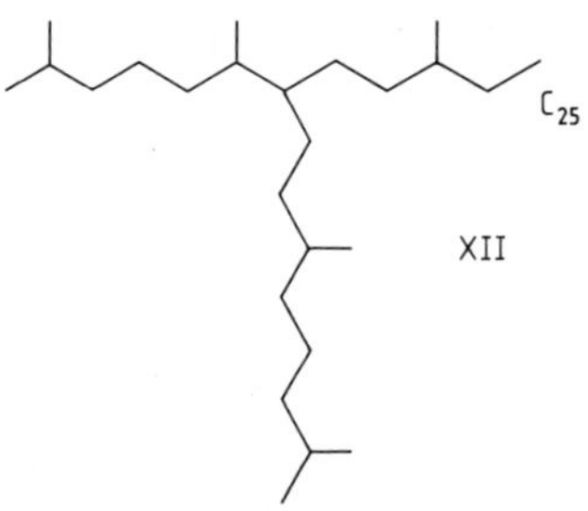

XII

Geochemical Significance of Isoprenoid Alkanes

The geochemical significance of isoprenoid alkanes lies mainly in their application to the genetic correlation between crudes, as well as between crudes and

Table 20. Retention indices of alkanes

Hydrocarbon	Retention index	Hydrocarbon	Retention index
Stationary phase squalane		2-Methylnonane	964
2,2-Dimethylbutane	530	3-Methylnonane	970
2,3-Dimethylbutane	558	2,4-Dimethylnonane	1012
2-Methylpentane	560	2,5-Dimethylnonane	1015
3-Methylpentane	580	2,6-Dimethylnonane	1020
2,2-Dimethylpentane	622	2,8-Dimethylnonane	1028
2,4-Dimethylpentane	624	2,7-Dimethylnonane	1033
2,2,3-Trimethylbutane	632	3,7-Dimethylnonane	1036
3,3-Dimethylpentane	654	5-Methyldecane	1054
2-Methylhexane	658	4-Methyldecane	1059
2,3-Dimethylpentane	667	2-Methyldecane	1063
3-Methylhexane	673	3-Methyldecane	1070
3-Ethylpentane	687	2,4-Dimethyldecane	1106
2,2-Dimethylhexane	716	2,5-Dimethyldecane	1112
2,4-Dimethylhexane	724	2,6-Dimethyldecane	1112
2,5-Dimethylhexane	728	2,7-Dimethyldecane	1121
2,2,3-Trimethylpentane	734	3,7-Dimethyldecane	1123
3,3-Dimethylhexane	740	2,9-Dimethyldecane	1126
2,3,4-Trimethylpentane	744	2,8-Dimethyldecane	1132
2,3-Dimethylhexane	756	3,8-Dimethyldecane	1138
2-Methyl-3-ethylpentane	759	6-Methylundecane	1151
2-Methylheptane	762	5-Methylundecane	1155
4-Methylheptane	764	4-Methylundecane	1159
3,4-Dimethylhexane	766	2-Methylundecane	1165
3-Methylheptane	769	3-Methylundecane	1170
3-Ethylhexane	769	Stationary phase Apiezon	
2,2,5-Trimethylhexane	773	2,6-Dimethylundecane	1206
2,3,5-Trimethylhexane	808	3,7-Dimethylundecane	1222
2,2-Dimethylheptane	814	6-Methyldodecane	1247
2,4-Dimethylheptane	818	5-Methyldodecane	1250
2-Methyl-4-ethylhexane	822	4-Methyldodecane	1255
2,6-Dimethylheptane	824	2-Methyldodecane	1263
2,5-Dimethylheptane	830	2,6,10-Trimethylundecane	1263
3,5-Dimethylheptane	831	3-Methylundecane	1269
3,3-Dimethylheptane	836	2,6-Dimethyldodecane	1308
2,3-Dimethylheptane	853	3,7-Dimethyldodecane	1310
4-Ethylheptane	858	6- and 7-Methyltridecane	1342
3,4-Dimethylheptane	858	5-Methyltridecane	1348
4-Methyloctane	861	4-Methyltridecane	1355
2-Methyloctane	862	2-Methyltridecane	1361
3-Ethylheptane	867	3-Methyltridecane	1370
3-Methyloctane	870	2,6,10-Trimethyldodecane	
2,4-Dimethyloctane	916	(Farnesane)	1368
2,5-Dimethyloctane	922	6- and 7-Methyltetradecane	1443
2-Methyl-5-ethylheptane	925	5-Methyltetra- and	
2,7-Dimethyloctane	929	2,6,10-Trimethyltridecane	1450
2,6-Dimethyloctane	933	4-Methyltetradecane	1454
2-Methyl-3-ethylheptane	937	2-Methyltetradecane	1461
4-Ethyloctane	952	3,7,11-Trimethyltridecane	1468
2,3-Dimethyloctane	954	3-Methyltetradecane	1470
5-Methylnonane	958	2,6,10-Trimethyltetradecane	1533
4-Methylnonane	961	7- and 8-Methylpentadecane	1539

Table 20 (continued)

Hydrocarbon	Retention index
6-Methylpentadecane	1542
5-Methylpentadecane	1546
4-Methylpentadecane	1554
2-Methylpentadecane	1562
3-Methylpentadecane	1569
2,6,10-Trimethylpentadecane	1633
2-Methylhexadecane	1659
3-Methylhexadecane	1668
2,6,10,14-Tetramethylpentadecane (Pristane)	1684
2,6,10-Trimethylhexadecane	1727
7-Methylheptadecane	1734
2-Methylheptadecane	1760
3-Methylheptadecane	1767
2,6,10,15-Tetramethylhexadecane	1788
2,6,10,14-Tetramethylhexadecane (Phytane)	1792
2,6,11,15-Tetramethylhexadecane (Crocetane)	1792
6- and 7-Methyloctadecane	1842
2-Methyloctadecane	1860
3-Methyloctadecane	1865
2,6,10,14-Tetramethylheptadecane	1867
2,6,10,15-Tetramethylheptadecane	1885
7- and 8-Methylnonadecane	1940
2,6,10,14-Tetramethyloctadecane	1957
2-Methylnonadecane	1963
3-Methylnonadecane	1968
2,6,10,15-Tetramethyloctadecane	1970
8- and 9-Methyleicosane	2038
2,6,10,14-Tetramethylnonadecane	2050
2-Methyleicosane	2063
3-Methyleicosane	2069
2,6,10,14,18-Pentamethylnonadecane	2105
9- and 10-Methylheneicosane	2135
2,6,10,14-Tetramethyleicosane	2140
2,6,10,15-Tetramethyleicosane	2155
2-Methylheneicosane	2163
3-Methylheneicosane	2169
2,6,10,14,18-Pentamethyleicosane	2210
2,6,10,14,19-Pentamethyleicosane	2226
2,6,10,15,19-Pentamethyleicosane	2229
10- and 11-Methyldocosane	2234
2-Methyldocosane	2263
3-Methyldocosane	2269
2,6,10,14,18-Pentamethylheneicosane	2293
2,6,10,14,19-Pentamethylheneicosane	2309
2,6,10,15,19-Pentamethylheneicosane	2319
11- and 12-Methyltricosane	2333
2,6,10,14,19-Pentamethyldocosane	2391
2,6,10,15,19-Pentamethyldocosane	2394
12-Methyltetracosane	2432
2,6,10,14,18-Pentamethyltricosane	2470
2,6,10,14,18,22-Hexamethyltricosane	2522
12- and 13-Methylpentacosane	2530
2,6,10,14,19-Pentamethyltetracosane	2569
2,6,10,15,19-Pentamethyltetracosane	2572
2,6,10,14,18,22-Hexamethyltetracosane	2627
12- and 13-Methylhexacosane	2628
2,6,10,15,19,23-Hexamethyltetracosane (Squalane)	2630
2,6,10,15,19,23-Hexamethyltetracosane	2632
2,6,10,14,18,22-Hexamethylpentacosane	2710
12- and 13-Methylheptacosane	2726
2,6,10,14,19,23-Hexamethylpentacosane	2728
2,6,10,14,19,23-Hexamethylhexacosane	2800
12- and 12-Methyloctacosane	2825
2,6,10,14,18,22,26-Heptamethylhexacosane	2870
12- and 13-Methylnonacosane	2924
2,6,10,14,18,22,26-Heptamethylheptacosane	2937
2,6,10,14,19,23-Hexamethyloctacosane	2979

Table 20 (continued)

Hydrocarbon	Retention index	Hydrocarbon	Retention index
2,6,10,14,18,22,26-Heptamethyloctacosane	3039	2,6,10,14,18,22,26,30-Octamethylhentriacontane	3350
2,6,10,14,19,23,27-Heptamethyloctacosane	3040	2,6,10,14,18,22,26,30-Octamethyldotriacontane	3444
2,6,10,14,18,22,26-Heptamethylnonacosane	3135	2,6,10,14,19,23,27,31-Octamethyldotriacontane (Lycopane)	3445
2,6,10,14,19,23,27-Heptamethylnonacosane	3139		
2,6,10,14,19,23,27-Heptamethyltriacontane	3211		
2,6,10,14,18,22,26-Heptamethylhentriacontane	3288		

possible source rocks. Any regularity in molecular distributions of branched alkanes may be used for correlation purposes. The pristane/phytane ratios acquired the widest popularity among isoprenoid alkanes (Safonova 1980). After extensive debates and misunderstandings, it was unequivocally established that phytane dominates in crudes originating from marine sediments deposits in an anoxic environment, while it is mostly pristane that emerges in crudes of continental origin from oxic depositional environments (Tissot and Welte 1981; Rashid 1979), although other factors, i.e., precursors and naturation, should not be excluded. However, it should be noted that under conditions of advanced catagenesis of dispersed organic material or kerogen and in related crudes, newly formed products are usually marked by a considerable predominance of pristane. From the genetic standpoint, such unique hydrocarbons as botryococcane, long-chain head-to-head isoprenoids and some others deserve special attention. As a rule, the presence of these compounds in crudes greatly facilitates the search for possible sources of petroleum (see Chap. 6).

Retention Indices of Alkanes

In conclusion, Table 20 presents retention indices of the hydrocarbons analyzed in this chapter. Retention indices are established for linear temperature programming. Capillary columns of 80 m length were used, and the carrier gas was hydrogen. The temperature programmed for columns with squalane was 50 °C→1°/min (end temperature, 150 °C); for columns with Apiezon, 100 °C→2°/min (end temperature, 320 °C).

CHAPTER 3

Cycloalkanes (Naphthenes)

Saturated cyclic hydrocarbons (naphthenes) comprise a most interesting, albeit complicated component of any crude. In effect, these hydrocarbons determine the special place of crude oils among natural organic compounds, as well as among fossil fuels.

The chemical structure of naphthenes varies to a great extent and differs primarily in the number of cycles per molecule. The increase of the molecular weight of naphthenes is accompanied by an increase of their content of polycylic molecules (Fig. 23). There is a maximum of five cycles in naphthenes, which are identifiable as individual compounds. However, there is circumstantial evidence that this number may be much greater (Ten Fu Yene and Chilingarian 1976).

Although investigations of naphthene structures are almost a 100-years-old and were initiated in the work of the founder of the chemistry of cyclic petroleum hydrocarbons, V. V. Markovnikov, it has only been in the last 10–12 years that we have gained a much better understanding of the structure of complex high-molecular-weight polycyclic compounds. Research at the molecular level has revealed that a most prominent role in the composition of polycyclic naphthenes is played by compounds of isoprenoid type.

All naphthenes may be tentatively subdivided into two large groups: mono- and polycyclic hydrocarbons. The composition and structure of hydrocarbons belonging to the first group, especially in the low-boiling range, have been analyzed extensively. Traditionally, monocyclic hydrocarbons are subdivided into groups of five- and six-membered naphthenes. Naphthenes with six-membered rings were studied with particular success by the well-known method of catalytic dehydrogenation. However, starting with C_8 and higher cyclanes, the concentrations of

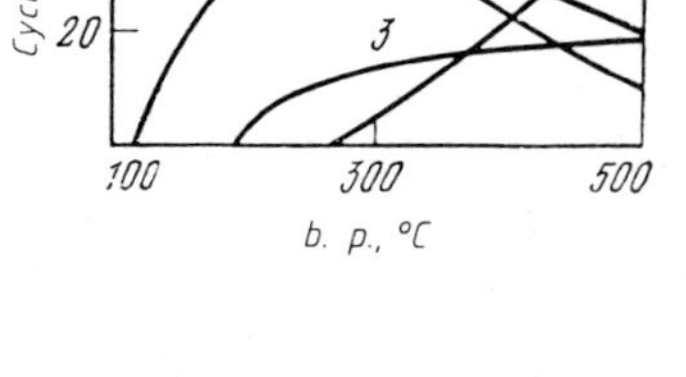

Fig. 23. Relative distribution according to structure types in Gryazevaya Sopka crude. Hydrocarbons: *1* monocyclic; *2* bicyclic; *3* tricyclic; *4* tetracyclic and higher

gem-substituted hydrocarbons in the cyclohexane series become appreciable, which disturb dehydrogenation results. Consequently, with regards to higher naphthenes, especially in view of large concentrations of polycyclic systems, catalytic dehydrogenation may be of limited use only. Gas chromatographymass-spectrometry (GC/MS) as well as the synthesis of standard compounds are preferable in establishing hydrocarbon structure.

Special classification principles should be applied to polycyclic hydrocarbons, because both five and six-membered rings may occur in a single molecule. Apparently, they should be classified according to the number of cycles per molecule, as well as the type of the polycyclic system: number of cycles, number of carbon atoms per cycle, type of ring junction, association with a certain class of natural compounds (such as steranes, triterpanes), etc.

The naphthenes analyzed below are grouped according to the number of cycles per molecule, beginning with monocyclic and ending with pentacyclic hydrocarbons.

C_7-C_{10} Monocyclic Naphthenes

Presently, a wealth of reliable information exists on the composition of various monocyclic naphthenes in crude oils. This information, gathered by means of capillary GC, provides an undistorted picture of the quantitative distribution of individual C_5-C_9 cyclanes in various oils. Appropriate experimental data are presented in Tables 21 and 22. Total contents numbers are represented by several diastereomers of a given cyclane. Distributions of some diastereomers are presented in Table 29 below.

Results of the analysis of light fractions presented in Tables 21 and 22 generally support the regularities suggested by F. Rossini (Petrov 1971; Rossini 1967):

1. All crudes contain an identical combination of hydrocarbons.
2. Crudes differ in relative concentrations of normal alkanes, branched alkanes, five- and six-membered naphthenes.
3. The Distribution of hydrocarbons within each of the identified groups appears constant at first approximation.

As compared to alkanes, the relative distribution of cyclanes is more homogeneous, but not without peculiarities, for example, in concentrations of *gem*-substituted hydrocarbons. Ratios of six- and five-membered cyclanes are of special significance. These are widely used in various geochemical correlations. Generally, a five-membered cycle is a strained system which is thermodynamically unstable at low temperatures. Moreover, this grouping has limited occurrence in those natural compounds which may be regarded as possible precursors of crude oil. Hence, relatively high concentrations of cyclopentane hydrocarbons in certain oils deserve considerable attention from the view point of their possible formation (for more details, see Chaps. 5 and 6).

A correlation between methylcyclohexane and the total sum of C_7 cyclopentanes in some crudes and gas condensates is presented in Table 23. As may be observed, this ratio ranges within 0.3 – 1.6 in crudes (with rare exceptions). However,

Table 21. Relative distribution of C_7-C_9 cyclopentanes in crudes (%)

Hydrocarbons	Deposit, Crude type				
	Grozne-nskoe, A[1]	Surgut, A[1]	Ekhabi-nskoe, A[2]	Gryazevaya Sopka, B[1]	Anastasi-yevsko-Troyit-skoe, B[2]
C_7 Composition[a]:					
Ethylcyclopentane	8	29.0	10	17	4.0
1,1-Dimethylcyclopentane	12	8.5	12	33	37.0
1,3-Dimethylcyclopentanes[b]	40	30.0	53	36	53.0
1,2-Dimethylcyclopentanes[b]	40	32.5	25	14	5.0
Σ C_7 Cyclopentanes	100	100	100	100	100
C_8 Composition:					
1,1,3-Trimethylcyclo-pentane	24.5	20.0	23.5	23.0	25.0
1,1,2-Trimethylcyclo-pentane	4.0	4.5	1.5	8.0	12.0
1,2,4-Trimethylcyclo-pentanes[b]	24.0	19.0	20.0	25.0	26.0
1,2,3-Trimethylcyclo-pentanes[b]	19.5	17.0	35.0	16.0	19.0
Σ Methylethylcyclo-pentanes	28.0	39.5	20.0	28.0	18.0
Σ C_8 Cyclopentanes	100	100	100	100	100
C_9 Composition:					
1,2,3,4-Tetramethylcyclo-pentanes[b]	10.0	18.0	32.0	6.5	22.0
1,1,2,4-Tetramethylcyclo-pentanes[b]	2.5	2.0	3.0	Trace	2.0
1,4-Dimethyl-2-ethyl-cyclopentanes[b]	33.0	21.0	14.0	20.0	12.0
1,2-Dimethyl-3-ethyl-cyclopentanes[b]	22.5	21.0	23.0	29.0	25.0
1,1-Dimethyl-3-ethyl-cyclopentane	8.0	2.5	6.0	3.0	3.0
Σ Di-substituted cyclo-pentanes	24.0	35.5	22.0	41.5	36.0
Σ C_9 Cyclopentanes	100	100	100	100	100

[a] Cyclopentane and methylcyclopentane are also found in all crudes (See Figs. 11 and 12).
[b] Total stereoisomers.

in gas condensates it increases to 3–5, which serves as a geochemical criterion for the phase characterization of reservoir fluids (Olenina and Petrov 1969; Tchakhmatchev et al. 1978). The reasons for such high concentrations of methylcyclohexane in primary gas condensates are yet unclear.

Table 22. Relative distribution of C_8-C_9 cyclohexanes in crudes (%)

Hydrocarbons	Deposit, Crude type				
	Grozne-nskoye, A^1	Surgut, A^1	Samotlor, A^1	Gryazevaya Sopka, B^1	Anastasi-yevsko-Troyit-skoe, B^2
C_8 Composition[a]:					
1,1-Dimethylcyclo-hexane	–	–	–	–	16
1,2-Dimethylcyclo-hexanes[b]	19	15	16	27.5	18
1,3-Dimethylcyclo-hexanes[b]	25	–	–	18.5	32
1,4-Dimethylcyclo-hexanes[b]	15	41.5	57	22.5	26
Ethylcyclohexane	41	44	27	31.5	8
$\sum$ C_8 Cyclohexanes	100	100	100	100	100
C_9 Composition:					
gem-substituted:					
1,1,2-Trimethylcyclo-hexane	3	–	7	15	10
1,1,3-Trimethylcyclo-hexane	85	–	63	74	30
1,1,4-Trimethylcyclo-hexane	12	–	30	11	60
$\sum$ *Gem*-substituted	100 (36)	100 (17)	100 (35)	100 (30)	100 (30)
Tri-substituted:					
1,2,3-Trimethylcyclo-hexanes[b]	17	30	20	21	10
1,2,4-Trimethylcyclo-hexanes[b]	80	61	80	79	88
1,3,5-Trimethylcyclo-hexanes[b]	3	9	Trace	Trace	2
$\sum$ Tri-substituted	100 (32)	100 (55)	100 (22)	100 (36)	100 (50)
Di-substituted:					
1-Methyl-3-ethyl-cyclohexanes[b]	61	68	50	58	51
1-Methyl-4-ethyl-cyclohexanes[b]	30	32	30	42	49
1-Methyl-2-ethyl-cyclohexanes[b]	9	–	20	–	–
$\sum$ Di-substituted	100 (32)	–	100 (20)	100 (34)	100 (20)
n-Propyl-cyclohexane	–	–	23	–	–

[a] Large amounts of cyclohexane and methylcyclohexane are usually present in crudes.
[b] Total stereoisomers.
Numbers in parentheses = sum of concentration of *Gem* %, Tri %, and Di-subst. %.
$\sum$ $C_9 = 100\%$.

Table 23. Correlation between methylhexane (M) and the total sum of C_7 cyclopentanes ($\sum$ C) in crudes and gas condensates

Deposit, Crude type	M/$\sum$ C	Deposit, Crude type	M/$\sum$ C	Deposit, Crude type	M/$\sum$ C
	Crudes			*Gas condensates*	
Samotlor, A[1]	1.02	Dagadjik, A[1]	1.8	Darvaza, B[1] (W. Turkmenistan)	3.2
Surgut, A[1]	1.5	Markovskoe, A[1]	2.0	Banka Yuzhnaya, A[1] (Azerbaijan)	3.0
Groznenskoe, A[1]	1.05	Norio, B[1]	1.14	Bitkovskiy, A[1] (Ukraine)	3.48
Romashkino, A[1]	1.57	Kos-Tchagyl	8.0	Shur-Tepe, A[1] (Turkmenistan)	3.35
Ekhabinskoe, A[2]	0.55	Mirzaani, A[1]	1.3	Maikopskiy, A[1] (Kuban)	5.2
Balakhanskoe, B[2]	0.55	E. Ekhabi	0.43	Berezanskiy, A[1] (Kuban)	4.95
Gryazevaya Sopka, B[1]	0.72	Rakushetchnoe, A[1] (Highly paraffinic)	3.2	Kanevskiy, A[1] (Kuban)	7.1
Anastasiyevsko-Troyitskoe, B[2]	0.34	Sartytchala, A[1]	1.95		
Paromay, A[1]	1.22				

Table 24. Equilibrium concentrations of C_7H_{14} cyclanes (%)

Hydrocarbons	295 K		500 K	
	Experiment	Calculation	Experiment	Calculation
1,1-Dimethylcyclopentane	23.4	26.5	12.7	16.8
1,3-Dimethylcyclopentanes[a]	36.6	34.5	41.8	37.4
1,2-Dimethylcyclopentanes	37.0	36.1	35.1	33.7
Ethylcyclopentane	3.0	2.9	10.4	12.1
$\sum$ C_7 Cyclopentanes	100 (6.9)	100 (3.5)	100 (41.8)	100 (42.6)
Methylcyclohexane	93.1	96.5	58.2	57.4

[a] Total stereoisomers.

In analyzing the isomeric composition of cyclanes the question of a potential equilibrium among isomers or of mean petroleum formation temperatures may arise again. In our opinion, there should be no complete equilibrium among structural isomers of cyclanes, i.e. the same applies as discussed before for the

Table 25. Equilibrium concentrations of C_8H_{16} cyclanes (%)

Hydrocarbons	295 K	400 K	500 K	600 K
1,1,3-Trimethylcyclopentane	28.5	23.2	16.3	15.8
1,2,4-Trimethylcyclopentanes [a]	30.0	28.0	24.5	24.9
1,2,3-Trimethylcyclopentanes [a]	29.3	23.0	20.6	19.4
1,1,2-Trimethylcyclopentane	12.2	8.1	7.0	5.5
1-Methyl-3-ethylcyclopentanes [a,b]		6.1	11.8	12.8
1-Methyl-2-ethylcyclopentanes [a,b]		7.2	14.0	15.7
1-Methyl-1-ethylcyclopentane [b]		2.5	2.4	1.2
Isopropylcyclopentane [b]		1.2	1.7	2.2
Propylcyclopentane [b]		0.7	1.7	2.5
$\sum$ C_8 Cyclopentanes	100 (0.7)	100 (9.1)	100 (31.0)	100 (40.0)
1,3-Dimethylcyclohexanes [a]	54.4	50.7	45.8	40.9
1,4-Dimethylcyclohexanes [a]	26.0	23.9	21.9	19.9
1,1-Dimethylcyclohexane	7.3	6.3	6.8	7.0
1,2-Dimethylcyclohexanes [a]	10.8	14.4	18.7	19.8
Ethylcyclohexane	1.5	4.7	6.8	12.4
$\sum$ C_8 Cyclohexanes	100 (99.3)	100 (90.9)	100 (69.0)	100 (60.0)

[a] Total stereoisomers.
[b] Equilibrium concentrations of hydrocarbons below the analytical detection limit (<0.1%).

Table 26. Equilibrium concentrations of C_9H_{18} cyclohexanes (%)

Hydrocarbons	300 K	500 K
1,1,3-Trimethylcyclohexane	13.5	10.6
1,1,4-Trimethylcyclohexane	9.1	9.7
1,3,5-Trimethylcyclohexanes [a]	28.1	24.3
1,2,4-Trimethylcyclohexanes [a]	36.8	29.6
1,2,3-Trimethylcyclohexanes	1.8	3.3
1,1,2-Trimethylcyclohexane	0.9	1.6
1-Methyl-3-ethylcyclohexanes [a]	5.7	11.6
1-Methyl-4-ethylcyclohexanes [a]	3.2	5.9
1-Methyl-2-ethylcyclohexanes [a]	0.4	1.8
1-Methyl-1-ethylcyclohexane	0.2	0.6
Isopropylcyclohexane	0.1	0.4
Propylcyclohexane	0.2	0.6
$\sum$ Tri-substituted (including geminal)	100 (90.2)	100 (79.1)
$\sum$ Geminal	23.7	22.5

[a] Total stereoisomers.

alkanes. We may only speak of a tendency towards the equilibrium state. Moreover, light petroleum hydrocarbons may be constantly replenished by destruction of high-molecular-weight alkanes and cyclanes. Tables 24–27 and Fig. 24 present some data on the thermodynamic equilibrium of C_7-C_{10} cyclanes (Petrov 1971).

Table 27. Equilibrium concentrations of $C_{10}H_{20}$ cyclohexanes

Hydrocarbons	300 K	600 K
1,1,3,5-Tetramethylcyclohexanes[a]	30.2	27.7
1,1,3,3-Tetramethylcyclohexane	0.5	0.5
1,1,4,4-Tetramethylcyclohexane	1.5	1.0
1,1,3,4-Tetramethylcyclohexanes[a]	17.1	15.6
1,2,4,5-Tetramethylcyclohexanes[a]	5.4	6.8
1,1,2,4-Tetramethylcyclohexanes[a]	5.6	5.8
1,1,2,5-Tetramethylcyclohexanes[a]	5.7	5.9
1,2,3,5-Tetramethylcyclohexanes[a]	8.2	9.3
1,1-Dimethyl-3-ethylcyclohexane	2.8	3.0
1,3-Dimethyl-5-ethylcyclohexanes[a]	10.2	6.6
1,1-Dimethyl-4-ethylcyclohexane	1.5	2.5
1,2,3,4-Tetramethylcyclohexanes[a]	0.2	0.2
1,4-Dimethyl-2-ethylcyclohexanes[a]	1.5	2.3
1,1,2,6-Tetramethylcyclohexanes[a]	0.6	0.6
1,1,2,3-Tetramethylcyclohexanes[a]	1.7	2.0
1,2-Dimethyl-4-ethylcyclohexanes[a]	2.6	3.4
1,3-Dimethyl-4-ethylcyclohexanes[a]	2.5	2.6
1,1-Dimethyl-2-ethylcyclohexane	0.8	0.5
1,3-Dimethyl-2-ethylcyclohexanes[a]	1.2	2.1
1,2-Dimethyl-2-ethylcyclohexanes[a]	0.6	1.7
Σ Cyclohexanes	100	100

[a] Total stereoisomers.

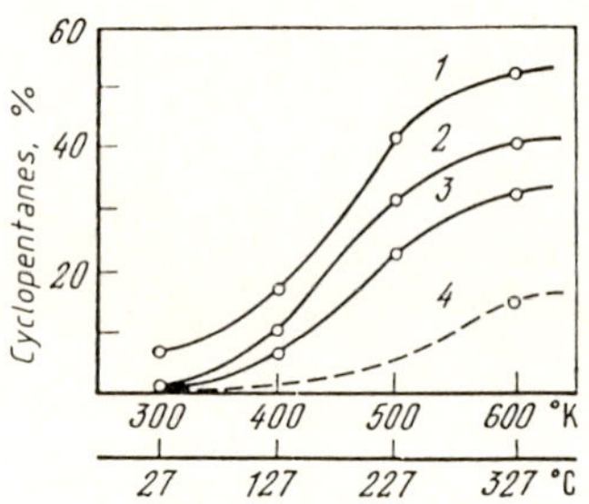

Fig. 24. Cyclopentanes in equilibrium mixtures at various temperatures. Hydrocarbons: *1* C_7; *2* C_8; *3* C_9; *4* C_{10} (approximate)

When comparing Tables 21 and 22 with Tables 24–27, we can conclude that there is no real equilibrium among structural isomers, hence, any attempt at establishing the temperature of petroleum formation must fail. Petroleum C_{10} cyclanes are also far removed from equilibrium. In view of the considerable methodological difficulties, the composition of these hydrocarbons could be elucidated only recently with the help of numerous standard hydrocarbons and GC-MS (Sokolova et al. 1981). On the whole, 87 hydrocarbons belonging mostly to six-membered naphthenes[1] were identified in the C_{10} (150°–175 °C) fraction of naphthenic crudes. The gas chromatogram of this fraction and the group distribution of the hydrocarbons are presented in Fig. 25 and Table 28.

[1] The fraction analyzed also contains some C_9 and C_{11} hydrocarbons.

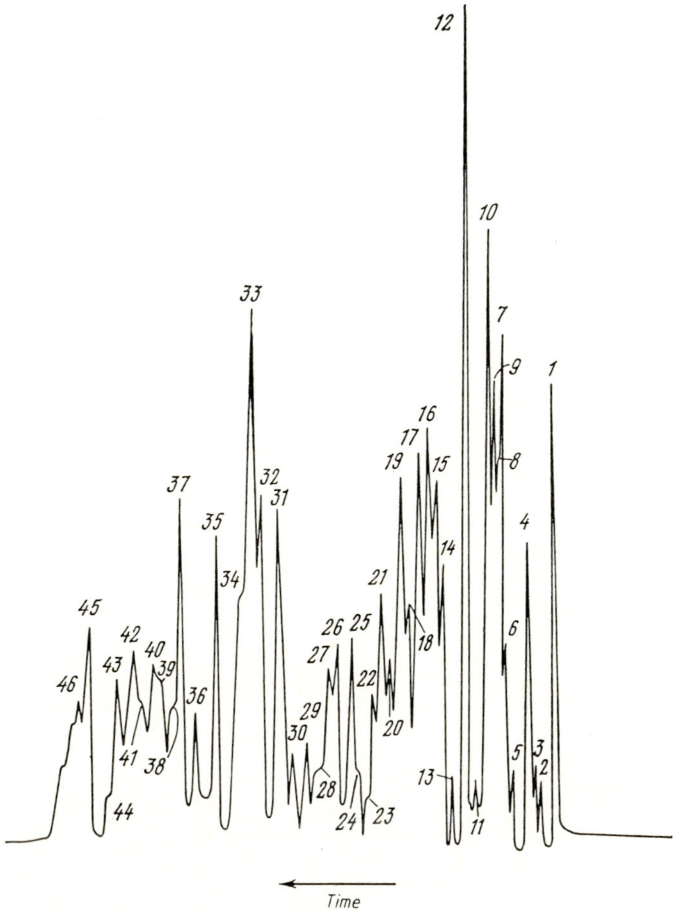

Fig. 25. Chromatogram of C_{10} saturated hydrocarbons in the 150°–175° fraction of a naphthenic crude (type B). Capillary column 50 m, squalane; 80 °C. For peak identifications, see Table 28

Data in the publication of Sokolova et al. (1981) reconfirm that the isomeric composition of petroleum hydrocarbons is far removed from equilibrium. In effect, over half of the hydrocarbons identified are thermodynamically unstable methylpropyl- and methylisopropylcyclohexanes. The concentration of dimethylethylcyclohexane is slightly inferior. At the same time, the concentrations of *gem*-substituted components, which are clearly prevalent under equilibrium conditions (see Table 27), are only 1–5% of the total. Certain C_{10} bicyclanes were also identified in the C_{10} fraction; hydrocarbons of this type will be thoroughly treated later.

1,1,3-Trimethylcyclohexane, which is the first representative of biological markers and an obvious decomposition product of higher-molecular-weight carotenoid systems, is present in the C_9 fraction. Its next homolog, 1,1,2,3-tetramethylcyclohexane, was found in the C_{10} fraction. These hydrocarbons, which

Table 28. Hydrocarbon composition of the C_{10} fraction (%)

Peak number in Fig. 25	Hydrocarbon	Deposit	
		Anastasiyevsko-Troyitskoe[a]	Norio[a]
1	1-Methyl-4-ethylcyclohexane, *trans*	4.1	3.5
2	n-Nonane	–	3.5
3	1,1,3,5-Tetramethylcyclohexane, *cis*	Trace	Trace
4	1-Methylbicyclo[3.2.1]octane	3.0	2.0
5	1,2,3-Trimethyl-4-ethylcyclopentane, *trans*, *trans*, *trans*	Trace	1.0
6	1-Methyl-3-ethylcyclohexane, *trans*	1.4	2.2
7	1,1,3,4-Tetramethylcyclohexane, *trans*	3.3	1.0
	1-Methyl-2-ethylcyclohexane, *trans*	Trace	3.0
8	2,4-Dimethyloctane	Trace	2.0
9	3-Methylbicyclo[3.3.0]octane, *endo*	2.4	1.8
10	2-Methylbicyclo[3.3.0]octane, *exo*	4.8	1.8
	1,1,3,5-Tetramethylcyclohexane, *trans*	Trace	–
11	1-Methyl-4-ethylcyclohexane, *cis*	0.3	2.1
	2,5-Dimethyloctane	–	1.8
12	Isopropylcyclohexane	9.4	7.0
13	3-Methylbicyclo[3.3.0]octane, *exo*	0.9	1.6
14	3-Methylbicyclo[3.2.1]octane, *exo*	1.1	3.0
	1,5-Dimethylbicyclo[3.2.1]octane	1.1	3.0
	2,6-Dimethyloctane	1.1	5.5
15	Propylcyclohexane	1.7	5.3
	1-Methyl-2-ethylcyclohexane, *cis*	–	Trace
16	3-Methyl-4-ethylheptane	2.3	2.0
17	2-Methyl-3-ethylheptane	2.9	1.7
	6-Methylbicyclo[3.2.1]octane, *exo*	–	Trace
18	1,3-Dimethylbicyclo[3.3.0]octane, *endo*	1.3	1.4
	1,1,3,4-Tetramethylcyclohexane, *cis*	–	Trace
19	1,1,2,4-Tetramethylcyclohexane, *cis*	3.7	1.6
	1,1,2,5-Tetramethylcyclohexane, *trans*	3.7	1.6
	1,3-Dimethyl-4-ethylcyclopentane, *trans*, *trans*	–	–
21	2-Methylbicyclo[3.3.0]octane, *endo*	1.7	–
	1,3-Dimethylbicyclo[3.3.0]octane, *exo*	1.7	–
	1,1-Dimethyl-3-ethylcyclohexane	0.8	3.2
	1,2,3,5-Tetramethylcyclohexane, *trans*, *trans*, *cis*	Trace	–
22	1,3-Dimethyl-1-ethylcyclohexane	1.0	0.8
24	3,4-Dimethyloctane (*α*), *trans*	0.8	1.6
25	3,4-Dimethyloctane (*β*), *cis*	0.8	1.6
	2,3-Dimethyloctane	0.8	1.6
	1,3-Dimethyl-5-ethylcyclohexane *cis*, *cis*	1.5	1.6
26	1,3-Dimethylbicyclo[3.2.1]octane, *exo*	1.8	–
	1,6-Dimethylbicyclo[3.2.1]octane, *exo*	1.8	–
	1,1-Dimethyl-4-ethylcyclohexane	Trace	0.6
27	Bicyclo[4.3.0]nonane, *trans*	0.4	0.4
	2-Methylbicyclo[3.2.1]octane, *exo*	0.4	–
	1,3-Dimethyl-5-ethylcyclohexane, *trans*, *cis*	0.4	0.5
	5-Methylnonane	–	0.5
28	4-Methylnonane	0.3	2.0

Table 28 (continued)

Peak number in Fig. 25	Hydrocarbon	Deposit	
		Anastasiyevsko-Troyitskoe[a]	Norio[a]
29	2,4-Dimethylbicyclo[3.3.0]octane, *exo*, *exo*	1.3	0.2
30	2-Methylnonane	1.2	1.2
31	1,3-Dimethyl-5-ethylcyclohexane, *trans*, *trans*	Trace	2.0
	1,5-Dimethylbicyclo[3.3.0]octane	5.6	Trace
	1,4-Dimethylbicyclo[3.3.0]octane, *endo*	5.6	Trace
	2,8-Dimethylbicyclo[3.3.0]octane, *exo*, *exo*	5.6	Trace
	1,1,2,3-Tetramethylcyclohexane, *trans*	Trace	4.2
	1,2,3,5-Tetramethylcyclohexane, *cis*, *trans*, *cis*	Trace	–
	1,2,3,4-Tetramethylcyclohexane, *trans*, *trans*, *trans*	Trace	–
32	1,4-Dimethyl-2-ethylcyclohexane, *trans*, *cis*	3.8	1.1
	2,7-Dimethylbicyclo[3.3.0]octane, *exo*, *endo*	Trace	–
	3-Methylnonane	–	1.1
33	1,2-Dimethyl-4-ethylcyclohexane, *trans*, *cis*	3.6	2.3
	2,3-Dimethylbicyclo[3.3.0]octane, *exo*, *endo*	3.6	2.3
	2,6-Dimethylbicyclo[3.3.0]octane, *exo*, *exo*	3.6	Trace
	3,7-Dimethylbicyclo[3.3.0]octane, *exo*, *endo*	3.6	Trace
34	1,4-Dimethylbicyclo[3.2.1]octane, *endo*	0.5	0.6
	2,7-Dimethylbicyclo[3.3.0]octane, *exo*, *exo*	0.5	0.5
35	1-Methyl-3-isopropylcyclohexane, *cis*	5.7	2.5
	1,3-Dimethylbicyclo[2.2.2]octane	5.7	2.5
	1,6-Dimethylbicyclo[3.2.1]octane, *endo*	Trace	Trace
	1,3,5-Trimethylbicyclo[3.2.1]octane	Trace	Trace
36	1,2-Dimethylbicyclo[3.3.0] octane, *exo*	0.9	0.3
	1,2,3,4-Tetramethylcyclohexane, *trans*, *cis*, *trans*	Trace	0.3
	1,4-Dimethyl-2-ethylcyclohexane, *cis*, *trans*	0.9	0.3
37	1-Methyl-4-isopropylcyclohexane, *trans*	5.6	2.4
	1,3-Dimethyl-2-ethylcyclohexane, *trans*, *trans*	Trace	–
38	Isobutylcyclohexane	0.5	1.9
39	Bicyclo[4.3.0]nonane, *cis*	0.9	–
40	1-Methyl-3-propylcyclohexane, *cis*	0.9	3.5
	1-Methyl-3-isopropylcyclohexane, *trans*	0.9	3.5
	tert.-Butylcyclohexane	0.9	3.5
41	1,2-Dimethyl-3-ethylcyclohexane, *trans*, *trans*	0.6	–
	1,2,3,4-Tetramethylcyclohexane, *cis*, *trans*, *cis*	Trace	–

Table 28 (continued)

Peak number in Fig. 25	Hydrocarbon	Deposit	
		Anastasiyevsko-Troyitskoe[a]	Norio[a]
42	1-Methyl-4-propylcyclohexane, *trans*	2.4	3.1
	1,2,3,4-Tetramethylcyclohexane, *trans, cis, cis*	Trace	–
43	1-Methyl-4-isopropylcyclohexane, *cis*	1.9	0.7
	1,1,3-Trimethyl-5-ethylcyclohexane, *cis*	0.9	0.7
44	1,2-Dimethyl-4-ethylcyclohexane, *cis, cis*	Trace	–
	1,2-Dimethyl-4-ethylcyclohexane, *cis, trans*	Trace	–
45	1-Methyl-2-propylcyclohexane, *trans*	3.7	1.7
46	n-Decane	–	0.9
	1,4-Diethylcyclohexane, *trans*	0.5	–
	1-Methyl-3-propylcyclohexane, *trans*	0.9	0.9

[a] Unidentified peaks (Nos. 20,23) comprise 3.4% and 2.3% of the total in the Anastasiyevsko-Troyitskoe and Norio crudes, respectively.

are important for understanding the origins of petroleum, are extensively analyzed below. The concentrations of methylisopropylcyclohexane, which have an obvious genetic link to terpenes, are also relatively high in the C_{10} fraction.

Stereoisomers belonging to the lowest boiling cyclanes are closest to the state of equilibrium. Some data on epimer pairs of C_7-C_8 cyclanes in equilibrium are presented in Table 29. In order to evaluate the temperatures of their formation, the following conditions should be met:

1. A thermodynamically controlled mechanism of epimer formation.
2. Absence of additional sources of these structure in reactions (such as decomposition) which are not subject to thermodynamic regularities (see Chaps. 5 and 6).

The values for the equilibrium constant between *trans*- and *cis*-1,2-dimethylcyclopentanes, ranging from 5.7 to 11.5, indicate petroleum formation temperatures of 130°–300 °C (160 °C on the average) (Fig. 26). The equilibrium constant of epimeric 1,3-dimethylcyclopentanes is independent of temperature variations and may not be used in calculations, although the relative abundance of these epimers in crudes corresponds to their equilibrium concentrations. Let us again recall that the thermodynamic properties of epimers thus analyzed were thoroughly treated in our monograph (Petrov 1981).

However, certain epimer pairs in the C_{10} cyclanes, such as 1,3-dimethyl-5-ethyl- and 1-methyl-3-propylcyclohexanes, are not present in equilibrium concentrations, because less stable epimers predominate (Sokolova et al. 1981). There is also no full equilibrium in epimeric steranes (see below). However, diastereomers (epimers) are closer to the state of equilibrium than any other isomer. Epimeriza-

Table 29. Relative distribution of cyclane epimers in crudes (%)

Hydrocarbons	Deposits						
	Grozne-nskoe	Dagad-jiks-koye	Ekhabi-nskoe	Gryaze-vaya Sopka	Balakha-nskoe	Samotlor	Norio
1,2-Dimethyl-cyclopentane, *trans/cis*	89:11	86:14	85:15	86:14	86:14	88:12	86:14
1,3-Dimethyl-cyclopentane, *cis/trans*	53:47	51:49	51:49	53:47	51:49	52:48	48:52
1,2,3-Trimethyl-cyclopentane, *trans, trans/ trans, cis*	76:24	75:25	81:19	81:19	89:11	83:17	84:16
Dimethylcyclo-hexanes (1,3-*cis* + 1,4-*trans*)/(1,3-*trans* + 1,4-*cis*)[a]	90:10	86:14	90:10	83:17	88:12	82:18	88:12

[a] Ratio between the totals of more stable and less stable epimers.

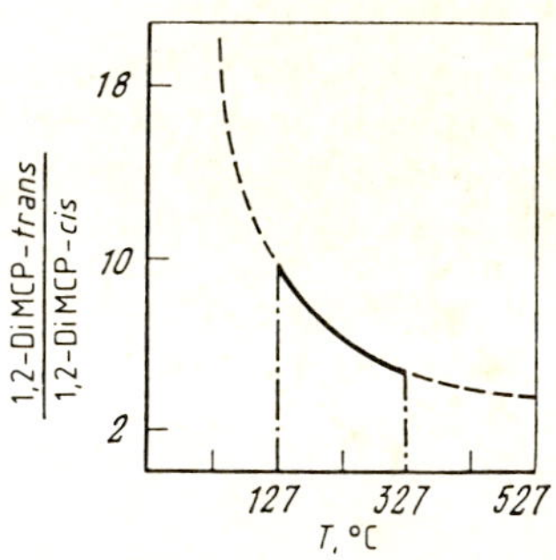

Fig. 26. Ratio of epimeric 1,2-dimethylcyclopentanes and possible upper temperature limit of petroleum formation. *Continuous line* represents the ratio of epimers in various crudes

tion is important for understanding the petroleum formation mechanisms as well as for petroleum exploration. "Stereochemical maturation" is especially important for evaluating catagenetic alterations of biogenic molecules reaching the maturity level of petroleum. These problems are treated later. At this point, it should be emphasized that precise determinations the ways of reciving equilibrium concentrations of stereoisomers should always be made.

Two main approaches are feasible: (1) preparation of epimeric equilibrium mixtures by way of their thermodynamically controlled formation from more labile sources, such as unsaturated hydrocarbons, alcohols, etc.; (2) preliminary formation of nonequilibrium mixtures of epimers in kinetically controlled reactions, with further stereoisomerization of original diastereomers to equilibrium.

Both procedures may occur in petroleum formation, although the former is the one favored.

We now move to cyclanes in higher boiling fractions.

$C_{12}-C_{25}$ Monocyclic Naphthenes

Relatively high concentrations of C_8-C_{12} monoalkylcyclohexanes and monoalkylcyclopentanes (see Figs. 11 and 16) led to the hypothesis that crudes also con-

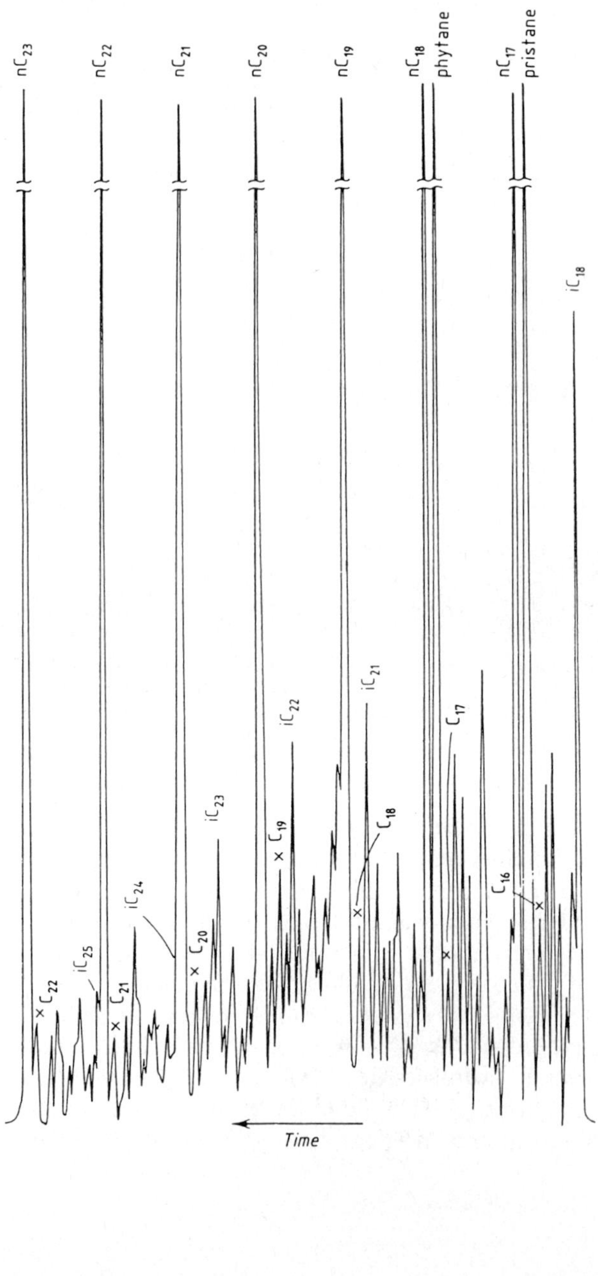

Fig. 27. Partial chromatogram of saturated hydrocarbon mixture of Samotlor crude. *X* indicates the elution points of $C_{16}-C_{22}$ n-alkylcyclohexanes. Capillary column 50 m, Apiezon; linear temperature programming, 200°→2°/min

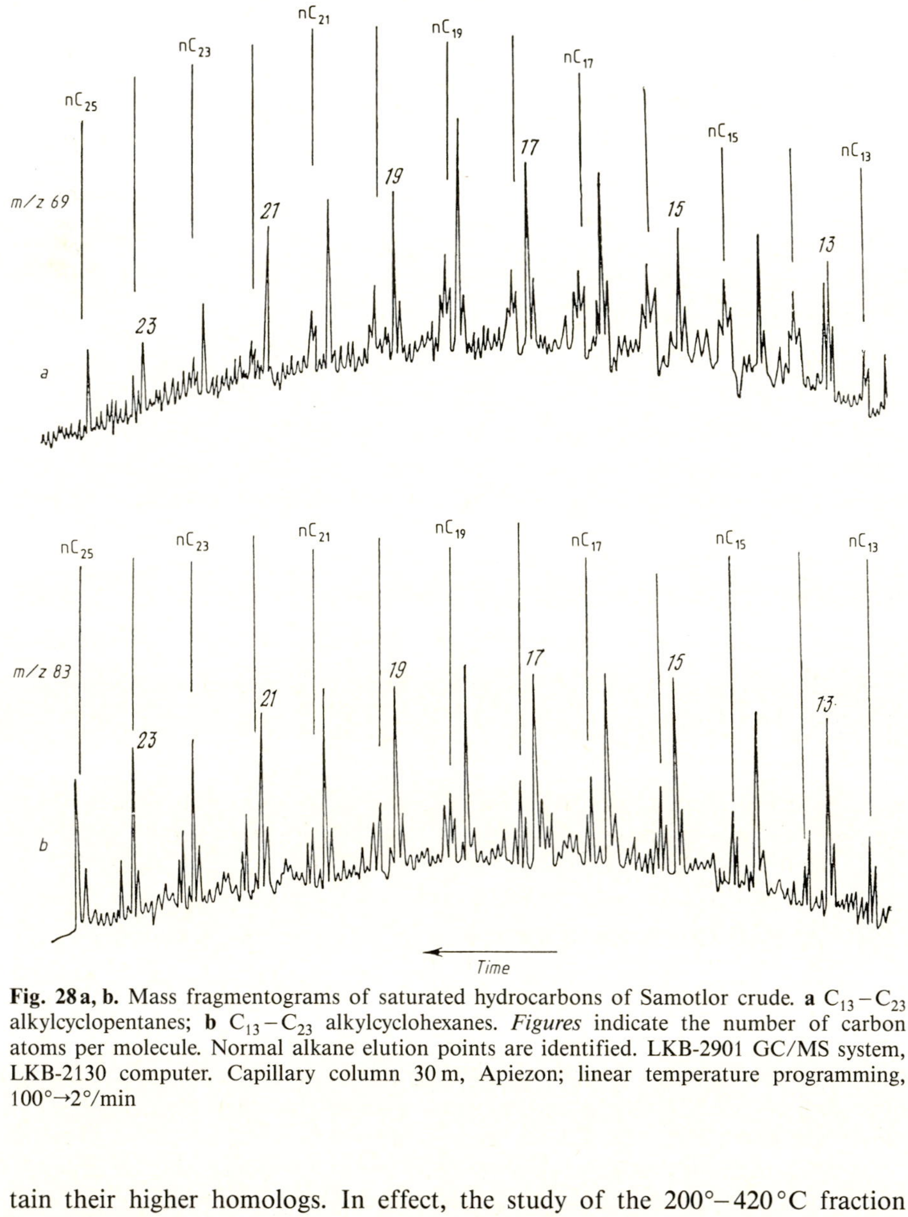

Fig. 28 a, b. Mass fragmentograms of saturated hydrocarbons of Samotlor crude. **a** $C_{13}-C_{23}$ alkylcyclopentanes; **b** $C_{13}-C_{23}$ alkylcyclohexanes. *Figures* indicate the number of carbon atoms per molecule. Normal alkane elution points are identified. LKB-2901 GC/MS system, LKB-2130 computer. Capillary column 30 m, Apiezon; linear temperature programming, 100°→2°/min

tain their higher homologs. In effect, the study of the 200°–420 °C fraction revealed the presence of the corresponding higher homologs. A partial gas chromatogram of the saturated hydrocarbons in the 200°–420 °C fraction ($C_{11}-C_{25}$) of the Samotlor crude is presented in Fig. 27. Peaks corresponding to normal alkylcyclohexanes are denoted by an "x". The hydrocarbons were identified by adding standard compounds easily produced in the cracking of higher-molecular-weight homologs, such as octadecyclcyclohexane. Under GC conditions with linear temperature programming, normal alkylcyclohexanes begin to elute progressively later, compared to normal alkanes with an equal number of carbon

Scheme 1

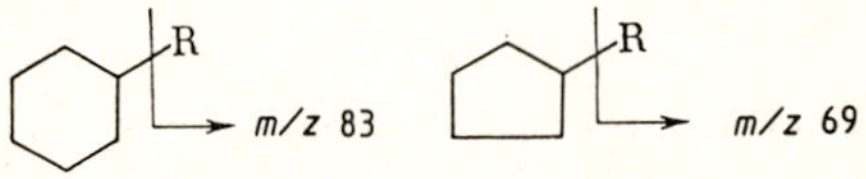

atoms per molecule. $C_{13}-C_{16}$ monocyclanes fall into the area of monomethylalkanes (with one extra carbon atom), monocyclanes $>C_{16}$ elute as separate peaks. Besides normal alkylcyclohexanes, a homologous series of normal alkylcyclopentanes is found in crude oils. N-alkylcyclopentanes elute slightly earlier than the isomeric n-alkylcyclohexanes; this difference becoming more apparent with the increase of the molecular mass.

The presence of a series of $C_{12}-C_{25}$ n-alkylcyclohexane and n-alkylcyclopentane homologs in the crude analyzed (which is a typical A^1 paraffinic crude) was proven by mass fragmentograms of the most characteristic fragment ions: *m/z* 69 for n-alkylcyclopentanes and *m/z* 83 for n-alkylcyclohexanes (Scheme 1).

The mass fragmentograms in Fig. 28 unequivocally point to the existence of hydrocarbons belonging to the respective homologous series in crudes. The total concentration of n-alkylcyclohexanes and n-alkylcyclopentanes amounts to $\approx 10\%$ of all n-alkanes in the same fraction. Recently, the presence of $C_{10}-C_{25}$ methylalkylcyclohexanes in crudes was demonstrated.

Trimethyl-Substituted Cyclohexanes with Isoprenoid Chains (Cyclohexanes of the Carotenoid Type)

Besides isoprenoid alkanes, a peculiar $C_{10}-C_{24}$ homologous series with an obviously relict structure was found in the afore-mentioned Karajanbas crude. These hydrocarbons belong to the 1,1,3-trimethyl-2-alkylcyclohexane series (Vorobyova et al. 1980) (structure type I):

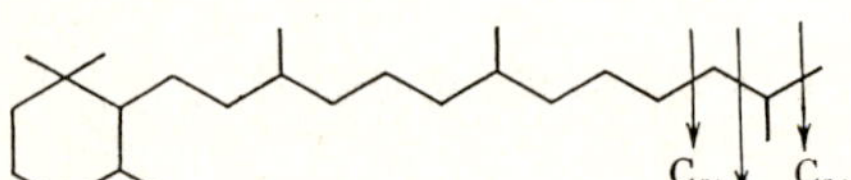

I

As expected, no C_{12} and C_{17} homologs were found. Figure 29 shows a chromatogram of saturated hydrocarbons boiling between 180° and 400 °C. The hydrocarbons analyzed are indicated by the letters M_{10}, M_{11}, etc. (i.e. C_{10}, C_{11} monocyclanes etc.).

A fairly high concentration of these hydrocarbons is apparent. The structure of monocyclanes was established by comparative synthesis of standard and GC-MS. The structure of the first homolog under analysis, *trans*-1,1,2,3-tetramethylcyclohexane was reconfirmed by ^{13}C NMR spectroscopy. The orientation of substituents at C-2 and C-3 is universally *trans*-(ee). The overall amount of these hydrocarbons in the fraction is $\approx 3\%$, or $\approx 1\%$ in petroleum.

The origin of these hydrocarbons if of interest. Both aliphatic isoprenoids with regular and irregular structures (with prior cyclization of one of the chain ends) and β-carotene may be considered as sources in this instance. The absence of the C_{17} cyclane and, in contrast, the presence of the C_{18} cyclane make the par-

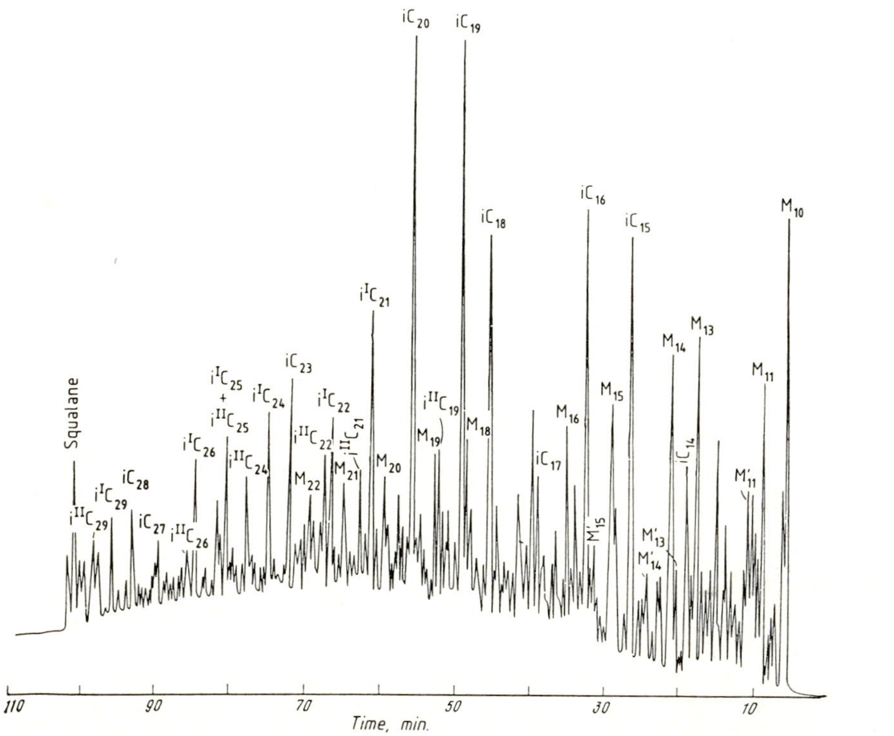

Fig. 29. Chromatogram of a mixture of isoprenoid alkanes and carotenoid-type monocyclanes in Karajanbas crude. i-C_{19}, i-C_{20} are isoprenoid alkanes (see also Fig. 22). M_{11}, $M_{13}-C_{11}$, C_{13} are monocyclanes with isoprenoid chains etc.; M'_{11} see text. *Indices I* and *II* identify regular and irregular structures, respectively. Capillary column 80 m, Apiezon; linear temperature programming, 100°→2°/min

ticipation of such an isoprenoid as squalane in the formation of these cyclanes doubtful. At the same time, the presence of the C_{22} monocyclane tends to exclude high-molecular-weight aliphatic isoprenoids of the regular type as possible precursors.

Thus, the information available seems to indicate that compounds of the carotenoid type participated in the formation of the monocyclanes under review. Indeed, this fact is surprising, taking into account the well-known instability of carotenoids, due to their numerous conjugated double bonds.

β-Carotane, which is is the saturated analog of β-carotene, was also discovered by us (with the help of GC) in the given crude, albeit in minute concentrations, much smaller than any of the analyzed carotenoid monocyclanes. In principle, the same monocyclane series may be produced by cyclization of one of the chain ends of lycopane.

Tetraalkyl-substituted cyclohexanes of apparently similar structure were found earlier in Green River shale, incidentally, having a relatively higher content of β-carotane (Anders and Robinson 1971). It is also beyond any doubt that 1,1,3-trimethylcyclohexane, identifiable in great quantities in gasolines of any crude, is the

Scheme 2

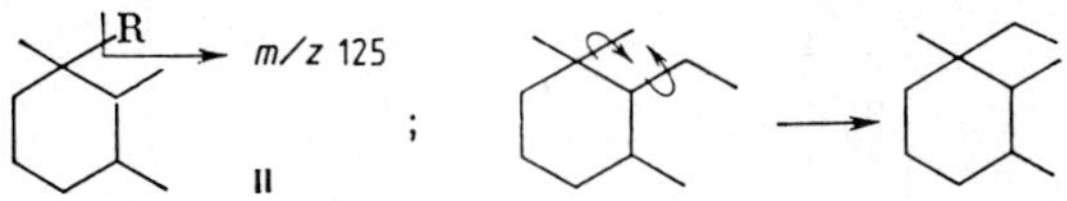

product of the loss of the aliphatic chain from the 1,1,3-trimethyl-2-alkylcyclohexanes under analysis.

Besides tetra-substituted monocyclanes of this type, a series of similarly structured $C_{11}-C_{15}$ hydrocarbons, however, belonging to a different homologous series, was identified in the Karajanbas crude oil.

According to their mass spectra, these hydrocarbons have a long aliphatic substituent connected to the quaternary carbon atom, which is cleaved in the mass spectrometer to form an intense fragment ion at *m/z* 125. Mass spectral information, as well as the absence of a C_{10} hydrocarbon of this type (identical to the above described 1,1,3-trimethyl-2-alkylcyclohexanes), lead to the assumption that their structural type is represented by formula II. These structures may be formed by the 1,2-shift of a methyl or alkyl radical as shown in Scheme 2.

The peaks corresponding to this type of hydrocarbon are identified by the letter M′ in Fig. 29.

Still another noteworthy representative of cyclanes with a C_{40} isoprenoid chain (type 3) is described in the publication of Chappe et al (1979):

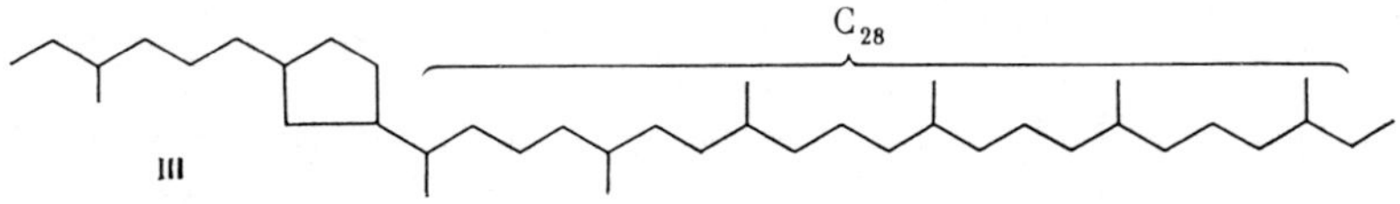

It is assumed that together with "head-to-head" aliphatic isoprenoids, this hydrocarbon is present in Archaebacterial lipids. There are indications that phytadiene dimers are present in crudes, such as a C_{40} hydrocarbon, formed by the Diels-Alder reaction. This hydrocarbon belongs to type IV:

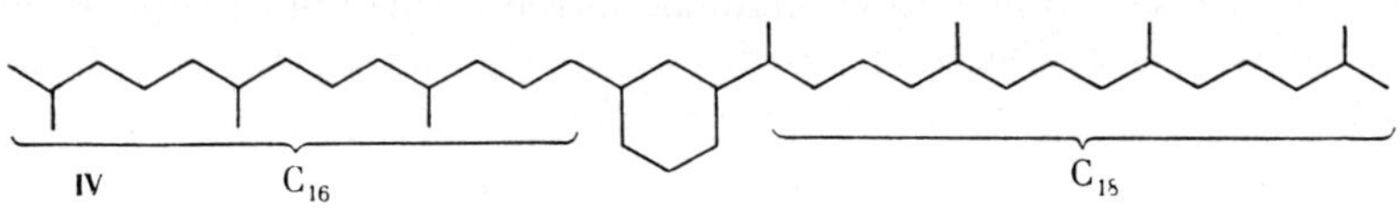

By decomposition of aliphatic chains in hydrocarbons III and IV, additional quantities of various isoprenoid alkanes may be generated. The formation of long aliphatic chains of normal and isoprenoid types has been demonstrated by the thermal cracking of high-molecular-weight petroleum cyclanes. Appropriate material is presented in Chapter 6.

Bicyclic Naphthenes

According to the type of relative arrangement of cyclic groupings, bicyclic hydrocarbons may be subdivided into hydrocarbons of the bridged (V), condensed (VI), connected (VII) and isolated (VIII) structure types:

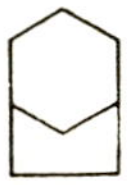
V

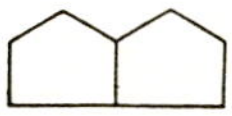
VI

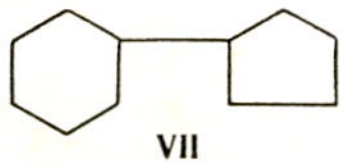
VII

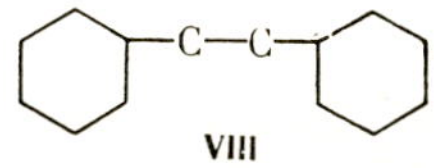

VIII

C_8-C_{10} Hydrocarbons

First allusions to bicyclic naphthenes in crudes, i.e. hydrindane and decalin, go back to the 1930's, when N.D. Zelinskii discovered indane and naphthalene among the products of the catalytic dehydrogenation of certain oil fractions.

In the research work of the American Petroleum Institute headed by Rossini, the presence of *trans*-decalin and *cis*-pentalane (bicyclo[3.3.0]octane) was established (Rossini 1967).

The publication of Lindeman and Tourneau (1963), which announced that norbornane and all four of its nearest methyl homologs as well as bicyclo[3.3.0]-octane, bicyclo[3.2.1]octane, bicyclo[2.2.2]octane and *trans*- and *cis*-bicyclo-[4.3.0]nonanes were separated and identified by thermal diffusion of a Californian crude, was a landmark in the study of relatively low boiling bicyclic naphthenes. This research gave an impetus to extensive studies in the chemistry of C_8-C_{10} bicyclanes, which led to their discovery in crudes (Petrov 1971, 1981; Bagrii et al. 1967).

At this point, the presence of the bicyclic hydrocarbons IX–XV, as well as their methyl and dimethyl homologs, has been established in a number of crudes (carbon atom numbers given to bicyclanes are used throughout):

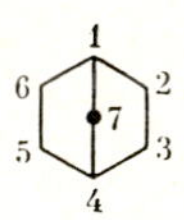
IX

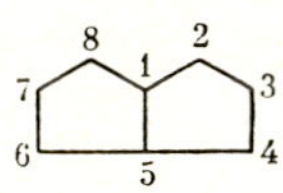
X

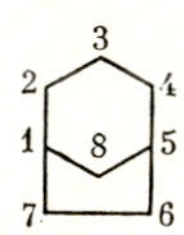
XI

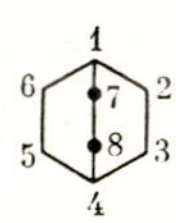
XII

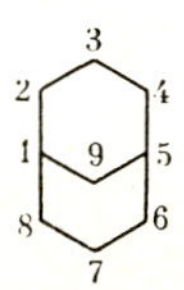
XIII

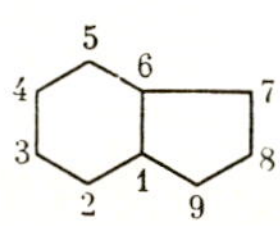
XIV

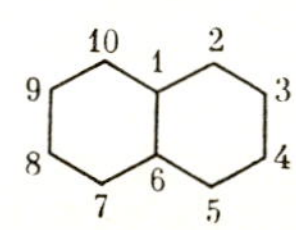
XV

IX. Bicyclo[2.2.1]heptane – norbornane;
X. Bicyclo[3.3.0]octane – pentalane; only the derivatives of the more stable *cis*-pentalane system are present in crudes;
XI. Bicyclo[3.2.1]octane;
XII. Bicyclo[2.2.2]octane;
XIII. Bicyclo[3.3.1]nonane;
XIV. Bicyclo[4.3.0]nonane – hydrindane; derivatives of *cis*- as well as *trans*-hydridane are present in crudes;
XV. Bicyclo[4.4.0]decane – decalin; the more stable hydrocarbons with *trans*-junction of the cycles predominate in crudes.

Table 30. C_8-C_9 Bicyclic naphthenes, identified in the Gryazevaya Sopka crude

Peak number in Fig. 30	Hydrocarbons	Content (%)	
		In petroleum	In the sum of all isomers
	C_8H_{14}:		
1	*exo*-2-Methylbicyclo[2.2.1]heptane	0.0001	0.4
2	*endo*-2-Methylbicyclo[2.2.1]heptane	0.00002	0.1
3	Bicyclo[3.3.0]octane	0.017	67.9
4	Bicyclo[3.2.1]octane	0.008	31.6
	Σ		100.0
	C_9H_{16}:		
6	1-Methylbicyclo[3.3.0]octane	0.032	16.4
8	1-Methylbicyclo[3.2.1]octane	0.027	13.8
9	1-Methylbicyclo[2.2.2]octane	0.015	7.6
10	*endo*-3-Methylbicyclo[3.3.0]octane	0.018	9.2
11	*exo*-2-Methylbicyclo[3.3.0]octane	0.052	26.6
12	*exo*-3-Methylbicyclo[3.2.1]octane	0.018	9.2
13	*exo*-3-Methylbicyclo[3.3.0]octane	0.007	3.6
14	*exo*-6-Methylbicyclo[3.2.1]octane	0.005	2.5
15	*endo*-2-Methylbicyclo[3.3.0]octane	0.006	3.1
16	*endo*-2-Methylbicyclo[3.2.1]octane	0.005	2.5
17	2-Methylbicyclo[2.2.2]octane	0.001	0.5
18	*endo*-6-Methylbicyclo[3.2.1]octane	0.002	1.0
19	*trans*-Bicyclo[4.3.0]nonane	0.007	3.5
20	*exo*-2-Methylbicyclo[3.2.1]octane	0.001	0.5
	Σ		100.0

Hydrocarbon IX and XI–XIII are compounds of the bridged-type. Hydrocarbons of the norbornane series are thermodynamically unstable. Their concentration in crudes is insignificant. As was pointed out, only norbornane proper and its methyl homologs were identified in crudes in the following relative concentrations: 1-methyl : 2-methyl (*exo*) : 2-methyl (*endo*): 7-methyl = 1 : 1 : 0.5 : 0.05 (Lindeman and Tourneau 1963).

Hydrocarbons of the bicyclo[3.2.1]octane and especially the bicyclo[3.3.0]-octane series have a much wider occurrence in crudes. Table 30 presents information on the content of C_8-C_9 bicyclanes in the Gryazevaya Sopka crude (type B^1, Apsheron, Tertiary sediments). Hydrocarbons were isolated with the help of thermal diffusion. A chromatogram of such a hydrocarbon concentrate is presented in Fig. 30. Special notice should be taken of the important role played by the pentalane bicyclanes, comprising over 50% of the C_9 bicyclanes. The same applies to C_{10} bicyclanes. It is noteworthy that the distribution of C_9H_{16} hydrocarbons in crudes is approximately the same as in an equilibrium mixture at $\approx 150°-200°$.

The relative distribution of bicyclic C_9H_{16} hydrocarbons in the equilibrium state at various temperatures is given in Fig. 31; bicyclo[2.2.2]octanes are ther-

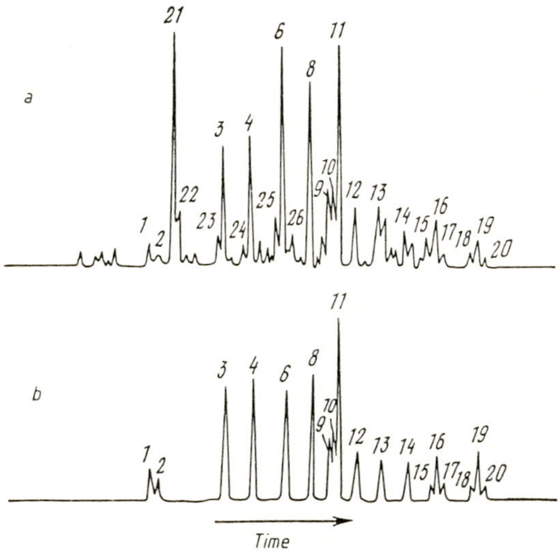

Fig. 30 a, b. Chromatograms of bicyclane mixtures. a Separated from Gryazevaya Sopka crude; b an equilibrium C_8 and C_9 mixture. Peaks *1–20* are identified in Table 30; peaks *21–26* belong to monocyclanes

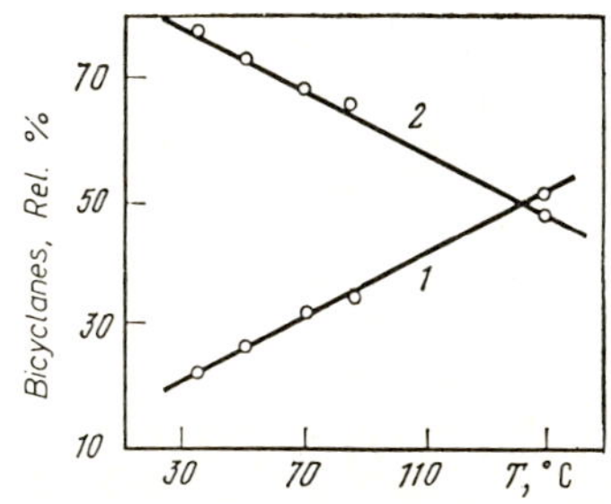

Fig. 31. Relative distribution of C_9 bicyclanes. Hydrocarbon series of: *1* bicyclo[3.3.0]octane; *2* bicyclo-[3.2.1]octane in equilibrium mixtures at various temperatures

Table 31. Equilibrium concentrations of C_9H_{16} bicyclic hydrocarbons at 323 K

Hydrocarbons	I[a]	II[b]
1-Methyl bicyclo[3.3.0]octanes	6.2	34.0
2-Methyl bicyclo[3.3.0]octanes	6.9	38.0
3-Methyl bicyclo[3.3.0]octanes	5.2	28.0
Σ Bicyclo[3.3.0]octanes	18.3	100.0
1-Methyl bicyclo[3.2.1]octanes	46.0	74.5
2-Methyl bicyclo[3.2.1]octanes	7.8	12.5
3-Methyl bicyclo[3.2.1]octanes	3.9	6.2
6-Methyl bicyclo[3.2.1]octanes	3.1	5.0
8-Methyl bicyclo[3.2.1]octanes	1.1	1.8
Σ Bicyclo[3.2.1]octanes	61.9	100.0
1-Methylbicyclo[2.2.2]octanes	6.3	86.5
2-Methylbicyclo[2.2.2]octanes	1.0	13.5
Σ Bicyclo[2.2.2]octanes	7.3	100.0
Bicyclo[4.3.0]nonanes	12.2	100.0
Bicyclo[3.2.2]nonane	0.3	–

[a] To the sum of all C_9 bicyclanes.
[b] Relative distribution.

modynamically unstable. The composition of an equilibrium mixture of all C_9 bicyclanes is given in Table 31. It should be stressed that compounds with substituents placed at the bridgehead in bridged-type hydrocarbons (IX, XI–XIII), are thermodynamically much more stable. Bicyclo[3.3.1]nonane (XIII) has a low thermodynamic stability because of the interaction of axial hydrogen atoms at C-3 and C-7.

A similar situation in the distribution of bicyclic hydrocarbon occurs in the study of the higher homologs. Thus, Table 32 contains data on the composition of some dimethylbicyclo-octanes in the Gryazevaya Sopka crude. The same bicyclanes were later found together with 1,2-dimethyl-, 2,3-dimethyl-, 2,6-dimethyl- and 3,7-dimethylbicyclo[3.3.0]octanes in crudes from the Norio and Anastasiyevsko-Troyitskoe oil fields (horizon IV) (Sokolova et al. 1981).

Table 33 and Fig. 32 present equilibrium concentrations of hydrocarbons belonging to different $C_{10}H_{18}$ series.

C_{10} bicyclanes are characterized by large amounts of thermodynamically stable *trans*-decalin (*trans*-isomer – 0.27%; calculated for the Gryazevaya Sopka crude, *cis*-isomer – 0.02%). Hydrocarbons of the decalin series comprise almost half of all C_{10}, C_{11} and C_{12} bicyclanes in crudes (Petrov 1971). The ratio of total decalins to other bicyclanes equals 57 : 43, 43 : 57 and 49 : 51 for C_{10}, C_{11} and C_{12} bicyclanes, respectively. *Cis*-3-methyl-*trans*-bicyclo[4.4.0]- and *trans*-2-methyl-*trans*-bicyclo[4.4.0]decanes in equal concentrations (0.15% calculated for petroleum) were reliably identified among the methyl decalins. The same C_{11} cyclanes

Table 32. Relative distribution of $C_{10}H_{18}$ bicyclanes in the Gryazevaya Sopka crude and in an equilibrium isomerizates

Hydrocarbon	Content (%)	
	In Petroleum[a]	In 450 K isomerizate
1,5-Dimethylbicyclo[3.2.1]octane	19.5	24.0
1,3-Dimethylbicyclo[3.2.1]octane, *exo*	9.0	20.5
1,6-Dimethylbicyclo[3,2,1]octane, *exo*	9.0	20.5
Σ Dimethylbicyclo[3.2.1]octanes	28.5	44.5
1,4-Dimethylbicyclo[3.3.0]octane, *exo*	7.0	10.0
1,3-Dimethylbicyclo[3.3.0]octane, *endo*	11.5	11.0
1,3-Dimethylbicyclo[3.3.0]octane, *exo*	9.5	10.0
2,4-Dimethylbicyclo[3,3,0]octane, *exo*, *exo*	6.0	5.0
1,5-Dimethylbicyclo[3.3.0]octane	4.5	1.5
1,4-Dimethylbicyclo[3.3.0]octane, *endo*	17.5	4.5
2,8-Dimethylbicyclo[3.3.0]octane, *exo*, *exo*	17.5	4.5
2,7-Dimethylbicyclo[3.3.0]octane, *exo*, *endo*	9.5	7.0
Σ Dimethylbicyclo[3.3.0]octanes	65.5	49.0
1,4-Dimethylbicyclo[2.2.2]octane	6.0	6.5
Σ	100.0	100.0

[a] Total content of all hydrocarbons amounted to 0.04% in petroleum.

Table 33. Equilibrium concentrations of C_{10} bicyclanes of various series (%)

Hydrocarbons	300 K	450 K	625 K
Bicyclo[3.2.1]octanes	5	16	19
	(55.9)	(31.7)	(28.6)
Bicyclo[3.3.0]octanes	2	20	31
	(22.3)	(39.6)	(46.7)
Bicyclo[2.2.2]octanes	1.3	2	2.5
	(14.4)	(4.0)	(3.7)
Bicyclo[4.3.0]nonanes	0.7	12.5	14
	(7.8)	(24.7)	(21.7)
Bicyclo[4.4.0]decanes	90	43	23
Others	1.0	6.5	10.6

Note: Concentrations relative to the sum of all bicyclanes, other than bicyclo[4.4.0]decanes, are in parentheses.

Table 34. Equilibrium concentrations of $C_{12}H_{22}$ bicyclo[4.4.0]decanes (%)

Hydrocarbons (*trans*-epimers)	Stereoisomers[a]	Equilibrium concentrations (%)		
		Experiment		Calculation (300 K)
		300 K	400 K	
3,3-Dimethylbicyclo[4.4.0]decane	–	6.5	5.0	4.0
3,9-Dimethylbicyclo[4.4.0]decane	*cis, cis*	26.0	16.5	30.5
3,8-Dimethylbicyclo[4.4.0]decane	*cis, cis*	26.0	16.5	30.5
2,9-Dimethylbicyclo[4.4.0]decane	*trans, cis*	8.5	10.0	7.0
2,4-Dimethylbicyclo[4.4.0]decane	*trans, cis*	8.0	10.0	7.0
2,8-Dimethylbicyclo[4.4.0]decane	*trans, cis*	–	–	7.0
3,9-Dimethylbicyclo[4.4.0]decane	*cis, trans*	14.0	18.5	1.5
3,8-Dimethylbicyclo[4.4.0]decane	*cis, trans*	–	–	1.5
3,4-Dimethylbicyclo[4.4.0]decane	*cis, cis*	6.5	8.0	7.0
2,5-Dimethylbicyclo[4.4.0]decane	*trans, trans*	1.5	3.0	1.5
2,3-Dimethylbicyclo[4.4.0]decane	*trans, cis*	1.5	2.0	1.5
3-Ethylbicyclo[4.4.0]decane[b]	*cis*	1.0	3.5	1.0
Others	–	0.5	7.0	–

[a] Orientation of the methyl groups; *trans* ring junction throughout.
[b] The stability of low 3-ethylbicyclo[4.4.0]decane is caused by the presence of only one substituent.

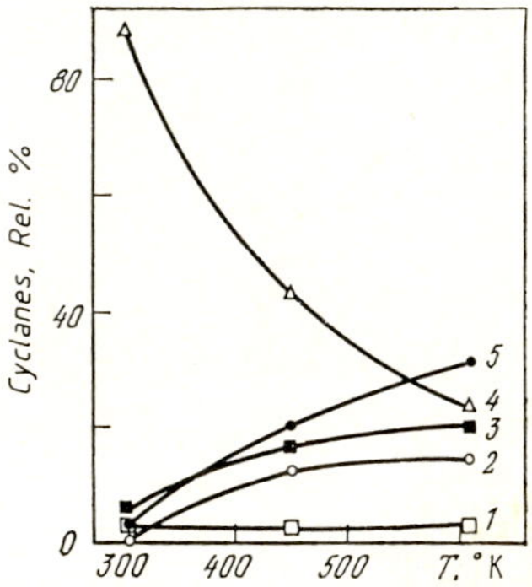

Fig. 32. Relative distribution of C_{10} bicyclanes in equilibrium mixtures at various temperatures. Hydrocarbon series of: *1* bicyclo[2.2.2]octane; *2* bicyclo[4.3.0]nonane; *3* bicyclo[3.2.1]octane; *4* bicyclo[4.4.0]decane; *5* bicyclo[3.3.0]-octane

were found in Surgut petroleum (see Fig. 16). Among the simplest bicyclanes, and besides those discussed above catalytic dehydrogenation allowed the identification of dicyclohexyl (Sevostyanova et al. 1967).

Let us underline that there are 81% 3-methylbicyclo[4.4.0]decanes, 18% 2-methylbicyclo[4.4.0]decanes and ≈1% 1-methylbicyclo[4.4.0]decanes (epimer mixtures throughout) in an equilibrium mixture of methylbicyclodecanes at 400 K.

Equilibrium concentrations of di- and trimethyldecalins are given in Tables 34 and 34a (Petrov 1971; Berman et al. 1973); a chromatogram of an equilibrium mixture of dimethylbicyclo[4.4.0]decanes is presented in Fig. 33.

Table 34a. Equilibrium concentrations of the most stable trimethylbicyclo[4.4.0]decanes at 300 K

Hydrocarbon[a]	Equilibrium concentrations[b] (%)	
	Experiment	Calculation
3,3,8-Trimethylbicyclo[4.4.0]decane, *cis*	28.5	12.5
3,3,9-Trimethylbicyclo[4.4.0]decane, *cis*	28.5	12.5
2,4,9-Trimethylbicyclo[4.4.0]decane, *trans, cis, cis*	29.0	25.0
2,4,8-Trimethylbicyclo[4.4.0]decane, *trans, cis, cis*	22.5	25.0
3,4,8-Trimethylbicyclo[4.4.0]decane, *cis, cis, cis*	20.5	25.0

[a] Orientation of the methyl substituents; *trans* ring junction throughout.
[b] Total amount of tabulated hydrocarbons is 80% of the equilibrium mixture.

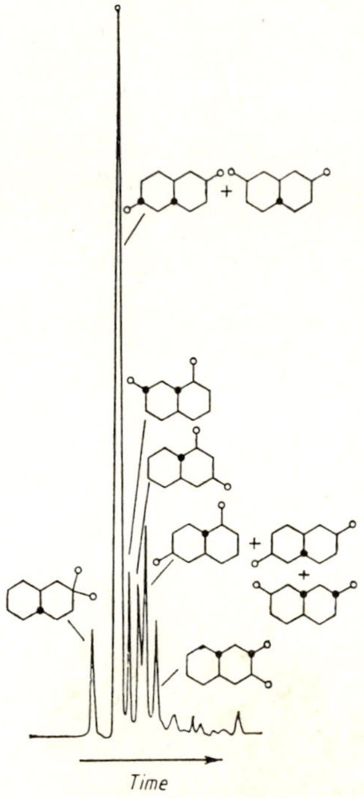

Fig. 33. Chromatogram of an equilibrium mixture (400 K) of dimethylbicyclo[4.4.0]decanes

$C_{14}-C_{16}$ Hydrocarbons Sesquiterpane Hydrocarbons

Polymethyl-substituted C_{15} decalins, i.e. hydrocarbons of the sesquiterpane type, were discovered in crudes of the Loma Novia and Anastasievsko-Troyitskoe deposits (Bendoraitis 1973; Kagramanova et al. 1976). The same hydrocarbons, together with their next highest homologs, were identified in Pre-Cambrian crude of the Sivinski deposit (Vorobyova et al. 1973). There are very high concentrations of these hydrocarbons in Sivinski crude. A chromatogram of a concentrate of these hydrocarbons (fraction 200°–300°C), consisting of 56% bicyclanes, is presented in Fig. 34. The structural formulae of these compounds (XVI–XXII) are obviously of relict character. The ring junction is always *trans*. Methyl (ethyl) substituents at C-2 and C-3 are always equatorial (applies only when one substituent at C-2 and C-3 is present):

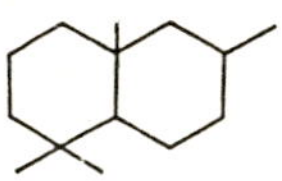
XVI

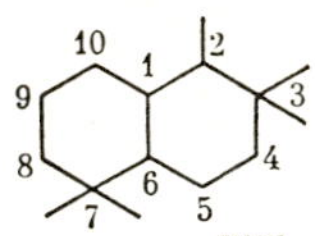

XVII

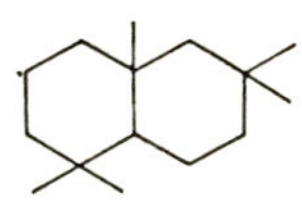
XVIII

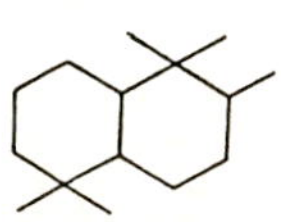
XIX

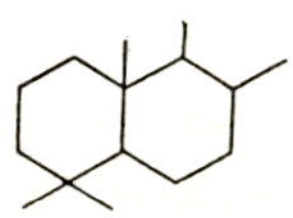
XX

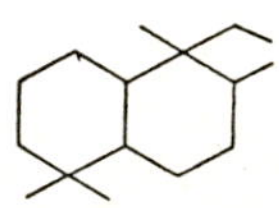
XXI

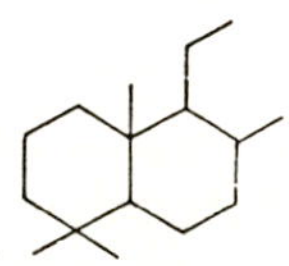
XXII

Structure XX was also found in Loma Novia crude, structures XVII and XIX, i.e. *trans*-2,3,3,7,7- and *cis*-2,2,3,7,7-pentamethylbicyclo[4.4.0]decanes were identified in Anastasievsko-Troyitskoe crude. At the same time, there is a considerable quantity of similarly structured hydrocarbons in Sivinski crude. Compounds XVII–XIX represent $C_{15}H_{28}$ sesquiterpanes. The structural similarities and stereochemical characteristics of these compounds are supposedly rooted in their common origin. Cyclization of squalene or some other higher aliphatic polyprene with subsequent chain degradation (see below) is a most probable way of their formation. In such a case, the first to appear is an XX compound, which may then be transformed into other structural isomers (XIX, XXIII, etc.) through subsequent 1,2-shifts of methyl groups.

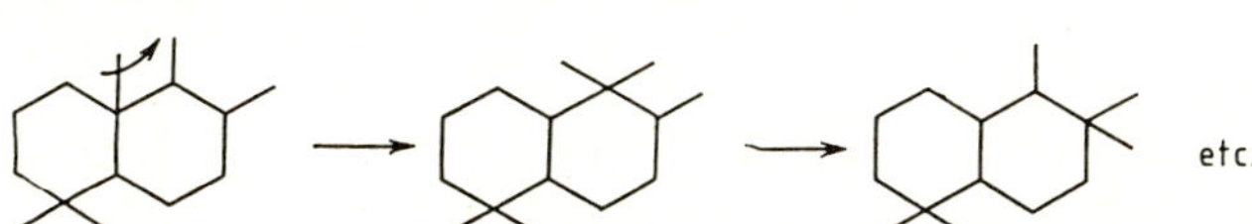

XX XIX XXIII

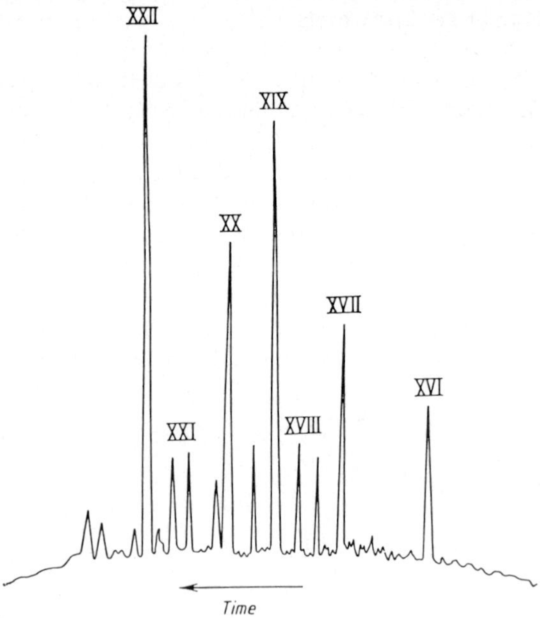

Fig. 34. Chromatogram of a $C_{14}-C_{16}$ bicyclane concentrate of Sivinski crude. For *XVI–XXII* peak identifications, see text. Capillary column 80 m, Apiezon

Hydrocarbon XX is called 8β (H)-drimane. A homologous series of these hydrocarbons with 15 to 24 carbon atoms and in isoprenoid chain at C-9 has been reported (Dimer et al. 1984). The numbering of carbon atoms is as adopted from natural product nomenclature.

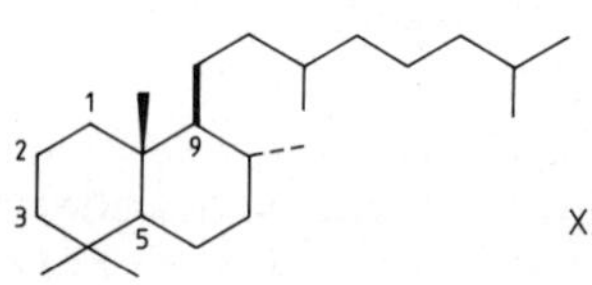

XXIII[a]

Besides drimane, the sesquiterpane eudesmane was isolated from crudes (Dimler et al. 1984):

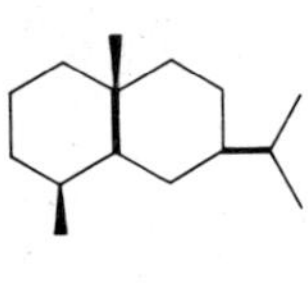

XXXIII[b]

To complete the study of bicyclic naphthenes, perhydrocarotane (XXIV), a peculiar hydrocarbon with a pair of isolated rings, discovered in Sivinski crude and in Green River shale will be analyzed (Ten Fu Yene and Chilingarian 1976; Petrov 1981):

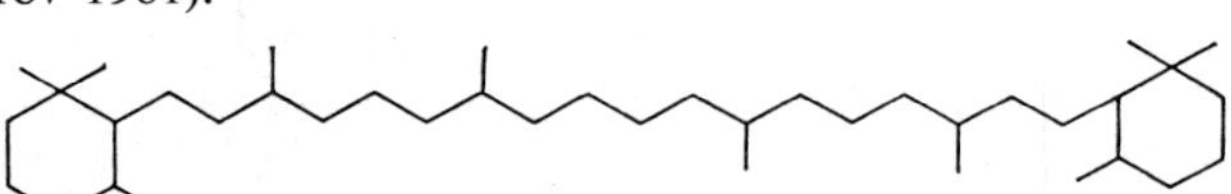

XXIV

As mentioned previously, it may be assumed that this hydrocarbon serves as a source of the 1,1,3-trimethyl-2-alkylcyclohexane hydrocarbon series.

Tricyclic Naphthenes

Depending on their relative ring arrangement, tricyclic petroleum hydrocarbons may be subdivided into three groups:

1. Hydrocarbons of a bridged structural type.
2. Hydrocarbons with a condensed ring system.
3. Hydrocarbons of a mixed structural type, i.e. hydrocarbons having both bridged and condensed ring junctions.

Hydrocarbons of the Adamantane Series

Adamantane, or tricyclo[3.3.1.1^{3,7}]decane (XXV), belongs to the purely bridged hydrocarbons:

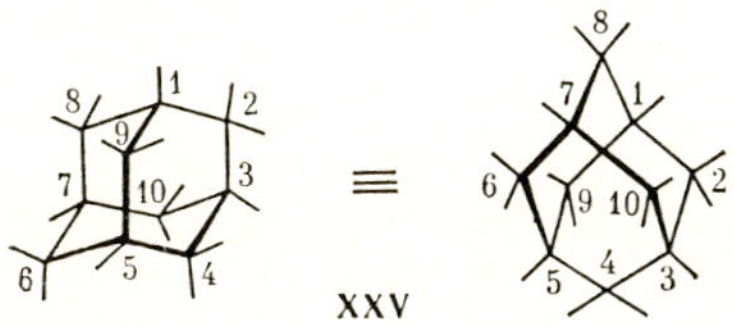

XXV

From a practical view point, adamantane, discovered in crudes by S. Landa in the early 1930's, has become a valuable cyclane in the last 2 decades. A new trend in science, the chemistry of polyhedranes, emerged on its basis.

Due to the development of fairly simple methods of the catalytic synthesis of adamantane and its homologs, the chemistry of adamantane occupies a prominent position in the chemistry of hydrocarbons and their derivatives.

Besides adamantane proper, many of its mono-, di- and trimethyl and -ethyl $C_{11}-C_{15}$ homologs were discovered in crudes (Petrov 1971). The qualitative and quantitative determination of adamantanes in crudes (200°–250 °C fraction) was made considerably more reliable with the invention of hydrocracking, a special process in which non-adamantane hydrocarbons undergo destruction, while adamantane and its homologs remain unchanged (Yakubson et al. 1973).

Table 35 presents data on the content of adamantane series hydrocarbons in various crudes (Yakubson et al. 1973; Sanin et al. 1974; Dididze et al. 1977). In analyzing the results of this research, the conclusion is easily reached that the ethyl substituent is the largest group connected with the nucleus of adamantane. Hydrocarbons having at least one substituent at the bridgehead have a wider occurrence.

At the same time, thermodynamically more stable hydrocarbons, e.g. 1-methyl-, 1,3-dimethyl-, 1,3,5-trimethyl- and 1,3,5,7-tetramethyl-adamantanes, although present in appreciable concentrations, do not belong to the main components of the petroleum mixtures.

Table 35. Relative distribution of adamantanes in crude (%)

Hydrocarbons[a]	Peak number in Fig. 35	Deposit, Crude type: Russkoe, B[1]	Anastasiievsko-Troyitskoe, B[2]	Gryazevaya Sopka, B[1]	Balakhanskoe, B[1]	Norio, B[1]	Sartytchaly, A[1]
C_{11}:							
1-Methyladamantane	2	48	23	13	55	39	45
2-Methyladamantane	5	52	77	87	45	61	55
C_{12}:							
1,3-Dimethyladamantane	3	20.6	15.4	12.0	33	20.4	18.6
1,4-Dimethyladamantane, *cis*	6	17.9	15.4	12.9	13	14.8	15.1
1,4-Dimethyladamantane, *trans*	7	17.5	15.4	12.9	13	15.5	18.6
1,2-Dimethyladamantane	9	18.5	22.6	30.2	18	22.5	21.3
1-Ethyladamantane	12	15.1	13.3	15.8	18	16.9	9.3
2-Ethyladamantane	15	10.4	17.9	16.2	5	9.9	17.1
C_{13}:							
1,3,5-Trimethyladamantane	4	10.2	8.4	4.4	–	14.1	11.2
1,3,6-Trimethyladamantane	8	23.4	16.0	11.8	–	10.6	17.8
1,3,4-Trimethyladamantane, *cis*	10	18.4	22.3	24.7	–	17.7	18.7
1,3,4-Trimethyladamantane, *trans*	11	22.4	22.3	24.7	–	17.8	18.7
1-Methyl-3-ethyladamantane	14	25.6	31.0	34.4	–	38.8	33.6

Note: Peak No. 1 in Fig. 35: adamantane (added); peak No. 13: 1,3,5,6-tetramethyladamantane.

[a] Overall adamantane content in crudes: approximately 0.02 – 0.04%.

Equilibrium mixtures of $C_{11} - C_{14}$ adamantanes comprise over 90% of the methyl-substituted isomers with methyl groups at the bridgehead, i.e. 1-methyl-, 1,3-dimethyl-, 1,3,5-trimethyl- and 1,3,5,7-tetramethyl adamantanes, respectively (Petrov 1971).

Detailed data on equilibrium mixtures of C_{12} adamantanes are listed in Table 36.

Table 36. Equilibrium concentrations of $C_{12}H_{20}$ adamantanes (%)

Isomer[a]	300 K		470 K	
	Calculated	Experiment	Calculated	Experiment
1,3-Dimethyladamantane	97.3	97.0	87.2	87.1
1,4-Dimethyladamantane, *cis*	1.0	1.0	4.7	3.9
1,4-Dimethyladamantane, *trans*	1.0	1.0	4.7	3.9
1,2-Dimethyladamantane	0.4	0.6	1.8	2.4
1-Ethyladamantane	0.3	0.4	1.6	2.7

[a] Only the most stable five of the eleven possible isomers are tabulated.

The elution order of alkyladamantanes (i.e. the sequence of their boiling points) is quite peculiar. Figure 35 shows a gas chromatogram of an adamantane hydrocarbon mixture isolated from Norio crude after hydrocracking. All methyl-substituted adamantanes (substituted at the bridgehead) have much lower boiling points than hydrocarbons with at least one of the substituents not situated at the bridgehead (2-methyl-, 1,2- and 1,4-dimethyladamantanes, etc.). The difference in the boiling points of these adamantanes is so large that 2-methyladamantane (C_{11}) elutes later than 1,3,5,7-tetramethyladamantane (C_{14}).

Special notice should be taken of large concentrations of ethyl- as well as methyl-substituted adamantanes in petroleums, in which one of the substituents is situated away from the bridgehead (1,2-dimethyl-, 1,3,4-trimethyladamantanes, etc.).

These thermodynamically unstable compounds are usually formed first in the isomerization of non-adamantane tricyclic hydrocarbons into adamantane series compounds (Petrov 1971).

They are kinetically favored among the adamantane hydrocarbons and are accumulated in the reaction products due to their isomerization rate. Apparently, similar processes occur in crudes.

Thus the study of the relationship between adamantane hydrocarbons and other $C_{11}-C_{14}$ tricyclanes reveals that non-adamantane tricyclanes comprise 80–90% of the entire amount of tricyclanes in crudes. Thus, an ample potential for the formation of adamantane series hydrocarbons exists, because any tricyclic saturated hydrocarbon is capable of being transformed (naturally in the presence of an appropriate catalyst) into such hydrocarbons.

From this viewpoint, adamantane hydrocarbons in crudes are not the only potential source of adamantane tricyclanes. In general, adamantanes are absent from original biogenic compounds. The presence of these hydrocarbons in crudes indicates the catalytic reactions possible in the presence of acidic catalysts (Ten Fu Yene and Chilingarian 1976). In natural environments the only real catalysts of this kind are aluminosilicates (clays).

Protoadamantane Tricyclanes[2]

Formation of hydrocarbons of the adamantane series in crudes is a multistage reaction which may be represented by the following scheme:

Initial tricyclic hydrocarbons (tricyclanes and, possibly, tricyclenes)

$\downarrow k_1$

Prototadamantane tricyclanes

$\downarrow k_2$

Kinetically favoured adamantanes (2-methyl-, 1,2-dimethyl-, 1-ethyl-adamantane, etc.)

$\downarrow k_3$

Thermodynamically stable adamantanes (1-methyl-, 1,3-dimethyl-, 1,3,5-trimethyl-adamantanes)

The rate constants k_1, k_2, k_3 vary usually with $k_1 > k_2 \gg k_3$. As revealed in analysis, petroleum $C_{11}-C_{12}$ tricyclanes are represented mostly by non-adamantane tricyclanes of chiefly protoadamantane structure (Vorobyova et al. 1975, 1977, 1979).

C_{11} Protoadamantane Hydrocarbons. As is shown in the publication of Vorobyova et al. (1975), in the course of isomerization of any C_{11} tricyclic hydrocarbon, for example, tricyclanes XXVI and XXVIII:

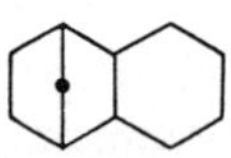

XXVI

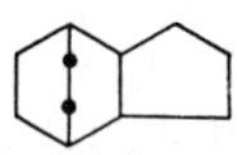

XXVII

which are saturated analogs of cyclopentadiene and cyclohexadiene codimers, a pseudoequilibrium mixture rapidly evolves, consisting of three hydrocarbons (XXVIII–XXX) in a 5 : 1 : 1 proportion:

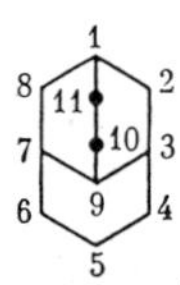

XXVIII

Tricyclo(5,3,1,$0^{3,8}$)undecane (homoisotwistane)

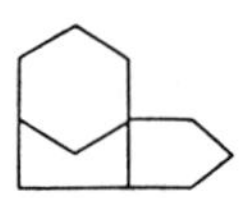

XXIX

Tricyclo(5,3,1,$0^{1,5}$)undecane (trisnormethylcedrane)

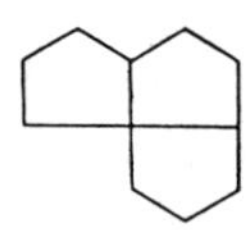

XXX

Tricyclo(6,3,0,$0^{1,5}$)undecane (1,2-trimethylenepentalane)

Later, the XXVIII+XXIX+XXX mixture gradually transforms into 1-methyl- and 2-methyladamantanes (in $\approx 1:1$ proportion).

[2] Protoadamantanes are the precursors of adamantanes, not to be confounded with protoadamantane, an unstable $C_{10}H_{16}$ hydrocarbon with seven- and five-membered rings.

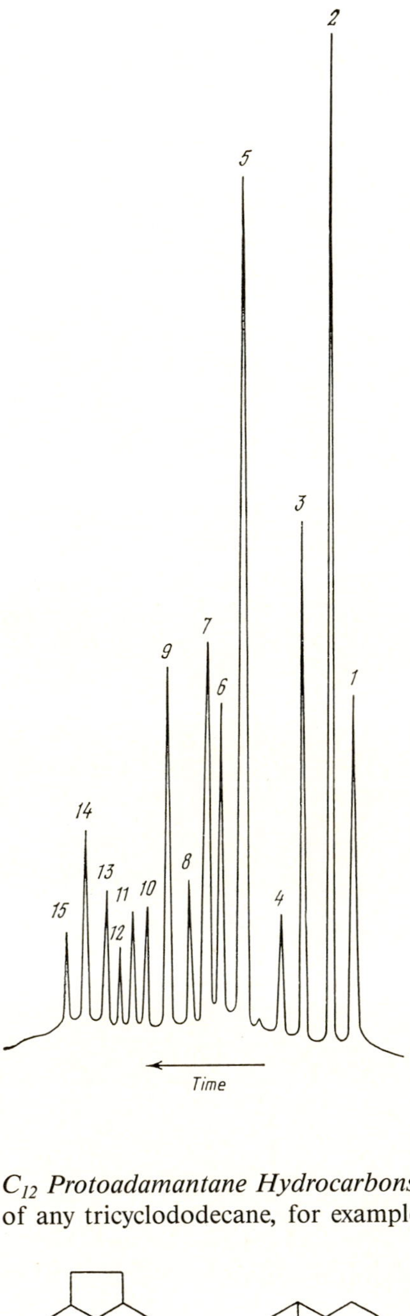

Fig. 35. Chromatogram of an alkyladamantane mixture isolated from Norio crude. Capillary column 80 m, Apiezon. *1* Adamantane addition. Other peak identifications, see Table 35

C_{12} Protoadamantane Hydrocarbons. Similarly, in the process of isomerization of any tricyclododecane, for example XXXI or XXXII,

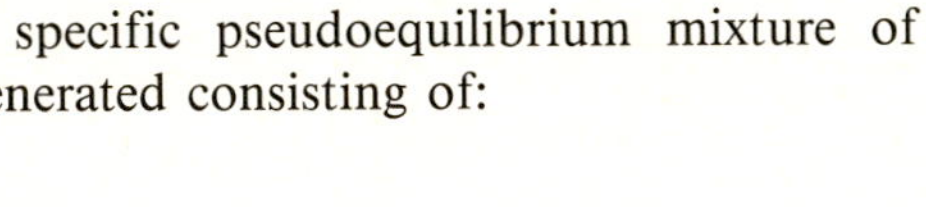

a specific pseudoequilibrium mixture of protoadamantane hydrocarbons is generated consisting of:

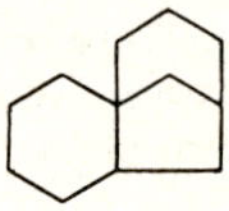

XXXIII

Tricyclo(6,3,1,$0^{1,6}$) dodecane, *endo*

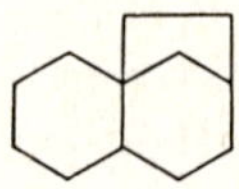

XXXIV

Tricyclo(7,2,1,$0^{1,6}$) dodecane, *endo*

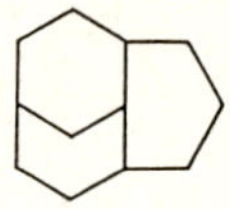

XXXV

Tricyclo(5,3,1,$1^{4,11}$) dodecane, *endo*

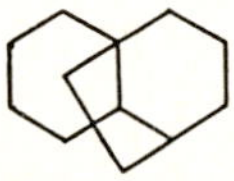

XXXVI

Tricyclo(5,3,2,$0^{1,6}$) dodecane, *anti*

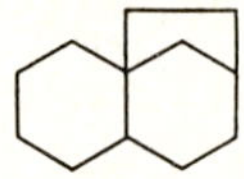

XXXVII

Tricyclo(7,2,1,$0^{1,6}$) dodecane, *exo*

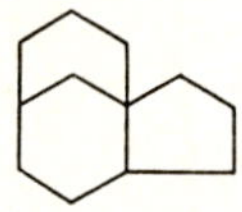

XXXVIII

Tricyclo(6,3,1,$0^{1,5}$) dodecane, *endo*

and 1-methylhomoisotwistane (XXXIX)

Compounds XXXIV, XXXV and XXXVII predominate in this mixture. Later, the XXXIII–XXXIX pseudoequilibrium mixture is slowly transformed into 1- and 2-ethyladamantanes.

Figure 36 shows chromatograms of pseudoequilibrium mixtures of C_{11} and C_{12} tricyclane, and a chromatogram of the concentrate of a tricyclane from the Gryazevaya Sopka crude. The similarity of the two mixtures is fairly obvious. All

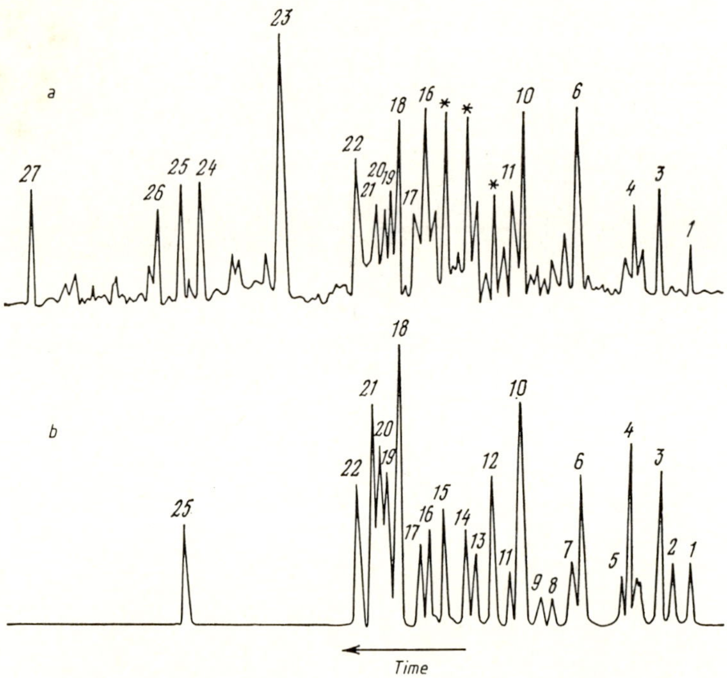

Fig. 36a, b. Chromatograms of (**a**) a C_{11} and C_{12} tricyclane concentrate of the Gryazevaya Sopka crude (200°–260 °C fraction) and (**b**) pseudoequilibrium mixtures of the same hydrocarbons. For peak identifications, see text. Methyltricycloundecanes denoted by *. Capillary column 100 m, Apiezon; 130 °C

the XXVIII–XXXIX hydrocarbons analyzed above, as well as a number of adamantanoid tetracyclic hydrocarbons to be described later, were identified in crude oils. High concentrations of different methyl-substituted tricyclo-undecanes of types XXVIII and XXX are found among the C_{12} tricyclanes in crudes and isomerizates.

Peak identifications for Fig. 36 are as follow:

1. 1-Methyladamantane;
2. 1,3-Dimethyladamantane;
3. Tricyclo[6.3.0.$0^{1,5}$]undecane;
4. Tricyclo[5.3.1.$0^{1,5}$]undecane;
5. Tricyclo[5.2.2.$0^{3,7}$]undecane;
6. 2-Methyladamantane;
7. *exo*-Tricyclo[5.3.1.$0^{2,6}$]undecane;
8, 9. 1,4-Dimethyladamantanes;
10. Tricyclo[5.3.1.$0^{3,8}$]undecane;
11. 1,2-Dimethyladamantane;
12. 1-Methyltricyclo[5.3.1.$0^{3,8}$]-undecane;
13. 1-Ethyladamantane;
14, 15. Methyl homologs of tricycloundecane;
16. 2-Ethyladamantane;
17. *exo*-Tricyclo[6.3.1.$0^{1,6}$]-dodecane;
18. *endo*-Tricyclo[7.2.1.$0^{1,6}$]-dodecane;
19. *anti*-Tricyclo[5.3.2.$0^{1,6}$]-dodecane;
20. *endo*-Tricyclo[6.3.1.$0^{1,6}$]-dodecane;
21. *exo*-Tricyclo[7.2.1.$0^{1,6}$]-dodecane;
22. *endo*-Tricyclo[6.3.1.$0^{1,6}$]-dodecane;
23. Tetracyclo[6.3.1.$0^{2,6}0^{5,10}$]do-decane;
24, 26. Methyltetracyclo-[6.3.1.$0^{2,6}$.$0^{5,10}$]dodecanes;
25. Tricyclo[5.3.1.$1^{4,11}$]dodecane (addition);
27. Tetracyclo[6.3.1.$1^{6,10}$.$0^{2,6}$]tri-decane

Apparently, the composition of the initial tricyclic compounds which produce pseudoequilibrium mixtures of petroleum tricyclanes as the result of isomerization, is more diverse than is assumed in the publication of Vorobyova et al. (1977). Specific isomerization conditions could probably be effective ($AlBr_3$, 30 °C, in laboratory experiments, and aluminosilicates, 100 °–150 °C, in the case of petroleum tricyclane formation in nature).

Formation of pseudoequilibrium mixtures is a characteristic feature of petroleum hydrocarbons. Knowledge of their composition may be helpful in the study of complex oil fractions. Thus, in analyzing isomeric transformations of various $C_{10}H_{18}$ bicyclic hydrocarbons, a pseudoequilibrium mixture of three kinetically stable hydrocarbons regularly emerged, namely: 1,4-dimethylbicyclo[3.2.1]octanes, *endo-* and *exo-* and 1,3-dimethylbicyclo[2.2.2]octane. The same hydrocarbons in identical proportions were later identified in certain crudes and condensates (Matveeva et al. 1975).

Hydrocarbons with Perhydrophenanthrene Structure ($C_{19}-C_{26}$ Pentamethylalkylperhydrophenanthrenes)

The publication of Vorobyova et al. (1973) contains a description of a peculiar homologous $C_{19}-C_{26}$ tricyclane series with a perhydrophenanthrene nucleus. These hydrocarbons were found in large amounts in Sivinski crude (Pre-Cambrian), but similar to the previously discussed polymethyldecalins, polymethylalkylperhydrophenanthrenes are present in many crudes and oil shales (Ten Fu Yene and Chilingarian 1976; Connan et al. 1979; Reed 1977a). Figure 37 shows a chromatogram of a concentrate of the saturated hydrocarbon (300 °–420 °C fraction) of the Sivinski crude. Gas chromatography-mass spectrometry helped in identifying a homologous series of tricyclic hydrocarbons (XL–XLVI) with a perhydrophenanthrene skeleton (since these compounds are regarded as fragments of triterpanes, the carbon atom numbering commonly used for steranes and cyclic triterpanes is also applied in their case):

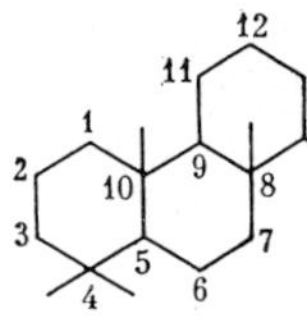

XL

Peak No. 1 (C_{19})

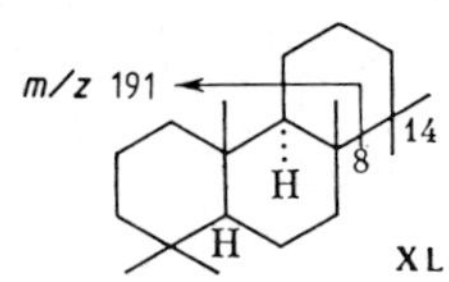

XLI

Peak No. 3 (C_{20})

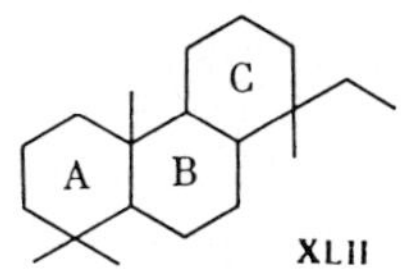

XLII

Peak No. 4 (C_{21})

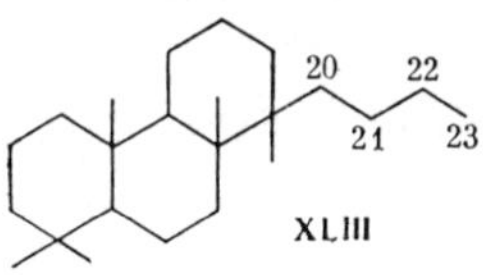

XLIII

Peak No. 6 (C_{22})

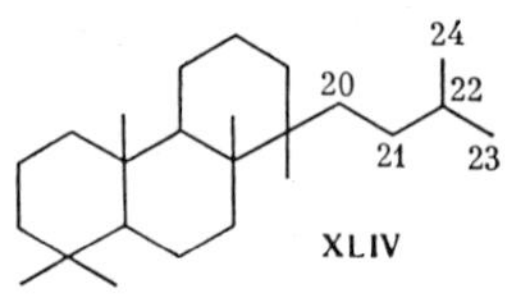

XLIV

Peak No. 7 (C_{24})

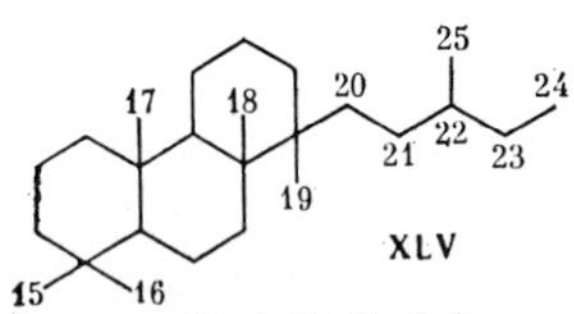

XLV

Peak No. 10 (C_{25})

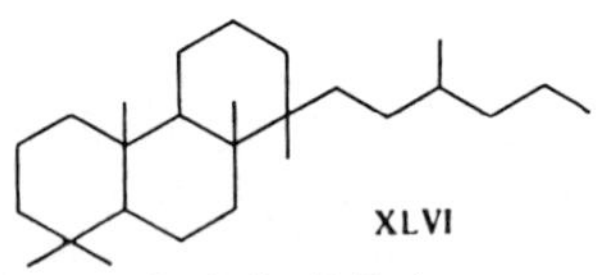

XLVI

Peak No. 11 (C_{26})

The C-22 atom is chiral in the analyzed of C_{25} hydrocarbons, and two epimers elute as a peak doublet (22S and 22R) in crudes. Mass spectra of these hydrocarbons (with the exception of the first homolog) are characterized by a base peak fragment ion at *m/z* 191, formed by the rupture of the C-8/C-14 bond. The rupture of the labile bond connecting two quaternary carbon atoms and the formation of an *m/z* 191 ion are typical of the spectra belonging to any hydrocarbon with this structural unit. It will be further demonstrated that the same fragment ion is the base peak of certain tetracyclanes and the majority of pentacyclanes as well.

All hydrocarbons under review have a *trans*-junction of cycles A/B and B/C, which corresponds to the junction of these rings in most important triterpenoids. Besides the main components (peak Nos. 1, 3, 4, 6, 7, 10, 11), other tricyclanes of similar structure were found in the fraction. Most probably, these hydrocarbons were formed through a possible 1,2-shift of certain methyl groups in the above structures.

A similar situation in the distribution of $C_{19}-C_{26}$ tricyclanes (as well as $C_{24}-C_{26}$ tetracyclanes) was registered in the publication of Connan et al. (1979),

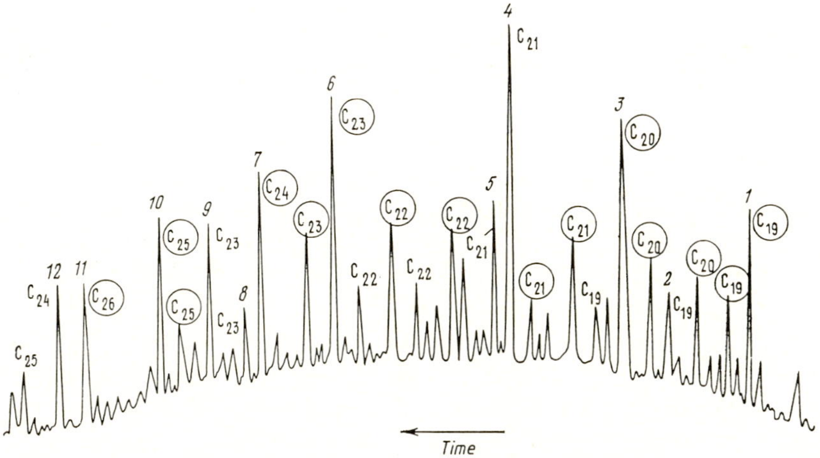

Fig. 37. Chromatogram of a mixture of $C_{19}-C_{25}$ tri- and tetracyclanes of the Sivinski crude (300°–420 °C fraction). Identified are the compositions of tricyclics (*in circles*) and tetracyclics (*without circles*). Peak (*1–12*) identifications are presented in the text (see p. 100, tricyclics; and p. 105, tetracyclics). Capillary column 80 m, Apiezon; linear temperature programming 100°→2°/min

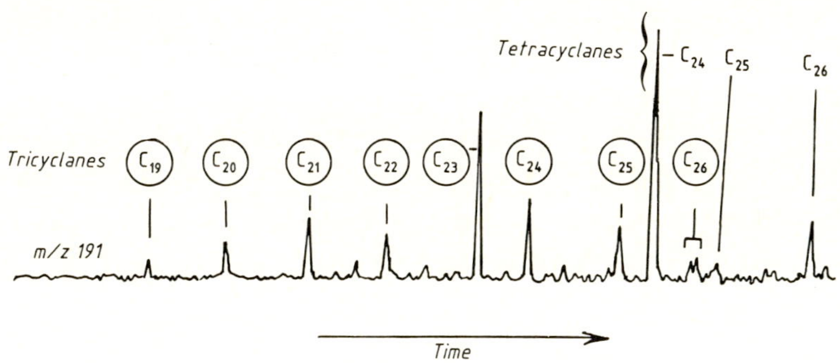

Fig. 38. Massfragmentogram (m/z 191) of a mixture of tri- and tetracyclanes of the Akvitanskaya crude. Peak identifications are the same as in Fig. 37

devoted to the analysis of Cretaceous crude oils in the Aquitaine basin. Figure 38 presents a mass fragmentogram of fractions containing the $C_{19}-C_{26}$ tri- and tetracyclic components of this crude.

The origin of this group of hydrocarbons attracts considerable attention. Different variations have been analyzed in the literature (Ten Fu Yene and Chilingarian 1976; Vorobyova et al. 1973). In our opinion, the formation of these hydrocarbons, as well as a series of other genetically related compounds, may be better explained by the cyclization of squalene or some other closely related aliphatic isoprenoid.

Figure 39 offers a possible scheme for the formation of bi-, tri-, tetra- and pentacyclic hydrocarbons from squalene. The formation of two bonds results in bicyclanes, that of three bonds in tricyclanes, etc. We are certainly not trying to

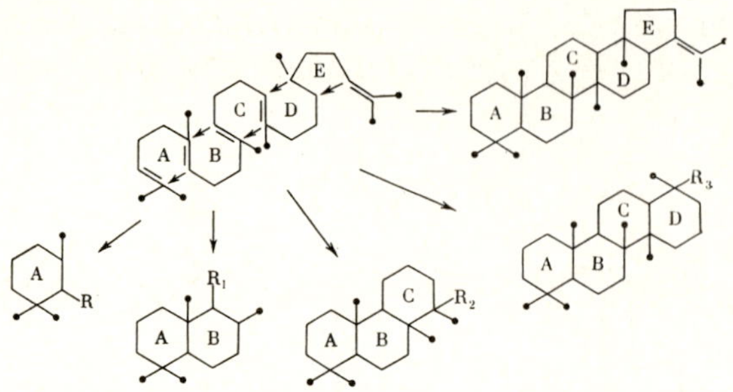

Fig. 39. Probable scheme of formation of mono-, bi-, tri-, tetra- and pentacyclanes of isoprenoid type in stepwise cyclization of squalene

assert that all the hydrocarbons tabulated were generated exclusively in this way. Our task is limited to demonstrating the genetic and structural resemblance of many petroleum polycyclanes. However, their possible sources as well as their ways of cyclization deserve special consideration.

Recently, these tricyclic hydrocarbons, called extended terpanes, acquired considerable geochemical importance. A homologous series of these compounds with 16 to 46 carbon atoms was described in the papers of Aquino Neto et al. (1981); Ekweozor and Strausz (1981); and Moldowan et al. (1983). A slightly different formula for these hydrocarbons, with a methyl group at C-13 rather than C-14, was suggested. Hydrocarbons of this type could be formed either by cyclization of regular aliphatic isoprenoids or, as proposed in the publication of Aquino Neto et al. (1981), on the basis of hexaprenol, a polyunsaturated isoprenoid C_{30} alcohol:

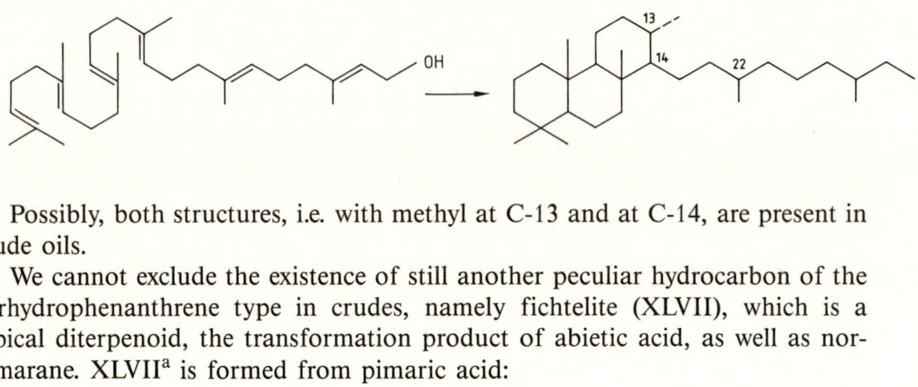

Possibly, both structures, i.e. with methyl at C-13 and at C-14, are present in crude oils.

We cannot exclude the existence of still another peculiar hydrocarbon of the perhydrophenanthrene type in crudes, namely fichtelite (XLVII), which is a typical diterpenoid, the transformation product of abietic acid, as well as norpimarane. XLVII[a] is formed from pimaric acid:

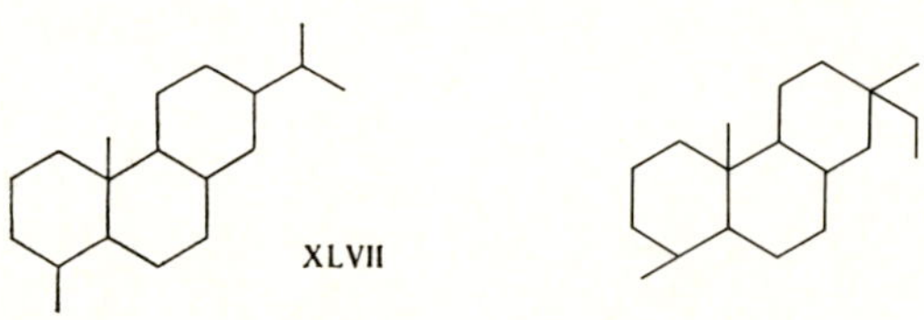

XLVII XLVII[a]

Tetracyclic Naphthenes

Tetracyclic saturated petroleum hydrocarbons represent an important group of biological marker compounds. Steranes are the most important and most widespread examples among these hydrocarbons. But let us first discuss adamantanoid hydrocarbons, as well as certain other compounds of bridged-type ring juction.

Bridged Hydrocarbons $C_{12}-C_{14}$ Adamantanoid Hydrocarbons

Four tetracyclic hydrocarbons (Vorobyova et al. 1979; Weidenhoffer and Hala 1971) of the adamantanoid type (XLVIII – L) were identified in various crude oils (see Fig. 36):

XLVIII
Peak No. 23

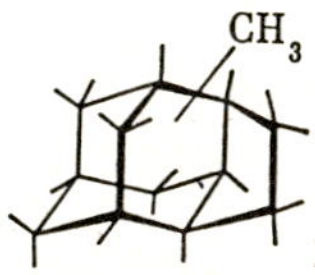
XLIX
Peak No. 24,26

L
Peak No. 27

Tetracyclo[6.3.1.$0^{2,6}$.$0^{5,10}$]dodecane, two methyl homologs of compound XLVIII, and tetracyclo[6.3.$1^{6,10}$.$1^{2,6}$]tridecane.

In addition, a tetradecane (cyclohexanoadamantane) with the following structure was identified:

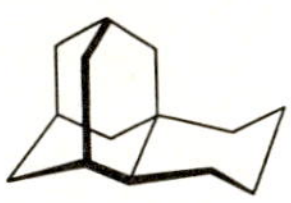
L^a

The presence of methyl homologs of type L^a cannot be excluded.

These hydrocarbons were again generated as the result of isomerization of tetracyclic non-adamantanoid hydrocarbons. Notice should be taken of the high packing degree of carbon atoms in a tetracyclic system. Only 12 – 14 carbon atoms were used here to create four rings. Viewed from this angle, the problem of possible precursors of these hydrocarbons arises. So far, no other tetracyclic systems were discovered in petroleum which are formed with that few carbon atoms. It is noteworthy that the first hydrocarbons discussed here could not be produced by isomerization of other isomeric tetracyclic systems. Adamantanoid petroleum hydrocarbons include a similarly structured pentacyclic hydrocarbon called congressane (LI, diamantane) (Petrov 1971):

LI
Congressane

Diterpenoids (Kauranes, Gibberellins)

Different natural products (plant resins, peat, brown coals, etc.) contain a wide variety of $C_{19}-C_{20}$ bridged tetracyclic diterpanes. Compounds with two similar structures are most common:

1. *Kauranes,* based on the perhydrophenanthrene and bicyclo[3.2.1]octane systems (LII, LIII):

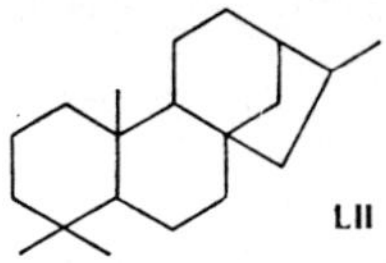
LII (C_{20})

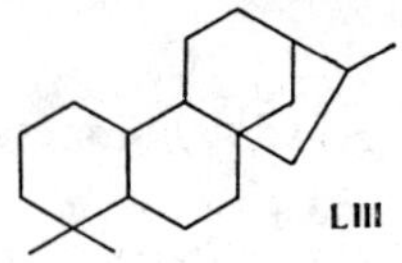
LIII (C_{19})

2. *Gibberellins,* based on the perhydrofluorene and bicyclo[3.2.1]octane (LIV, LV):

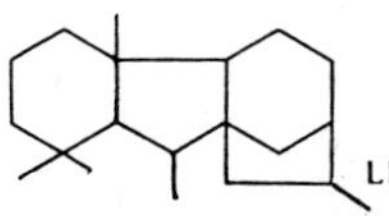
LIV (C_{20})

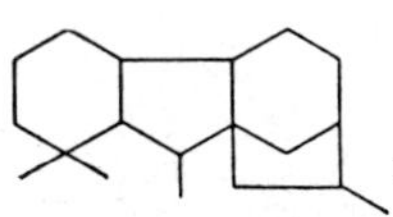
LV (C_{19})

It took time to obtain clear-cut evidence of the presence of these compounds in petroleums, although a group of very light-boiling $C_{19}H_{32}$ tetracyclanes (eluting in the range of nonadecane, was registered by gas chromatography-mass

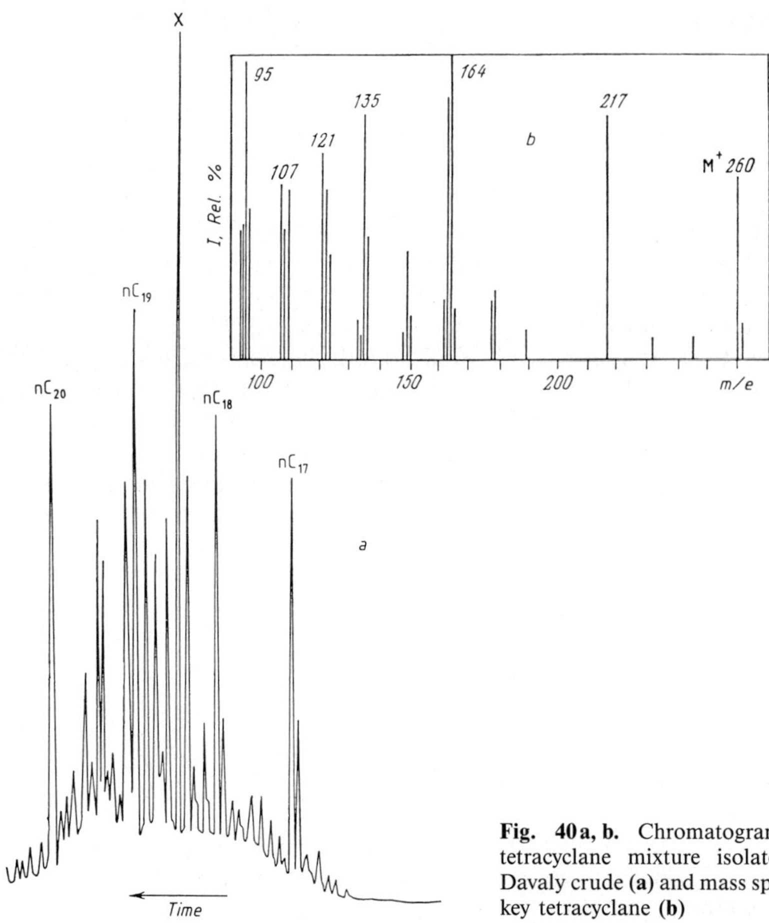

Fig. 40a, b. Chromatogram of a C_{19} tetracyclane mixture isolated from the Davaly crude (**a**) and mass spectrum of the key tetracyclane (**b**)

spectrometry in Davaly crude (type A[1], West Turkmenistan). According to their mass spectra, these hydrocarbons may belong to the just reviewed kauranes and gibberellins.

A gas chromatogram of tetracyclic hydrocarbons in the n-C_{19} range and the mass spectrum of a basic tetracyclane are presented in Fig. 40.

As recent investigations have shown using ^{13}C and ^{1}H NMR spectroscopy, the structure of the main hydrocarbon peak (X) is much more complex and may be represented by formula LV^a (Vorobyova et al. 1986):

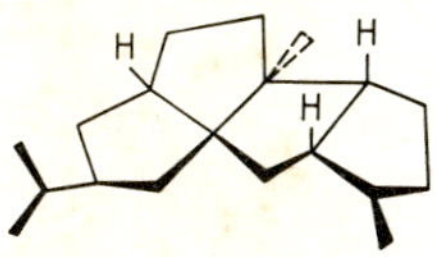

LV^a

Structure LV^a with four cyclopentane rings helps elucidate the genesis of hydrocarbons of the pentalane and tricycloundecane series (XXX). The formation mechanism of the hydrocarbon LV^a is evidently connected with the fermentative C_5 cyclization of monocyclic isoprenoids (Kaneda et al. 1972). At the same time, kaurane and its stereo- and structural isomers – dihydrophyllocladene and beyerane – were finally discovered in Australian crudes (Noble et al. 1985).

Hydrocarbons with a Perhydrochrysene Structure

Certain tetracyclic saturated hydrocarbons having a perhydrochrysene structure (LVI and LVII) were identified in crude oils according to the publications of Vorobyova et al. (1979) and Connan et al. (1979). Previously, these hydrocarbons were identified in Green River shales (Ten Fu Yene and Chilingarian 1976). For the structures and gas chromatographic retention times of these hydrocarbons see Fig. 37 and 38.

The ring junction in the analyzed tetracyclic perhydrochrysene system is *trans* throughout:

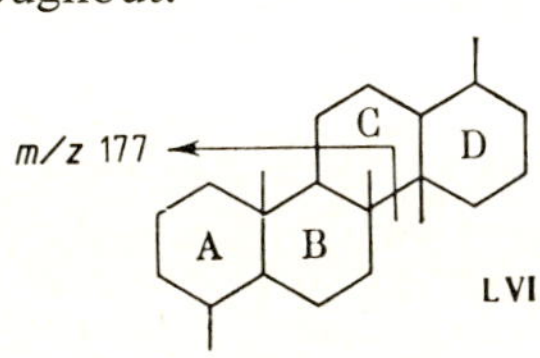

LVI

Peak No. 8 (C_{23})

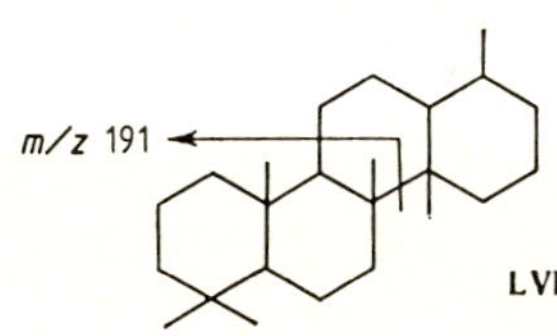

LVII

Peak No. 12 (C_{24})

A description of a homologous series of hydrocarbons with structure LVII is given in the publication of Aquino et al. (1981). These compounds are considered to be seco-(17,21)-hopanes (see below):

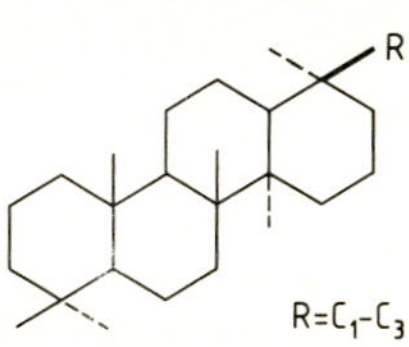

LVII^a

Other tetracyclic $C_{19}-C_{23}$ compounds were discovered in the same Sivinski crude, but their mass spectra suggest a different structure. The tetracyclic nucleus is based on the cyclopentanoperhydrophenanthrene system, i.e. these compounds already belong to the sterane hydrocarbons.

Steranes

Together with hopanes, steranes are the most important, biological marker petroleum hydrocarbons. Although over 15 years have elapsed since the discovery of steranes in geological samples (see: Ten Fu Yene and Chilingarian 1976), the chemistry of these hydrocarbons and their geochemical significance became prominent only in recent years. Significant research was undertaken primarily in the stereochemistry of sterane hydrocarbons, which is a demanding task, because in contrast to natural biological compounds, petroleum steranes proved to be stereochemically and structurally altered to a considerable extent. Through the joint effort of chemists and geochemists from France, Great Britain, the USA and the Soviet Union, this research led to a comprehensive explanation of the composition and stereochemistry of petroleum steranes (Petrov 1981; Eusminger et al. 1978; Mulheirn and Ryback 1975a; Seifert and Moldowan 1979; Pekhk et al. 1982). Overall, there are a few dozen studies devoted to petroleum steranes and especially steranes in dispersed organic material. The importance of these works on the stereochemistry of petroleum steranes can hardly be overestimated, because during dia- and catagenesis in sediments, these compounds undergo complicated, gradual configurational transformations of a number of chiral centres. As will be demonstrated below, it is this epimerization that is a criterion in evaluating the catagenetic maturation of biogenic molecules towards the level of petroleum, and accordingly, in forecasting petroleum depositions in specific regions.

Regular Steranes

A structural formula of regular, i.e. unaltered, steranes (LVIII) with details of their sterical configuration and numbering of carbon atoms is given below (Petrov 1981):

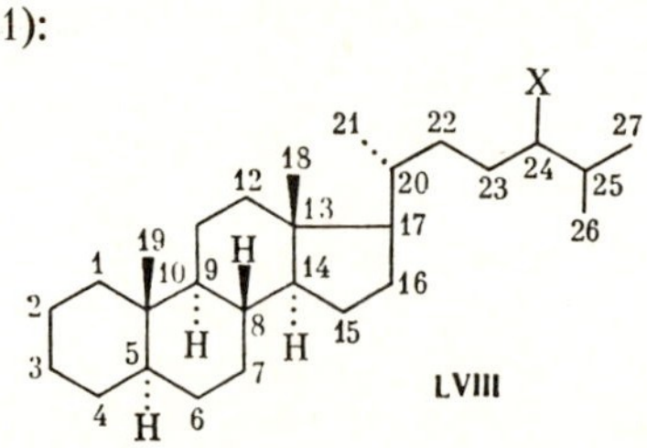

LVIII

$(X=H, CH_3, C_2H_5)$

The chiral centres at C-5, C-14, C-17, C-20 and C-24 (for C_{28} and C_{29} steranes) are possible locations for epimerization. The above structure corresponds to the natural α-cholestane, having a 5α(H), 14α(H), 17α(H), 20 *R* configuration of the respective chiral centres. (According to the Physer nomenclature, the denotations α and β, with the exception of special cases, in the following will indicate the orientation of hydrogen atoms at the corresponding chiral centres. Hence, the symbol *H* will be omitted.) In a nomenclature which is more familiar to petroleum chemists, α-cholestane has *trans*-connections of all rings (A/B, B/C and C/D), and a relative *cis*-orientation of bonds 17–20 and 13–18. The basic fragmentation as a result of electron impact is indicated in Scheme 1.

Scheme 1

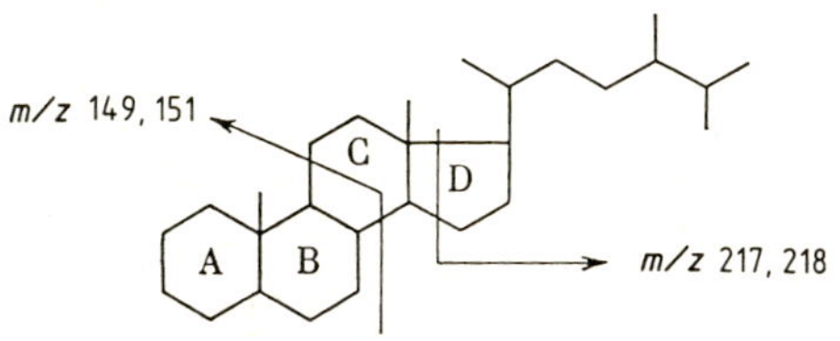

α-Steranes, such as α-cholestane, 24-methyl-α-cholestanes (ergostane and campestane), 24-ethyl-α-cholestanes (sitostane, poriferastane) were the first compounds identified in crudes due to their ability to easily enter into a complex with thiourea. Besides 5α-steranes, there are 5β-steranes in natural samples, i.e. hydrocarbons differing from α-steranes only in the A/B ring junction (*cis*-).

Although α-steranes are usually present in all crudes, the concentrations of stereochemically transformed hydrocarbons are significant, because the sterane configuration derived from the biosynthesis of living organisms, is thermodynamically less stable. The *trans*-junction of the C/D rings is especially unfavorable, because hydrocarbons with *cis*-junction are much more stable in angular-substituted hydrindanes.

Therefore, configurational transformations of the original molecules proceed in sediments under the catalytic influence of minerals. The time and place of these configurational alterations (diagenesis, catagenesis) are still heavily debated. Besides the configurational transformation at C-14, resulting in the appearance of molecules having a *cis*-connection of the C/D rings (14β), epimerization at C-17 and C-20 is also feasible in petroleum (geological) steranes. In the course of such transformations, the so-called isosteranes (two isomers) are generated with the following configuration: 5α,14β,17β,20*R* and 20*S*. These two hydrocarbons are usually easily separable by GC, and are well identifiable in *m/z* 217 mass chromatograms of petroleum hydrocarbons. Their sterical structure was established in an unusual way. Isosteranes, identical to those found in crudes, were produced by catalytic isomerization of 5α-cholestane by the author as described in his works (Petrov et al. 1976; Pustilnikova et al. 1980). Based on their mass spectra, a *cis*-C/D-, i.e. 14β-configuration attributed to these compounds. Analogous steranes were described in crudes by Ryback (Mulheim and Ryback 1975). Later, in the publication of Seifert and Moldowan (1979a), it was established that isosteranes have the 17β-configuration and the two main peaks are 14β,17β,20*S* and 20*R* epimers. However, the final elution order of the 20*S* and 20*R* epimers as well as the unequivocal confirmation of the structure of the most important configurational sterane isomers (e.g. cholestane), were done independently in the works of Pekh et al. (1982), MacKenzie et al. (1980), and Moldowan et al. (1980).

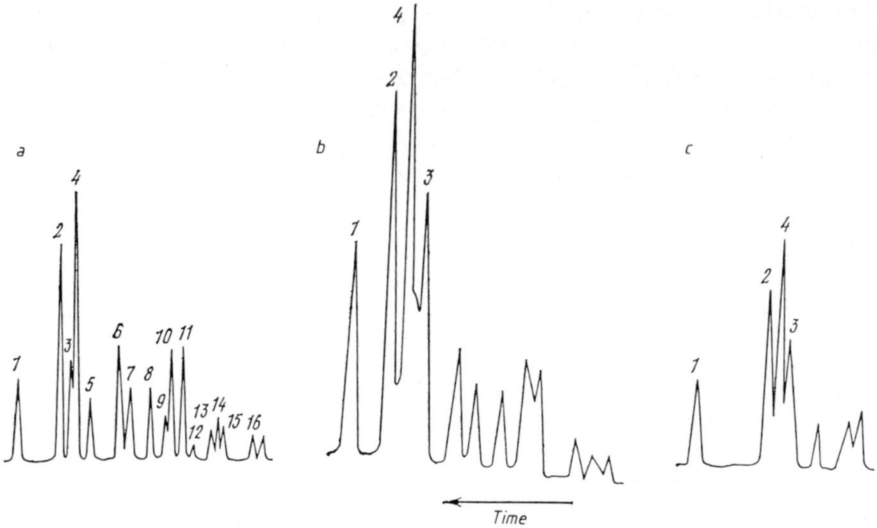

Fig. 41 a – c. Chromatograms of equilibrium mixtures of various sterane epimers. **a** Cholestanes, capillary column 80 m, Apiezon, 300 °C; **b** cholestanes, capillary column, Dexil, 270 °C; **c** 24-ethyl-cholestanes, Apiezon, 280 °C. *1* 5α,20R; *2* 5α,14β,17β,20S; *3* 5α,20S; *4* 5α,14β,17β,20R. For complete peak identifications, see Table 37. Peak numbers start at the end of the chromatogram

Besides the three stereoisomers under review, geochemistry uses the primary epimerization product of biosteranes, at the chiral centre C-20, i.e. 5α, 14α, 17α, 20*S* isomers (for brevity these isomers will be identified as 5α, 20*R*; 5α,20*S*; 14β,20*R* and 14β,20*S*).

Gas chromatograms of cholestane (C_{27}) and sitostane (C_{29}) equilibrium mixtures are presented in Fig. 41. As expected, the equilibrium mixtures of these hydrocarbons are similar to each other. Table 37 presents the composition of an equilibrium mixture of cholestane.

As can be easily observed in Fig. 41 as well as in Table 37, isosteranes (5α,14β,17β,20*S*, and 20*R* epimers) are the main components in an equilibrium mixture consisting of 16 compounds, while the total concentration of epimers with a *cis*-junction of the C/D rings (14β) in the isomerizate reaches 75%). For geochemical purposes an isosterane/α-sterane relationship is often used, i.e. the 14β,17β,20*R*+20*S*/5α,20*R* ratio, and is sometimes called the aging (maturation) coefficient (Petrov et al. 1976; Seifert and Moldowan 1981), as well as the 5α,20*S*/5α,20*R* relationship, called the migration coefficient (Seifert and Moldowan 1981). A $C_{27}:C_{28}:C_{29}$ sterane carbon number ratio is of certain genetic interest, although it is more or less the same in most crudes. Mass spectrometry provides clear-cut readings of the junction of the C/D ring in steranes. For all epimers with *cis*-C/D-connection, the intensity of the *m/z* 218 ion is greater than that of the *m/z* 217 ion.

Figure 42 presents a typical gas chromatogram of a mixture of petroleum steranes. Since the elution range of epimeric steranes of the same mass is fairly large, the above coefficients should preferably by determined for epimeric

Table 37. Equilibrium concentrations of cholestane isomers

Isomers	Type of ring junction A/B	Type of ring junction C/D	Orientation of the substituent at C-17	Equilibrium concentration (%) Calculation 400 K	Calculation 570 K	Experiment 570 K	Peak number in Fig. 41
5α,14β,17β,20R[a]	*trans*	*cis*	*trans*	21.1	15.6	22.0	2
5α,14β,17β,20S[a]	*trans*	*cis*	*trans*	21.1	15.6	23.1	4
5α,14β,17α,20R	*trans*	*cis*	*cis*	7.5	7.7	4.4	7
5α,14β,17α,20S	*trans*	*cis*	*cis*	7.5	7.7	7.1	6
5α,14α,17β,20R	*trans*	*trans*	*trans*	4.6	5.5	1.6	13
5α,14α,17β,20S	*trans*	*trans*	*trans*	4.6	5.5	3.0	9
5α,14α,17α,20R[b]	*trans*	*trans*	*cis*	4.6	5.4	6.4	1
5α,14α,17α,20S	*trans*	*trans*	*cis*	4.6	5.4	7.3	3
5β,14β,17β,20R	*cis*	*cis*	*trans*	6.8	7.1	6.5	10
5β,14β,17β,20S	*cis*	*cis*	*trans*	6.8	7.1	6.7	11
5β,14β,17α,20R	*cis*	*cis*	*cis*	2.5	3.6	1.6	15
5β,14β,17α,20S	*cis*	*cis*	*cis*	2.5	3.6	2.4	14
5β,14α,17β,20R	*cis*	*trans*	*trans*	1.5	2.6	1.1	16
5β,14α,17β,20S	*cis*	*trans*	*trans*	1.5	2.6	2.2	12
5β,14α,17α,20R[c]	*cis*	*trans*	*cis*	1.4	2.5	1.9	5
5β,14α,17α,20S	*cis*	*trans*	*cis*	1.4	2.5	2.7	8

Note: Unusual numbering of peaks is explained by the fact that the most important and better identifiable epimers elute at the end of the chromatogram.
[a] Isosteranes.
[b] α-Cholestane.
[c] Coprostane.

sitostanes (C_{29}), because these hydrocarbons are least affected by coeluting sterane admixtures of differing molecular mass.

More precise results (less background interference in the chromatograms) can be obtained using mass fragmentograms on the basis of the most characteristic sterane fragment ion, *m/z* 217 or the molecular ions. For appropriate examples, see Fig. 43. Mass fragmentograms show the enhanced chromatographic resolution by computer reconstructions based on different ions, although quantitative correlations are not always preserved.

Similar $C_{mat.}$ readings have been reported by Seifert (Seifert and Moldowan 1981). Apparently, the main role here is being played not by the age of the oil-bearing rocks, but the burial depth, hence, the thermal conditions in the layers. It is also noteworthy from the genetic point of view that Pre-Cambrian crudes contain only C_{29} steranes (Petrov et al. 1976; Arefyer et al. 1980).

Maturation coefficient values, calculated to concentrations of 5α,14β,17β steranes (isosteranes) are widely used in the evaluation of the maturity of dispersed organic matter in prospective oil regions.

Figure 44 reproduces the change of isosterane concentrations with increasing depth of Toarcian shales, which are potential source rocks in the Paris Basin (MacKenzie et al. 1980).

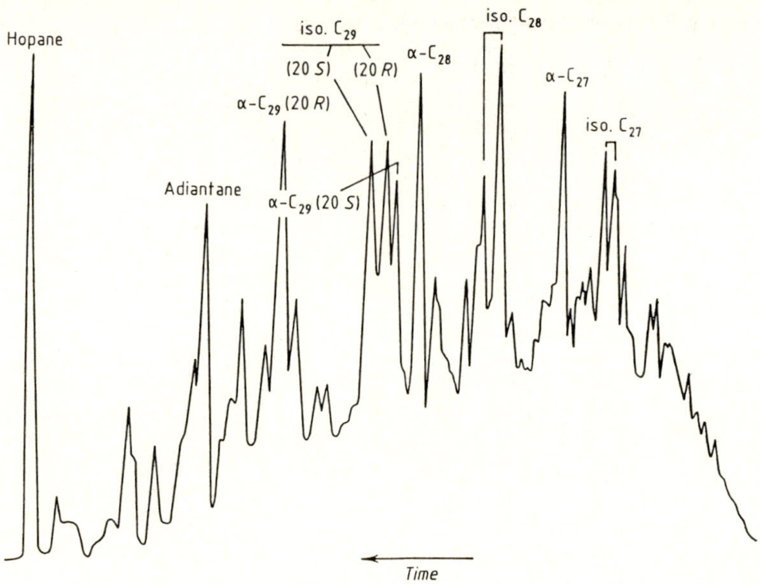

Fig. 42. Chromatogram of a sterane (hopane) concentrate in the Gurgyanskoe crude (420°–500 °C fraction). Identified are the most important $C_{27}-C_{29}$ sterane peaks and their partial stereochemical configurations. Capillary column 80 m, Apiezon; 290 °C

Another coefficient – that of migration – was suggested by Seifert (Seifert and Moldowan 1981). It is calculated on the basis of the 5*a*, 20*S*/5*a*, 20*R* ratio[3], i.e. it is connected to the epimerization of the C-20 chiral centre in initial *a*-steranes. The maximum value of this coefficient may be 1–1.2 (Petrov et al. 1976).

Table 38 presents data on the relative concentrations of steranes in crude oils from different deposits, as well as the values for the maturation (aging) coefficient of petroleums, computed according to the ratio:

$$k \text{ maturation} = \frac{14\beta,20R + 14\beta,20S}{5a,20R} = \frac{\text{iso}}{a}.$$

Table 38 clearly indicates the gradual increase of the maturation coefficient with the age of the oil-bearing rocks (though some exceptions do occur). Maximum (equilibrium) values for the coefficient derived from computed and experimental data range between 8 and 6.5 for temperatures of 400–570 K (Petrov 1981).

The theoretical basis for this coefficient has been previously published (Mulheim and Ryback 1975b; Ryback 1976; Kayukova et al. 1981a), which demonstrated the feasibility of liquid chromatographic separation of 20*R* and 20*S* epimers on aluminium oxide and zeolites 10X. However, the ratio of 20*S* and 20*R* epimers is also influenced by catagenesis, because in geological situations an

[3] The 20*S*/(20*S* + 20*R*) ratio is sometimes used. At maximum transformation it reaches a value of 0.55–0.60 in this case.

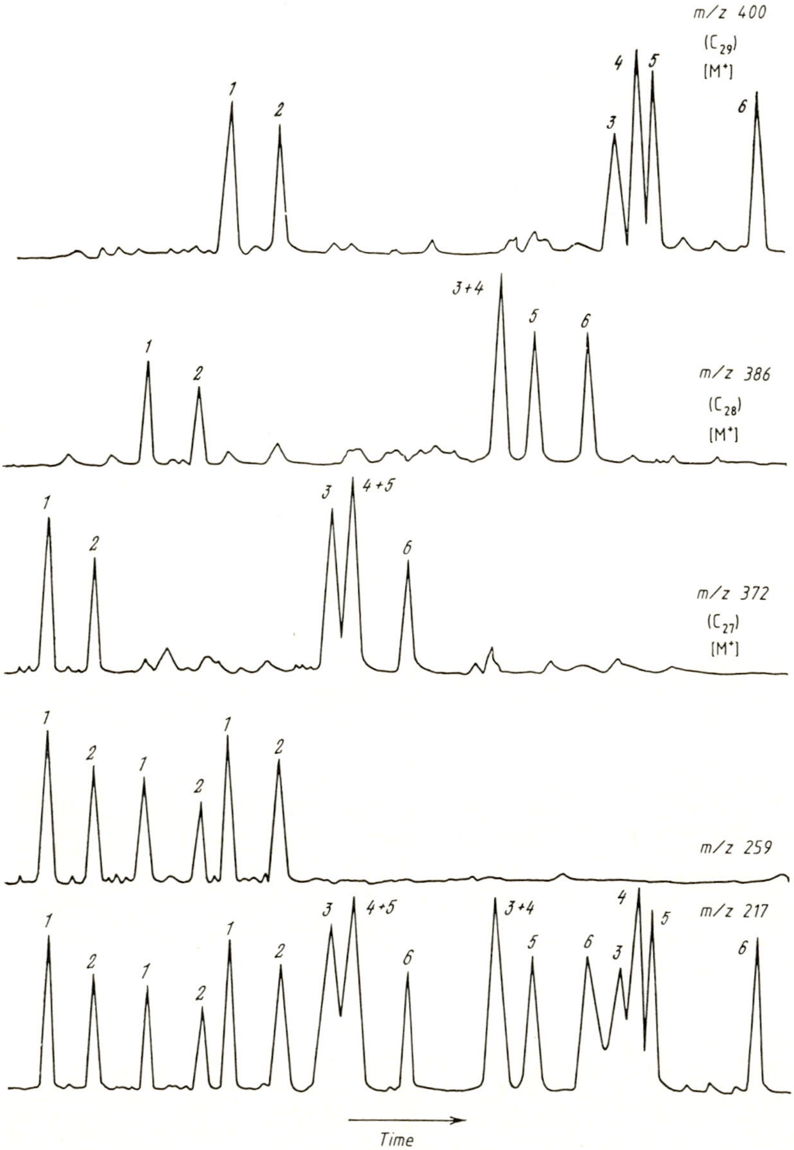

Fig. 43. Computer reconstruction (mass fragmentogram) of a sterane mixture in the Samotlor crude. Molecular ions for C_{29}, C_{28} and C_{27} steranes: m/z 400, 386 and 372; basic fragment ion m/z 259 for rearranged steranes. Fragment ion m/z 217 is the key ion for all steranes. *1, 2* Rearranged steranes (diasteranes-I) 20S and 20R, respectively; *3* 5α,20S; *4* 5α,14β,17β20R; *5* 5α,14β,17β,20S; *6* 5α,20R; LKB-2090 GC/MS system: LKB-2130 computer (peak heights may differ from those in regular chromatograms). Capillary column 30 m, Apiezon

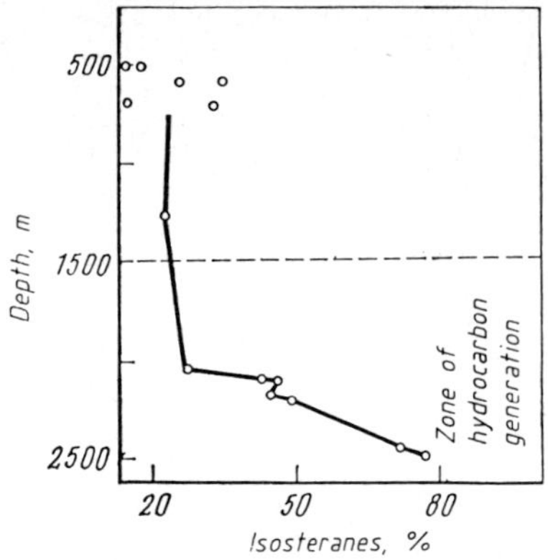

Fig. 44. Changes in isosterane (5α,14β,17β,20R and 20S) concentrations with an increase of burial depth (Toarcian shales). Initial hydrocarbon: 5α,14α,17α,20R

Table 38. Relative distribution of steranes in crudes (%)

Deposit	Cholestanes (C_{27})			Ergostanes (C_{28})			Sitostanes (C_{29})			Maturation K
	ISO	α	Σ ISO + α	ISO	α	Σ ISO + α	ISO	α	Σ ISO + α	
Balakhanskoe (Cenozoic)	19.2	9.4	28.6	24.0	9.1	33.1	24.1	14.2	38.3	1.7
Banka Darvina (Cenozoic)	14.2	8.2	22.4	14.2	9.3	23.5	23.5	14.1	37.6	1.7
Gryazevaya Sopka (Cenozoic)	14.5	11.3	21.8	25.6	13.4	39.0	24.8	14.4	39.2	1.7
Gyurgyanskoe (Cenozoic)	23.6	9.9	33.5	17.4	8.7	26.1	26.4	14.0	40.4	1.9
Sartytchala (Cenozoic)	36.1	5.7	40.8	26.9	6.2	33.1	20.4	5.7	26.1	3.6
Ostasakovitchi (Paleozoic)	–	–	–	–	–	–	–	–	–	3.6
Pripyatskoe (Paleozoic)	–	–	–	–	–	–	–	–	–	3.6
Stepnoozerskoe (Paleozoic)	28.7	5.4	34.1	18.5	5.6	24.1	33.2	8.6	41.8	3.8
Samotlor (Mesozoic)	17.4	5.8	23.2	33.3	10.0	33.3	32.1	11.4	43.5	2.9
Yaraktinskoe (Pre-Cambrian)	–	–	–	–	–	–	82	18	100	4.6
Vanavarskoe (Pre-Cambrian)	–	–	–	–	–	–	82	18	100	4.6
Sivinski Pre-Cambrian)	–	–	–	–	–	–	83	17	100	4.9

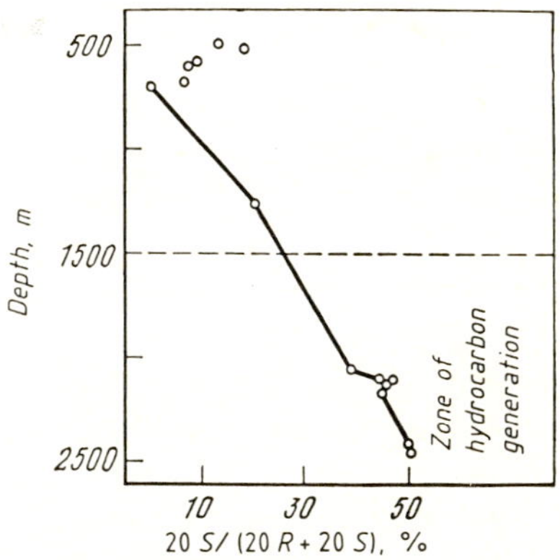

Fig. 45. Changes in 5*a*,20S epimer concentration with an increase of burial depth (Toarcian shales)

Table 39. Configurational transformations of chiral centres in steranes upon heating of Green River shales

Heating time Days (260 °C)	5*a*,20*S*/5*a*,20*R*	14*β*,20*R*/5*a*,20*R*	Heating time Days (260 °C)	5*a*,20*S*/5*a*,20*R*	14*β*,20*R*/5*a*,20*R*
0 (Original sample)	0.29	0.05	171	0.44	0.07
10	0.29	0.05	242	0.59	0.14
27	0.33	0.05	348	0.59	0.12

epimerization at C-20 occurs, which leads to the generation of large quantities of 20*S* isomers.

Figure 45 is vivid proof of the dependence of the relative concentrations of the 20*S* epimer of C_{29} 5*a*-steranes and depth for the Toarcian shale. These changes were also successfully demonstrated by Seifert (Seifert and Moldowan 1979b). Heating of Green River shales led to a gradual epimerization at C-20 and C-14 (as well as C-17) and the formation of the thermodynamically more stable epimers (Table 39).

Characteristically, C-14 epimerization and the formation of isosterane proceed much slower than the epimerization at C-20. The 14*β*,17*β* isosteranes are sometimes called geosteranes, while the original 5*a*,20*R* steranes are called biosteranes.

Hence, the ratio of the 20*S*/20*R* epimers should only be used cautiously in establishing the distance of petroleum migration. Since 5*a*,20*S* epimers are eluted first in liquid chromatography, and the 5*a*,20*R* epimer is adsorbed more intensely, the 20*S*/20*R* ratio may only be used in the study of secondary migration processes, i.e. the migration of mature crude oils through porous media (adsorbents). The original mixture should contain equal concentrations of 20*R* and 20*S* epimers in this case, so that migration and adsorption lead to a higher 20*S* epimer concentration in the migrated products. We should also note that the best GC separation of 5*a*,20*S* and 14*β*,20*R* epimers which elute close to each other is obtained on capillary columns with silicone phases (Seifert and Moldowan 1979b) (see Fig. 41).

So far we have been dealing with the results of the epimerization of the chiral centres at C-14, C-17, and C-20. However, the centres C-5 and C-24 (the latter in the case of C_{28} and C_{29} steranes) can also epimerize. It is true, however, that epimerization at C-5 in steranes (formation of 5β-hydrocarbons) is limited, due to the lower thermodynamic stability of hydrocarbons with a *cis*-junction of the A/B rings. The presence of β-steranes, i.e. 5β,14α,17α,20*R* epimers, has been mentioned in a number of publications. The corresponding C_{27} sterane, known as "coprostane", may be regarded as a biosterane. For the characteristics of other 5β-steranes, see Table 37.

A different situation emerges with the possible epimerization of the chiral centre at C-24. As has been frequently observed, 24*S* and 24*R* epimers cannot be separated on commonly used (non-polar) GC phases, hence, it was concluded that all C_{28} and C_{29} steranes consist of equal amounts of both diastereomers (their thermodynamic stabilities should be similar). However, as reported by MacKenzie et al. (1980), on Toarcian shales, where these epimers could be separated. C_{28} steranes consist of equal amounts of 24*R* and 24*S* isomers even in early stages (and at low depths) regardless of varying categenetic transformations (see Figs. 44 and 45). This is explained by fast chiral centre epimerization in the aliphatic chain, similar to that in pristane. An alternative explanation is also feasible. In fact, zooplankton and phytoplankton sterols and stanols, sources of steranes, already have different configurations of the C-24 centre. Accordingly, ergosterol (C_{28}) is characterized by a 24*S*-configuration, whereas campesterol (C_{28}) has a 24*R*-configuration. Sitosterol (C_{29} has a 24*R*-configuration, whereas poriferasterol (C_{29}) has a 24*S*-configuration.

Thus the ratio of the 24*R* and 24*S* epimers may serve both as a maturation criterion and a genetic indicator. Hence, the catagenetic meaning of this ratio in crude oils is seriously limited, especially because petroleums are stereochemically mature products. At the same time an obvious predominance of 24*R* or 24*S* epimers may be used for the determination of sterane sources (genetic indicator) in dispersed organic matter at an early categenetic transformation stage:

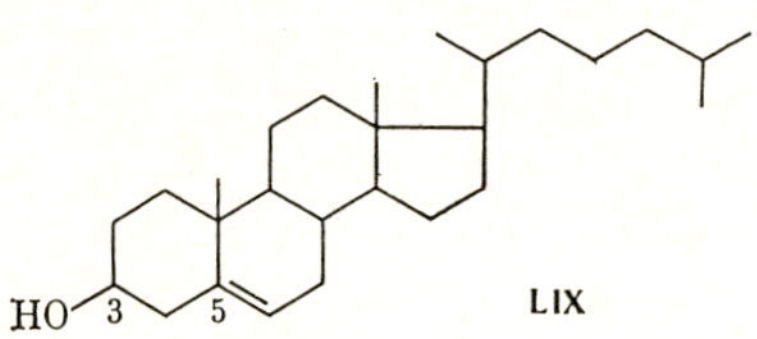

LIX

Cholesterol (C_{27})

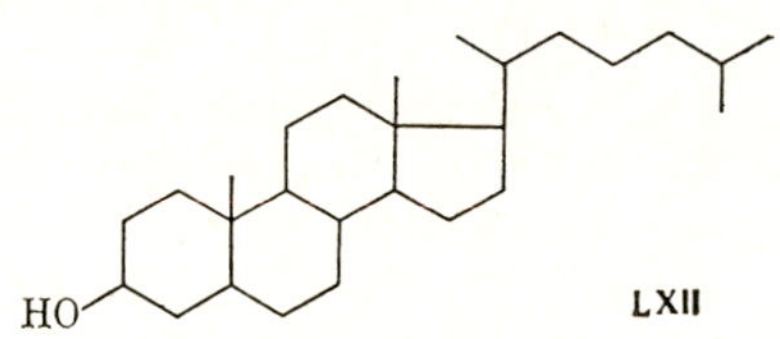

LXII

Cholestanol (C_{27})

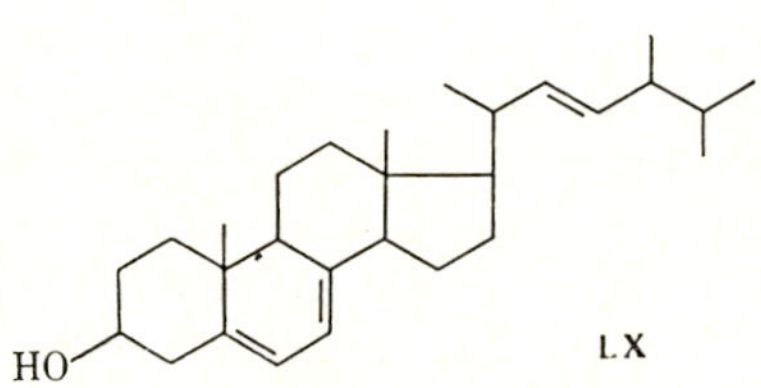

LX

Ergosterol (C_{28})

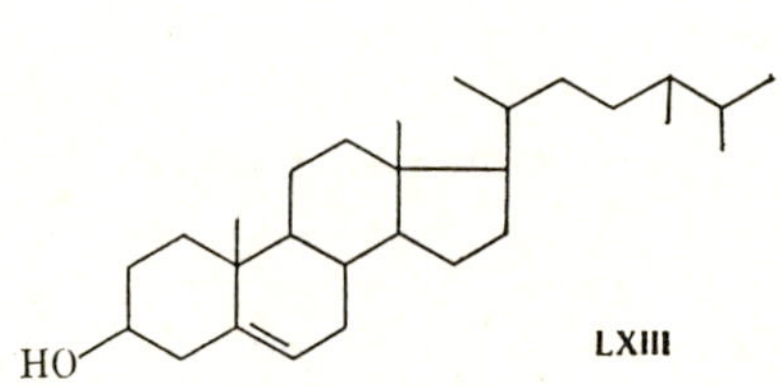

LXIII

Campasterol (C_{28})

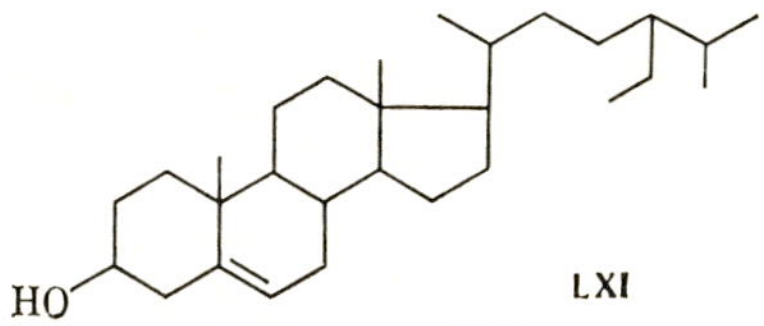

LXI

β-Sitosterol (C_{29})

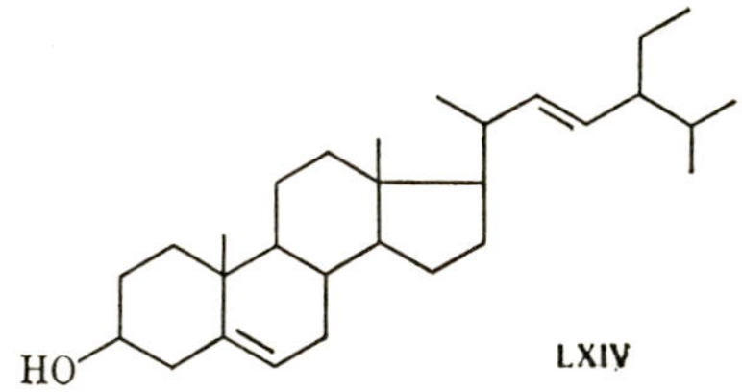

LXIV

Poriferastanol (C_{29})

Since the configuration of the C-24 chiral centre in petroleum steranes is usually not determined, it is preferable to call C_{28} and C_{29} steranes "24-methylcholestane" and "24-ethylcholestane" (rather than "ergostane" and "sitostane").

Structural formulae of the most important natural steroids, possible sources of petroleum steranes, are given above [these include unsaturated alcohols, e.g. sterols or stenols, which are the same (LIX–LXI, LXIII, LXIV), saturated alcohols, e.g. stanols (LXII), as well as saturated ketones, e.g. stanones, (the aliphatic side chain often contains double bonds)]. For a more detailed analysis of the sources and formation mechanisms of petroleum sterane, see Chapter 5.

Besides ordinary (regular) steranes, their 4-methyl homologs, i.e. C_{28}, C_{29}, C_{30} hydrocarbons, have been identified in crudes (Ensminger et al. 1978; Kimble et al. 1974; Rubinstein and Albrecht 1975). These compounds have the basic structure LXV:

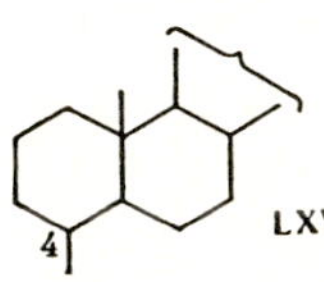

LXV

Their molecular masses would be 386, 400 and 414 respectively. Key fragment ions are *m/z* 231 and 232. All of the above stereochemical properties and the conclusions with respect to regular steranes apply to 4-methylsteranes almost entirely. The properties of 4-methylsteranes are comprehensively discussed in the publication of Kayukova et al. (1980a).

The molecular ion $M^+ = m/z$ 414 is characteristic not only of 4-methyl-24-ethylcholestane. Other C_{30} steranes have been identified in crudes in recent years (Moldowan 1984). The presence of the key fragment ion at *m/z* 217 in the mass spectra of these hydrocarbons indicates an additional carbon atom in the aliphatic chain and not in the cyclic system. Those hydrocarbons, possibly, are epimeric 24-propylcholestanes.

Figure 46 presents a chromatogram of an equilibrium mixture of 4-methylcholestane. 4-Methylsteranes are found in smaller quantities in petroleums, compared to regular steranes, and amount ot approximately one-third of the latter. Besides 4-methylsteranes, the presence of 4,4-dimethylsteranes with molecular masses of 400, 414 or 428 is possible. Key fragment ions in this case are *m/z* 245, 246, 315 and 316. The properties of 4,4-dimethylsterane are analyzed in the work of Kayukova (1980). Reference to the presence of 4,4-dimethylsteranes in geological samples may be found in the publication of Pryce (1971).

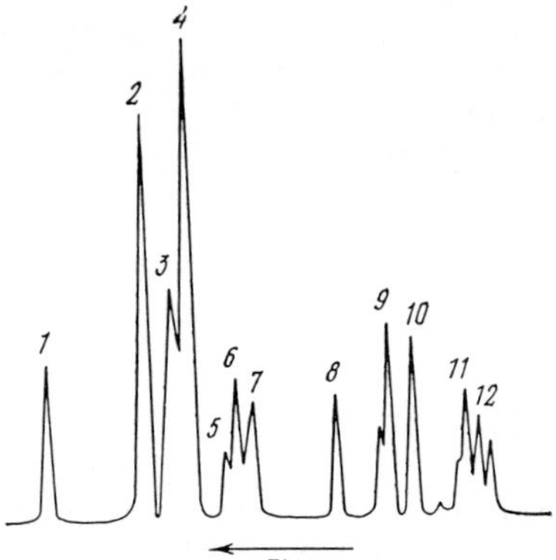

Fig. 46. Chromatogram of an equilibrium mixture of 4-methylcholestane epimers. Compare to the chromatogram of an equilibrium mixture of epimeric cholestanes in Fig. 41 a and data in Table 37

Rearranged Steranes

Rearranged (structurally modified) steranes have a different structure (LXVI), occurring as the result of carbonium-ionic rearrangement of sterenes:

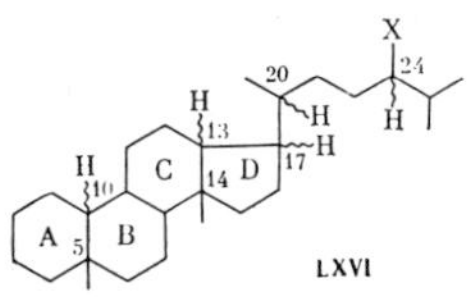

LXVI

($X = H, CH_3, C_2H_5$)

This reaction proceeds also in the presence of aluminosilicates, because such compounds have been identified in a number of crudes (Ensminger et al. 1978; Seifert and Moldowan 1979a; Kayukova et al. 1981; Sieskind et al. 1979).

The C-H bonds of chiral centres, actually capable of epimerization, are marked by a wavey line.

As opposed to regular (ordinary) steranes, the migration of a methyl group occurred from C-10 to C-5 and from C-13 to C-14 in the rearranged steranes. All epimers discussed below have trans-junctions of the A/B and B/C rings. The C/D ring junction may vary. Since rearranged steranes are not biological products, but hydrocarbons, which were produced by acidic rearrangement of steranes, the determination of the sterical structure of the most stable epimers acquires special importance. The equilibrium of the epimer has been studied by Kayukova et al. (1980a), while their configuration was established by Seifert and Moldowan (1979a) and Pekhk et al. (1982). A chromatogram of an equilibrium mixture of epimeric rearranged cholestanes is shown in Fig. 47.

The four most stable epimers of rearranged steranes should have a $10\alpha,13\beta,17\alpha,20R$, and $20S$ configuration, as well as a $10\alpha,13\alpha,17\beta$, $20R$, and $20S$ configuration. These two pairs of stereoisomers were called "diasteranes I and II" by Seifert and Moldowan (1979a). Diasteranes I have a $13\beta,17\alpha$-configuration, $\beta\alpha$ in the abbreviated form, while diasteranes II have an $\alpha\beta$-configuration. The C/D ring junction is *cis* (more stable) in the first pair, while it is *trans* in the second.

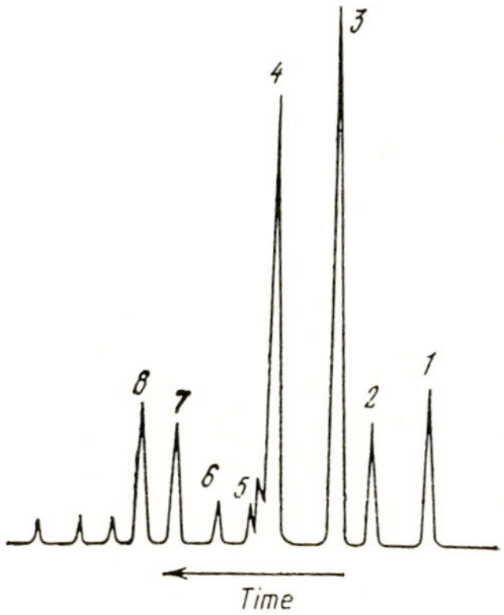

Fig. 47. Chromatogram of an equilibrium mixture of rearranged cholestane epimers. *1, 2* 10β,13β,17α,20S and 20R; *3, 4* 10α,13β,17α,20S and 20R (diasteranes I); *5, 6* 10β,13α,17β,20S and 20R; *7, 8* 10α,13α,17β,20S and 20R (diasteranes II)

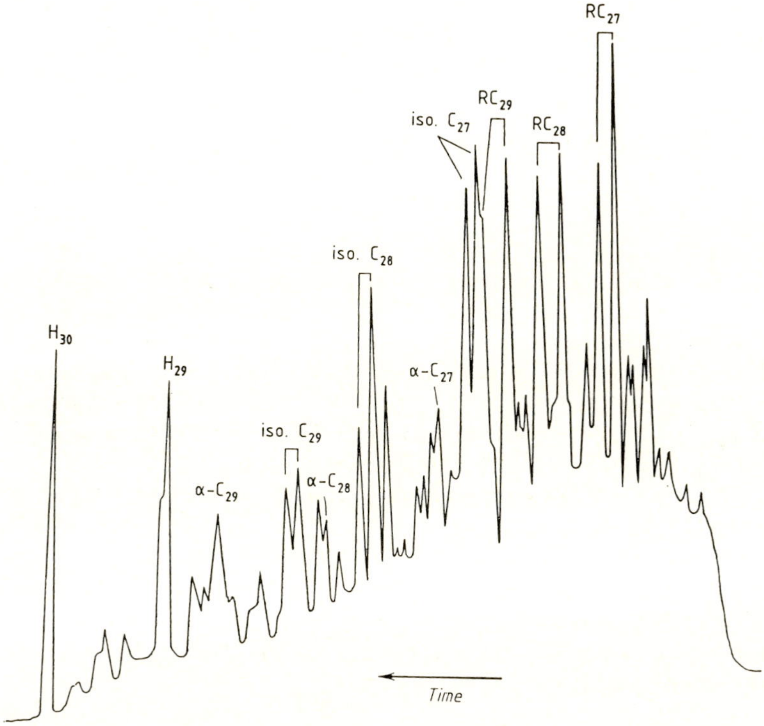

Fig. 48. Chromatograms of a sterane mixture in the Samotlor crude. The Main peaks are identified: *R* Rearranged steranes 20S and 20R (diasteranes I); isosteranes (14β,17β,20R and 20S); α-steranes (5α,20R); H_{29} and H_{30}: adiantane and hopane. Capillary column 80 m, Apiezon; 290 °C

The orientation of a substituent aliphatic radical is *trans* everywhere. Diasteranes I and II are found in crude oils.

Figure 48 presents a chromatogram of the sterane fraction (400°–470 °C) of Samotlor crude. Peak doublets of 20*S* and 20*R* epimer pairs, belonging to rearranged cholestane, 24-methylcholestane and 24-ethylcholestane (*βα*-diasteranes throughout) are easily identifiable. It is worthwhile to note that rearranged steranes (Δb.p. 20 °C) have much lover boiling points than regular steranes. Thus, for example, the epimers of the rearranged 24-ethyldiacholestane elute together with $5\alpha,14\beta,17\beta$, $20R$ cholestane.

In our opinion, diasteranes I are more suitable for determining petroleum migration than regular steranes, since $20R$ and $20S$ epimers in equilibrium concentrations appear immediately after the formation of rearranged steranes. This is especially true in view of the possibility of liquid chromatographic separation of $20S$ and $20R$ epimers of diasteranes I on aluminium oxide, as was proven in the publication of Kayukova et al. (1980a).

Schema 2 presents a structural formula of diasteranes I, and identifies the most typical points of bond cleavage under electron impact.

Scheme 2

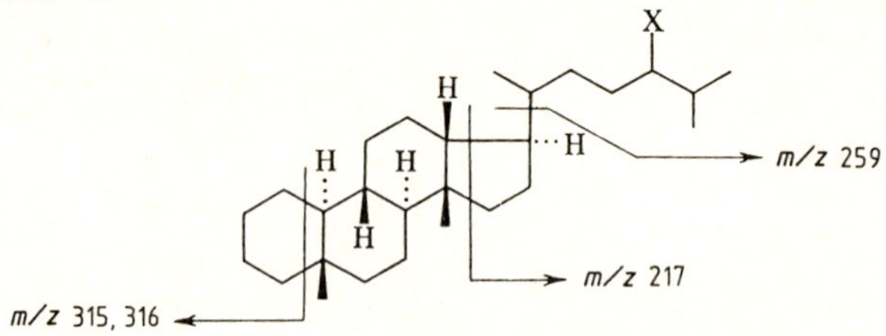

As compared to regular steranes, the mass spectra of rearranged steranes are characterized by an intensive peak at m/z 259 due to the loss of the side chain, as well as an m/z 189 peak. Both ions are typical of rearranged steranes. At the same time, the m/z 218 peak, characteristic of regular steranes, is insignificant in the spectra of rearranged steranes.

It appears that besides rearranged steranes, petroleums contain their 4-methyl homologs, which is proven by the presence of m/z 203, 273 and 315. The properties of rearranged 4-methylcholestane are described by Kayukova et al. (1981a). The sequence of relative thermodynamic stability, in this case, is the same as in regular rearranged steranes.

Besides the above analyzed steranes, crude oils may also contain 17-methyl-D-homosteranes, which are also formed by acid rearrangement of steranes. D-homosteranes possess a six-membered D-ring, and can thus be included with the homologs of perhydrochrysene:

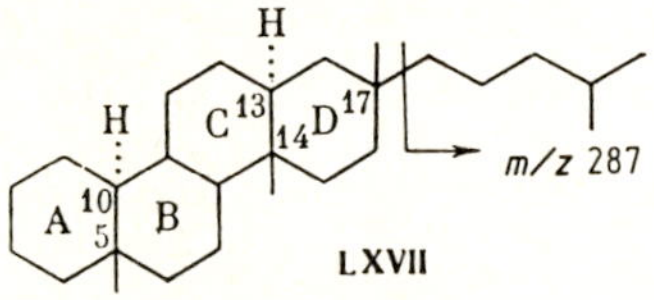

LXVII

17-Methyl-D-homocholestane (LXVII) consists of two epimers with $17\alpha CH_3$ and $17\beta CH_3$ configurations. D-homocholestanes have a characteristic m/z 287 (100%) fragment ion due to the loss of an alkyl radical (Pustilnikova et al. 1981).

Besides C_{27}-C_{30} steranes, lower molecular weight steranes have also been found in crudes. They are formed by the loss (complete or partial) of the aliphatic side chains of regular or rearranged steranes. Two such hydrocarbons (LXVIII and LXIX) have been identified in Sivinski crude. These isomers correspond to peaks Nos. 2 and 5 in Fig. 37:

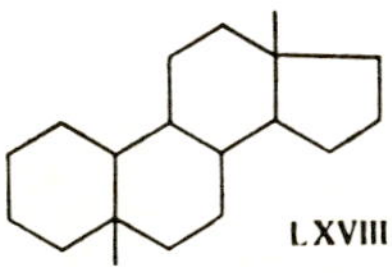
LXVIII

Peak No. 2 (C_{19})

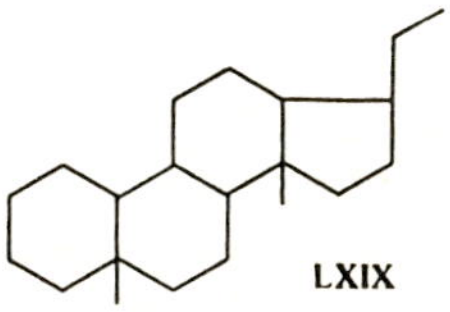
LXIX

Peak No. 5 (C_{21})

These hydrocarbons represent rearranged androstane and pregnane structures. The properties of regular androstane and pregnane epimers are described in the publication of Pustilnikova et al. (1981).

As opposed to cholestanes, there are only eight epimers (the C-20 chiral centre is missing) in an equilibrium mixture of pregnanes, and only four epimers (the C-17 chiral centre missing) in an equilibrium mixture of androstanes. The formulae of androstane (LXX) and pregnane (LXXI) are presented below. Figure 49 shows a chromatogram of an equilibrium mixture of androstane epimers:

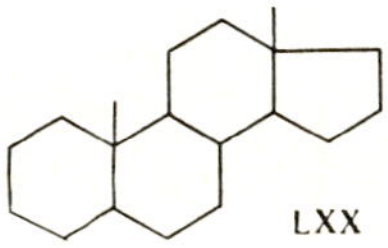
LXX

Androstane (C_{19})

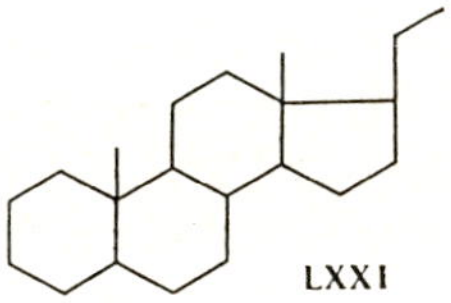
LXXI

Pregnane (C_{21})

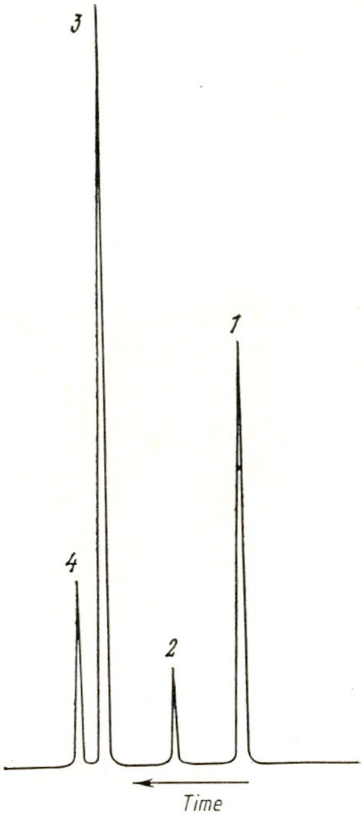

Fig. 49. Chromatogram of an equilibrium mixture of androstane epimers. *1* 5β,14β; *2* 5β,14α; *3* 5α,14β; *4* 5α,14α

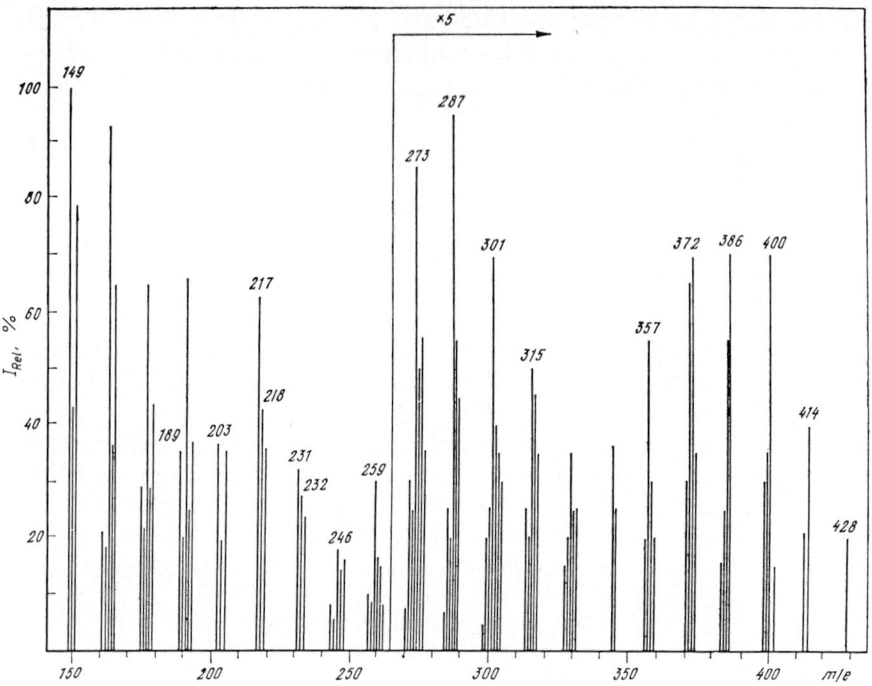

Fig. 50. Mass spectrum of a sterane concentrate of the Balakhanski crude. Identified are key fragment and molecular ions for $C_{27}-C_{31}$ steranes

A group of regular and rearranged $C_{19}-C_{22}$ steranes was found in Naftalan crude (Azerbaijan) (Kuliev et al. 1984).

The mass spectrum of a sterane concentrate from the Balakhansky crude is presented in Fig. 50. Numerous representatives of sterane hydrocarbons may be easily identified with the help of characteristic molecular and fragment ions.

Thus, the molecular ions $M^+ m/z$ 372, 386, 400, 414 indicate the presence of $C_{27}-C_{30}$ steranes; m/z 218 fragment ions, isosteranes; m/z 189, 259 ions, rearranged steranes; m/z 231, 232, 4-methylsteranes, etc.

In conclusion, we can provide a general scheme of the genetic ties between biosteranes and the most important geosteranes (Scheme 3) (Seifert and Moldowan 1981). Experimental data on the possibilities and conditions of these reactions will be discussed in Chapter 5. The concentrations of the sterane hydrocarbon in crudes do not exceed 0.3–0.5%, which does not affect their considerable geochemical importance. Moreover, these figures relate only to steranes eluting as separate peaks. A "hump" is often observed in the area where steranes elute, consisting of a large number of tetracyclic hydrocarbons, possibly steranes with slightly modified structures. Thus, potentially, steroid compounds play an important role in the formation of petroleum hydrocarbons.

In addition, steranes, as well as other tetracyclic hydrocarbons, e.g. perhydrophenanthrene derivatives, are representatives of cyclic isoprenoid hydrocarbons, although they are found only in some crude oils. They are absent from light

Scheme 3

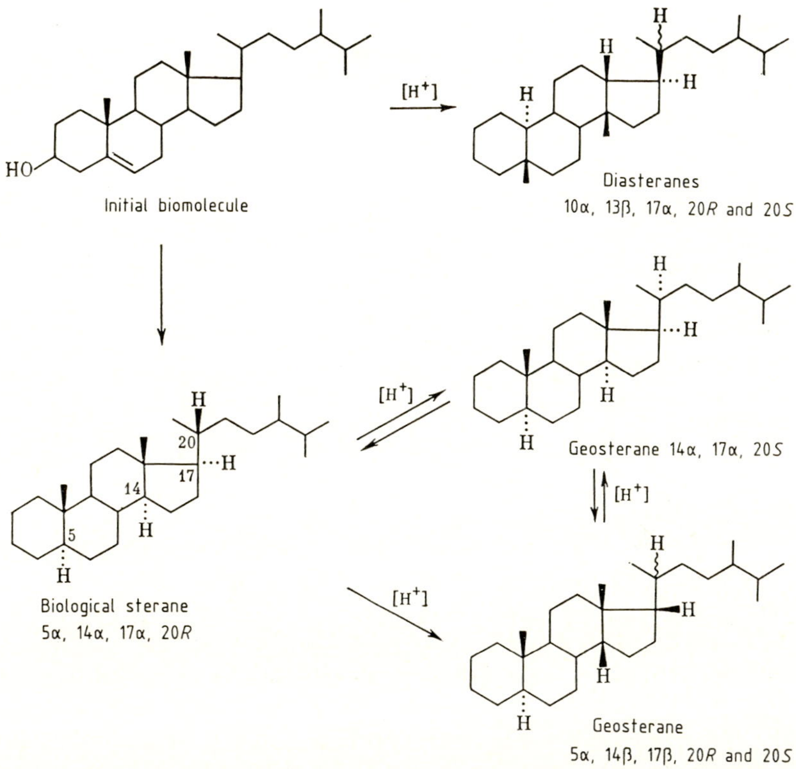

crudes containing no higher fractions. Only insignificant sterane concentrations exist in crudes (rich in other biomarkers) from the Anastasievsko-Troitskoe, Romashkinskoe, Karajanbas, Katangli and some other deposits. This is obviously due to the absence of appropriate sterols in the initial biomass, or their selective degradation by microorganisms.

Triterpenoid Hydrocarbons. Hopanes

Together with steranes, triterpanes belong to the most important petroleum hydrocarbons that retained the characteristic structure (but no the sterical configuration) of the original biological compounds.

Due to obvious connections with their biological precursors, these hydrocarbons are also called "biological markers" or "chemofossils". Most of these hydrocarbons are based on perhydropicene (LXXII) and cyclopentanoperhydrochrysene (LXXIII) structures,

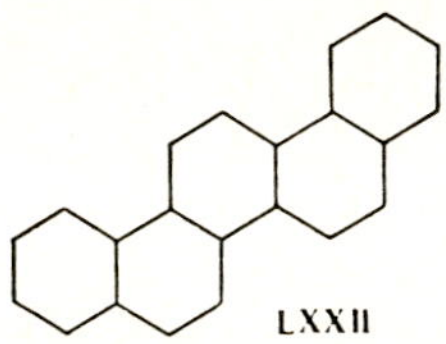
LXXII

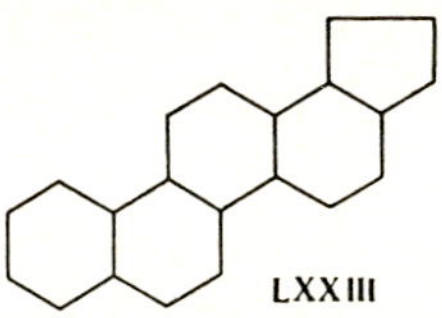
LXXIII

with compounds related to (LXXVIII) having much wider occurrence. Hydrocarbons of the hopane series are of special importance in petrochemistry since they are present in crude oils, not only as hopane, a $C_{30}H_{52}$ hydrocarbon, but also in the form of a homologous $C_{27}-C_{35}$ series (LXXIV–LXXVIII). The hopane hydrocarbons possess a common polycyclic ring system, but differ in the length of the alkyl side chain:

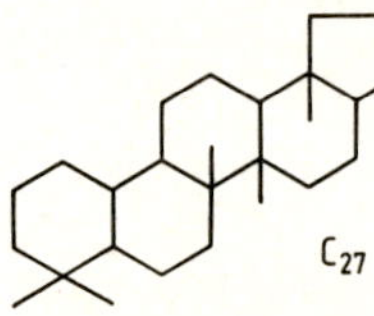

LXXIV

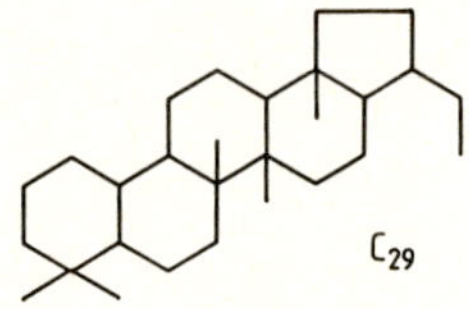

LXXV

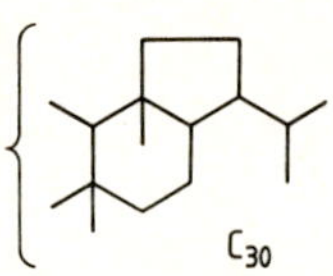

LXXVI

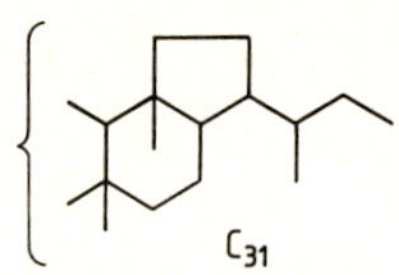

LXXVII

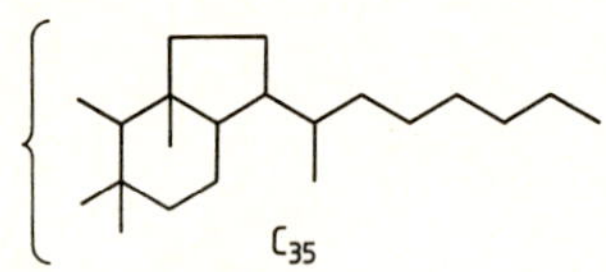

LXXVIII

Although saturated pentacyclic petroleum hydrocarbons were discovered in the early 1960's (Eglinton and Murphy 1974), their structure and stereochemical properties were finally established only in 1973 (Whitehead 1973; Ensminger et al. 1973). An outstanding role in determining the structure and stereochemistry of hopanes was played by French chemists, headed by Prof. Ourisson (Strasbourg, France). The school of Prof. Eglinton (Bristol, Great Britain) made a similarly important contribution to the chemistry of isoprenoid chemofossils and organic geochemistry.

Immediately after these first publications, numerous identifications of hopane hydrocarbons in crudes, coals, shales and dispersed organic matter in sediments followed. As a result, it was shown that the hydrocarbons of this series are ubiquitous (Van Dorsselaer et al. 1974).

Hopanes in Soviet crude oils were originally described in the publications of Petrov et al. (1976); Pustilnikova et al. (1976); and Ushakova et al. (1975). Structural formulae and sterical properties of hopanes (LXXIX) are discussed below (hopane, the parent compound of this homologous series, is a $C_{30}H_{52}$ hydrocarbon with $R = CH_3$ in LXXIX):

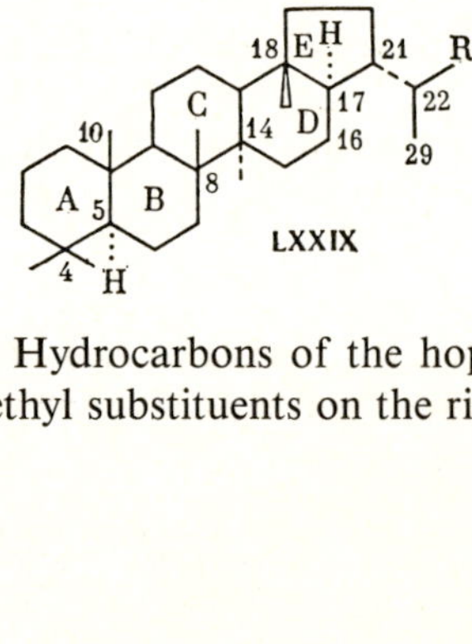

LXXIX

$(C_{29}-C_{35})$ $(R = H, CH_3-C_6H_{13})$

$17\alpha, 21\beta$

Hydrocarbons of the hopane series have 21 carbon atoms in their cycles, 6 methyl substituents on the ring system, of which 4 are angular. Rings A/B, B/C and C/D have *trans*-junctions. Rings D/E have a *trans*-junction in biohopane and a *cis*-junction in petroleum hopane. Thus, two main types of hopanes exist:

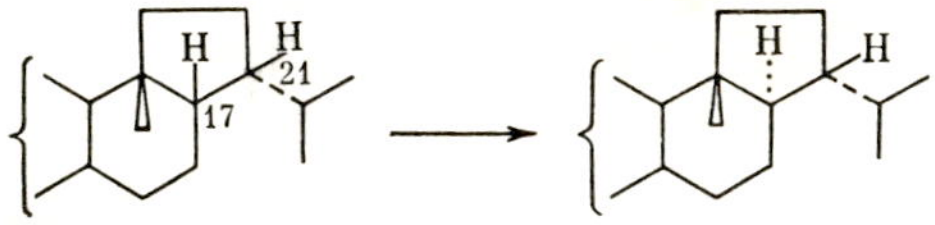

17β,21β 17α,21β

17*α*(H), 21*β*(H) (in petroleum) and 17*β*(H),21*β*(H), (the biological configuration). Stereochemically more labile are the junctions of the D/E ring and the substituent configuration at C-21. Biological hopane is thermodynamically unstable due to the instability of the D/E *trans*-junction and the staggered 21 – 22 and 17 – 16 bonds (*cis*-orientation of the substituents). Epimerization at C-17 removes both of these tensions.

Besides 17*α*,21*β*- and 17*β*,21*β* hopanes, small concentrations of 17*β*,21*α* hydrocarbon, called "moretane", are found in crude oils. Moretane has a *trans*-junction of the D/E rings and a *trans*-orientation of the side chain. The expected equilibrium of the hopane stereoisomers is (%): 17*α*,21*β*-86,17*β*,21*α*-11,17*β*,21*β*-3 (calculated, 400 K).

As a rule, only traces of hopanes with the "biological" configuration (17*β*H, 21*β*H) can be identified in crude oils. The bulk of hopanes is represented by the 17*α*,21*β* hydrocarbons. Unstable biohopanes were found only in peat, brown coals, immature shales and dispersed organic matter at the initial stage of catagenesis.

Configurational variations of hopanes are demonstrated by their mass spectra. The main direction of the decomposition of hopane under electron impact is the rupture of the 8 – 14 bond with the formation of two fragments, a common fragment A and a variable fragment B (which depends on the initial molecular mass) (Scheme 4). Moreover, hopanes have the following key ions: 370, 384, 398, 412, 426, 440, etc. (molecular ions) as well as $M^+ - 15$ and $M^+ - R$ (*m/z* 369).

Fragment A is more intense than B in petroleum (17*α*) hopanes. In contrast, fragment B is more intense than A in biohopanes (Whitehead 1973).

The higher boiling points of biohopanes should also be noted. In addition, they elute much later in GC than the stereoisomeric petroleum hopanes (thus, the C_{30} biohopane elutes in the area of the C_{32} petroleum hopane).

In fact, the name "hopane" should be used only for the C_{30} hydrocarbon. The C_{29} hydrocarbon is called nor-hopane or adiantane, C_{27} (with the whole side chain missing at C-21) is called trisnorhopane. C_{31}, C_{32}, etc. hydrocarbons are correspondingly called homohopane, bishomohopane, etc.

C_{31} and higher hopane hydrocarbons, i.e. when the C-22 atom becomes chiral, may have two epimers with an *R*- or *S*-configuration at C-22. In the original C_{31} and higher biohopanes, C-22 has a strictly defined *R*-configuration. At the same time, a transfer to petroleum hopanes is accompanied, not only by the already noted C-17 epimerization, but also by epimerization at C-22, which results in a number of doublet peaks, prominent in gas chromatograms of petroleum hopane mixtures. The 22*S* epimer is known for its slightly higher stability. In equilibrium situations, as well as in crude oils, the ratio of 22*S* to 22*R*

Scheme 4

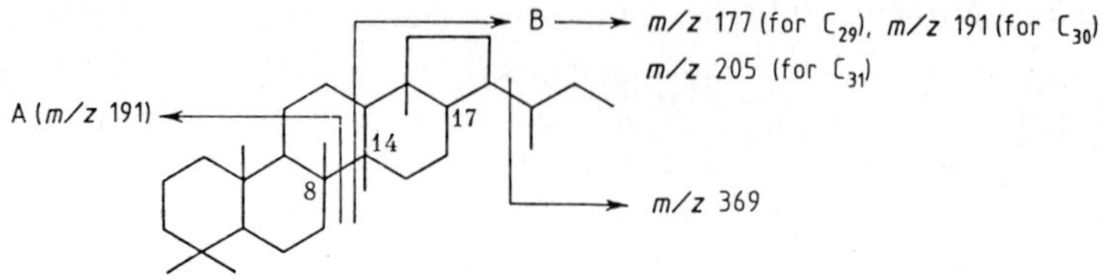

epimers is usually 1.2–14. C-22 isomers have similar properties as the analogous sterane C-20 epimers as in $\beta\alpha$-diasteranes in particular.

Figure 51 shows a typical chromatogram of a petroleum (17α) hopane mixture with obvious doublet peaks characteristic of C_{31} and higher hydrocarbons. The concentrations of C_{34} and C_{35} hopanes are usually lower than those of C_{29}, C_{30} and C_{31} hopanes. Therefore, many publications deal only with the relative distributions of $C_{27}-C_{33}$ hopanes. However, we discovered a Tertiary crude of the Shakalyk-Astana deposit (Middle Asia) and a Devonian crude of the

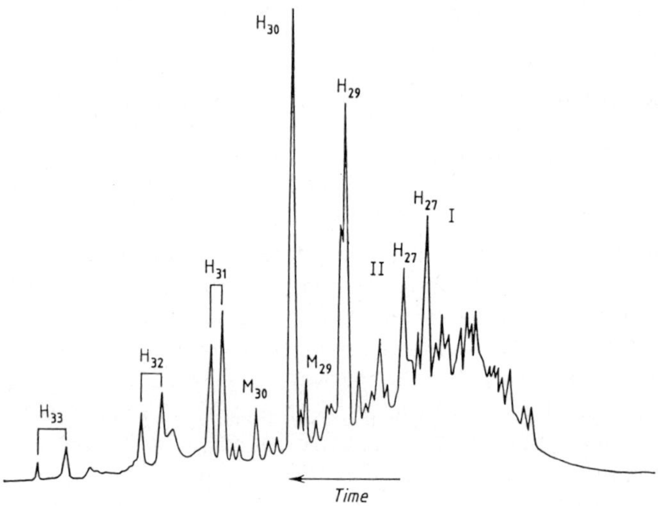

Fig. 51. Chromatogram of a hopane concentrate obtained by thermodiffusion from the Anastasievsko-Troitskoe crude. $H_{27}-H_{33}$: $C_{27}-C_{33}$ hopanes ($17\alpha,21\beta$). M_{29}, M_{30}: C_{29}, C_{30} moretanes ($17\beta,21\alpha$). Doublet peaks belong to 22S and 22R epimers respectively in C_{31} and higher hopanes

Table 40. Relative distribution of $C_{29}-C_{35}$ hopanes in crudes (%)

Deposit	Relative concentration (%)						
	C_{29}	C_{30}	C_{31}	C_{32}	C_{33}	C_{34}	C_{35}
Shakalyk-	11.8	12.2	16	16.2	12	15.7	15.7
Astana	–	–	(65:35)	(60:40)	(53:47)	(57:43)	(57:43)
Ostashkovi-	6.9	12.3	14.4	16.2	9.1	18.5	22.6
tchi	–	–	(69:31)	(60:40)	(57:43)	(65:35)	(62:38)

Note: The ratios of 22*S* and 22*R* epimers are in parentheses.

Ostashkovitchi deposit (Pripyat), with an unusual distribution of hopanes, which is reproduced in Table 40.

Computer processing of chromatograms may involve their reconstruction to a hopane base peak of *m/z* 191. Figure 52 shows such a reconstruction for a naphthenic crude oil from a Tertiary reservoir in California (Seifert and Moldowan 1979). Ion *m/z* 191 is so typical of hopanes, that Prof. Ourisson aptly called such fragmentograms hopanograms (Ourisson et al. 1979).

Besides ordinary (regular) $C_{27}-C_{33}$ hopanes, Figs. 51 and 52 contain a peak (No. 1) belonging to a structurally modified hopane, [17α-methyl-22,29-30-tris-nor-18α(H)-hopane (LXXX)]:

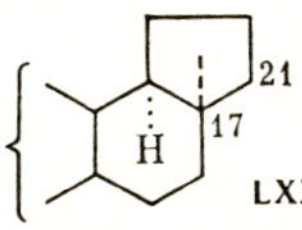

LXXX

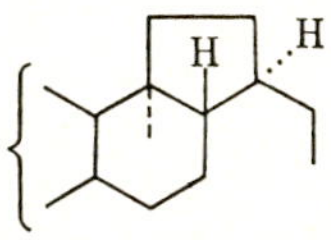

LXXXI

This rearrangement is similar to the formation of rearranged steranes and occurs only in the case of the C_{27} hopane, i.e. a hydrocarbon with no substituent at C-21.

Peaks belonging to C_{29} and C_{30} moretanes (17β,21α) (LXXXI), hydrocarbons of considerable geochemical importance, which will be demonstrated later, can also be seen in Figs. 51 and 52.

Since the relative distribution of hopanes is regarded as a characteristic "fingerprint" of crudes from a given area (Ourisson et al. 1979), and possibly a given sedimentary basin, data from various deposits worldwide (Petrov et al. 1976; Dididze et al. 1979; Pum et al. 1975) are of interest. Corresponding material is included in Table 41.

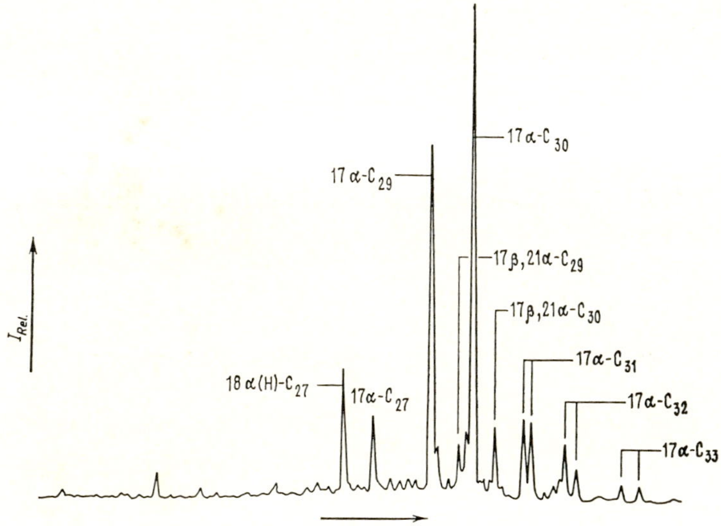

Fig. 52. Masschromatogram of a mixture of hopanes and moretanes in a Californian crude, reconstructed m/z 191 fragment ion

Table 41. Relative distribution of hopanes in crudes (%)

Deposit	Relative concentration (%)						
	C_{27}	C_{28}	C_{29}	C_{30}	C_{31} [a]	C_{32} [a]	C_{33} [a]
Balakhanskoe (Apsheron)	5.7	6.4	14.5	28.5	20.2	15.2	9.5
Banka Darvina (Apsheron)	4.3	5.1	13.8	29	21.0	15.8	10.6
Gryazevaya Sopka (Apsheron)	4.7	3.8	15.3	29.7	21.2	13.8	11.5
Gyurgyany-more (Apsheron)	3.3	3.7	12.6	32.9	24.5	16.2	6.8
Anastasievskoe-Troitskoe							
IV Horizon	6.6	2.7	18.1	30.4	23.9	12.4	5.9
VI Horizon	2.8	1.8	8.8	25.5	25.7	21.9	13.5
Groznenskoe	3.2	3.2	16.8	24.3	19.2	20.4	12.9
Sivinski (Volga-Urals)	3.3	22.9	8.9	18.8	18.6	16.8	10.7
Samotlor (West Siberia)	5.7	4.9	15.4	23.7	25.4	13.7	11.2
Romashkino (Tatariya-Devon.)	3.4	4.3	12.6	13.3	27.8	23.6	15.0
Makarovskoe (Tatariya Carbon.)	7.4	–	20.7	17.6	26.7	15.6	12.0
Ashaltchinskoe (Tatariya Perm.)	7.5	–	20.5	16.6	26.9	17.2	11.3
Norio (Georgia)	2.5	–	11.2	31.2	25.8	16.7	12.8
Sartychala (Georgia)	7.7	–	12.3	33.3	23.3	13.6	9.8
Burgan (Kuwait, Cret.)	11.2	–	30.3	18.1	27.5	13.9	–
Obigbo (Nigeria)	16.2	–	26.4	29.2	18.9	9.8	–
Prudhoe-Bay (Alaska)	10.1	–	17.7	23.8	28.2	20.2	–
North Sea	9.3	–	18.4	25.7	27.1	19.5	–

[a] The ratio of the 22*S* and 22*R* epimers is 1.2 – 1.4 throughout.

The ratios of adiantane (C_{29}) and hopane (C_{30}) are the most commonly used. Sometimes ratios of "primary" hopanes ($C_{29}+C_{30}$) to the C_{27} degraded products are used, as well as that of 17αH- and 18αH-trisnorhopanes (Tm, Ts).

As can be seen from Table 41, the concentration of hopane is higher than that of adiantane in Soviet crude oils. Crudes from Tatariya are an exception: regardless of the age (Devonian-Carboniferous-Permian) and chemical type of the sediments, adiantane is present in clearly higher concentrations than hopane, with the C_{31} hopanes having the highest relative concentrations. In the foreign crudes, adiantane predominates in some Middle Eastern oils (Ourisson et al. 1979).

Structurally Modified Hopanes

A representative of these hopanes (17-methyl-trisnor-18α(H)-hopane) was discussed earlier. However, structurally modified hopanes are much more numerous. A

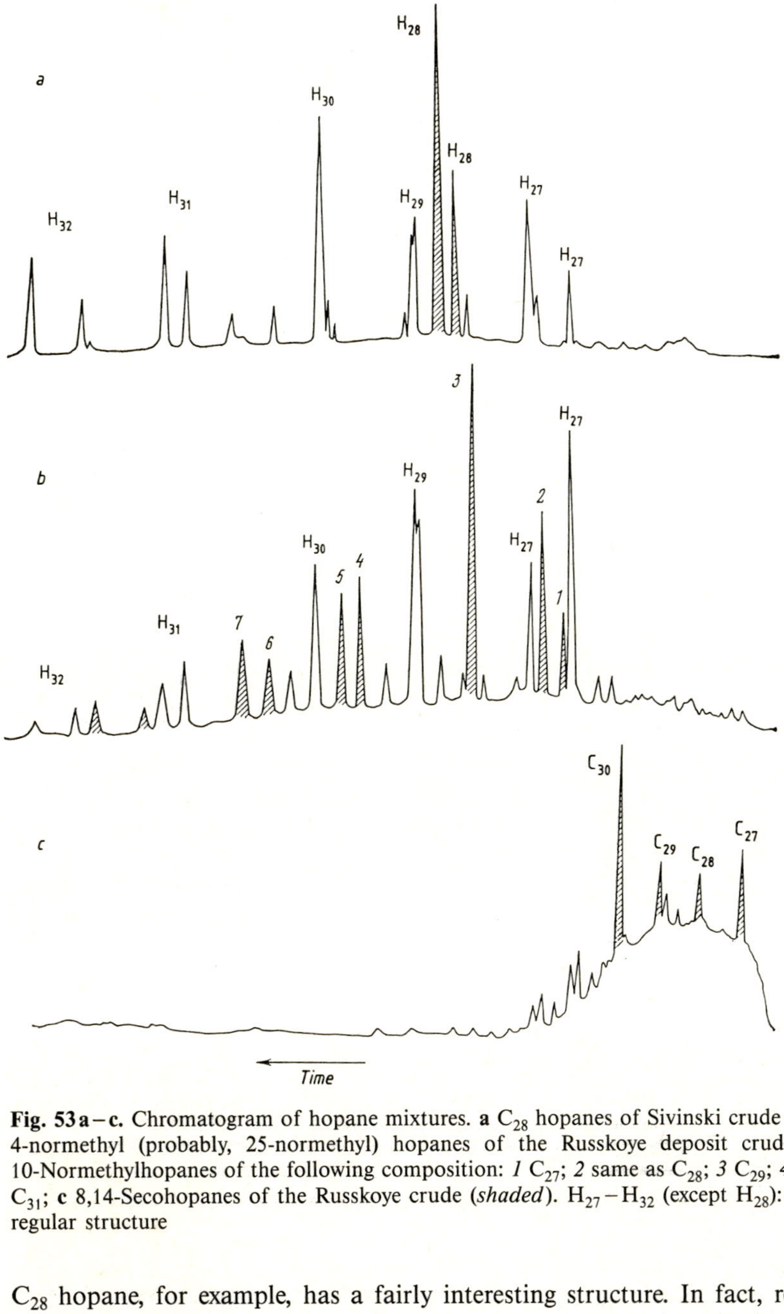

Fig. 53a–c. Chromatogram of hopane mixtures. **a** C_{28} hopanes of Sivinski crude (*shaded*); **b** 4-normethyl (probably, 25-normethyl) hopanes of the Russkoye deposit crude (*shaded*). 10-Normethylhopanes of the following composition: *1* C_{27}; *2* same as C_{28}; *3* C_{29}; *4, 5* C_{30}; *6, 7* C_{31}; **c** 8,14-Secohopanes of the Russkoye crude (*shaded*). $H_{27}-H_{32}$ (except H_{28}): hopanes of regular structure

C_{28} hopane, for example, has a fairly interesting structure. In fact, regular C_{28} hopanes (18-methyl) do not occur by the rupture of the aliphatic side chains of higher homologs as in the case of C_{17} isoprenoid. First found in Sivinski crude (Petrov et al. 1976), the C_{28} hopane was later identified in a number of other crudes and shales. The structure and stereochemistry of this compound were finally established through X-ray analysis by Seifert et al. (1978).

A gas chromatogram of a Sivinski crude hopane mixture is presented in Fig. 53a, featuring two peaks (H-28) with identical mass spectra. The structure (LXXXII) of the main (second) component is as follows:

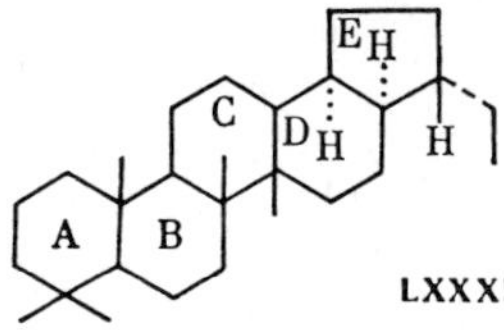

LXXXII

As can be observed, compared to ordinary (regular) hopanes, the C_{28} hopane has lost one of the angular methyl groups and has become 18-nor-17α,21β-adiantane. The D/E ring junction is *cis.* Such a selective loss of a methyl group remains to be explained. The most plausible explanation may be the microbial degradation of ordinary hopanes (adiatane). Different explanations can not be excluded.

Other modified hopanes (25,30-bisnorhopanes) are described in the publication of Moldowan et al. (1984).

As mentioned earlier, besides regular 17α- and 17β-C_{27} hopanes, a 17-methyl-trisnor-18α(H)-hopane is also found in crudes. Moreover, a C_{27} hopane, genetically related to C_{28} hopane, was discovered in North Sea crudes. It may be regarded as C_{28} hopane with a methyl group lost at C-10 (Grantham et al. 1979). This hydrocarbon was early identified as 17β,21α,24,28,30-trisnormethylmoretane by Grantham et al. (1979). Lost methyl groups in normethyl derivatives are sometimes unnumbered depending on the carbon atoms to which they were connected; rather, these lost methyl carbon atoms are numbered in accordance to the hopane nomenclature.

There are indications in the publication of Rulkötter and Wendisch (1982) that this hydrocarbon is not 24,28,30- but 25,28,30-trisnormoretane, i.e. that the methyl group (in ring A) is lost not at C-4, but at C-10. Probably the same is true for a 4-normethylhopane series, described below, which accordingly turns into that of 10-normethylhopanes:

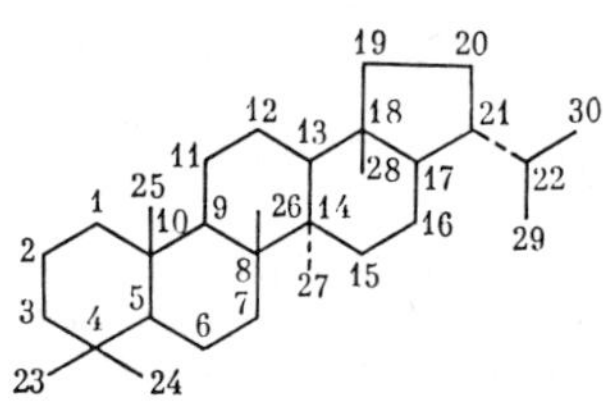

Hopane

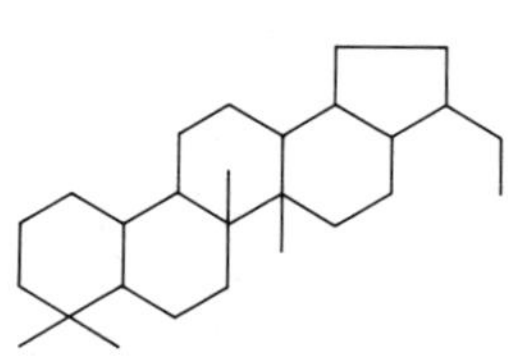

25,28,30-Trisnormethylhopane

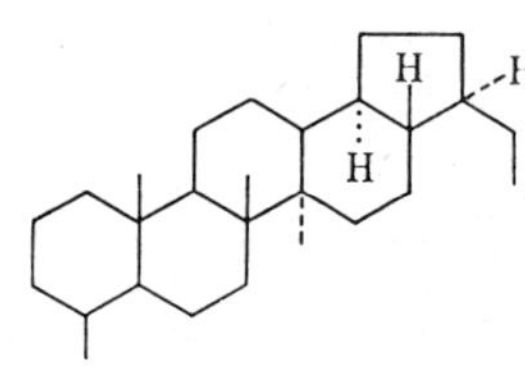

24,28,30-Trisnormethylmoretane

A series of 4-normethylhopanes (or, even more probable, 25-normethylhopanes), which is structurally similar to regular hopanes, was identified in various crude oils (Kimble et al. 1974; Reed 1977b; Kayukova et al 1981b). It is assumed that these hydrocarbons are present in strongly biodegraded crudes. In effect, 4-normethylhopanes are generally present in crudes with insignificant concentrations of regular hopanes. The structure of 4-normethyl-hopanes is presented in Scheme 5.

Scheme 5

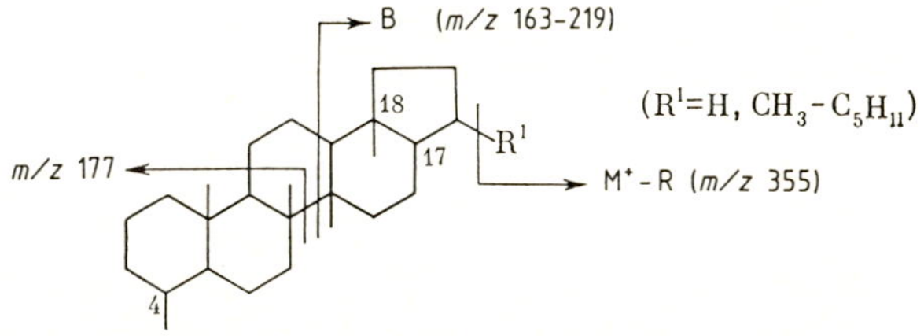

The base peak of *m/z* 177 corresponds to the *m/z* 191 fragment of regular hopanes. The $C_{27}-C_{31}$ 4-normethylhopanes were identified in the Russkoye field crude oil (type B^1, western Siberia) (Fig. 53b). The C_{30} and C_{31} 4-normethylhopanes are also present in the form of two diastereomers (22*S* and 22*R*).[4]

Besides the homologous 4-normethylhopanes, an interesting series of $C_{27}-C_{30}$ 8,14-secohopanes was isolated from the same crude (Fig. 53c). Secohopanes are genetically, clearly related to hopanes and may be generated from them by various catalytic transformations (Kayukova et al. 1981). Scheme 6 shows the structure of secohopanes and the main directions of fragmentation under electron impact. The most characteristic fragment peak is at *m/z* 123.

Seifert (Seifert and Moldovan 1978) described another homologous series of hopanes, differing from the regular ones in that they have an additional methyl group in position 3.

Scheme 6

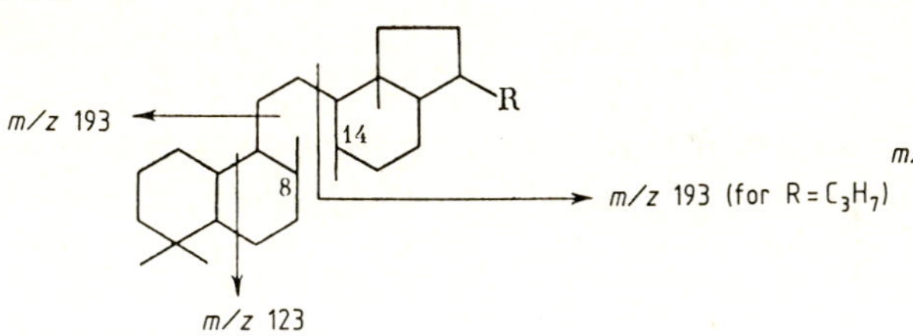

Scheme 7

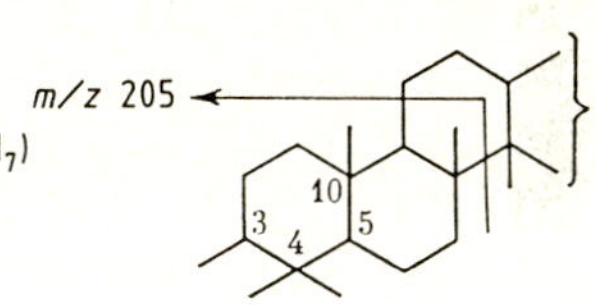

The most characteristic fragment ion in these hydrocarbons is at *m/z* 205 (Scheme 7). Chirality of C-22 in these homologs must occur starting with C_{32}.

Thus far we have centred our analysis on pentacyclic hydrocarbons of the hopane series. This structure is undoubtedly the most important for triterpanes in any crude. It is geochemically significant that it is the hopanes, with their skeletons created by the simplest procaryote bacterial cells or blue-green algae, that occupy such a prominent position in crudes (Ensminger et al. 1973; Dididze et al. 1979). Supposedly, hopane series hydrocarbons were produced by bacteria, and were included, among other compounds, into the lipids of their cell membrane, i.e. hopane formation occurred during the early diagenesis of organic sediments.

Several hypotheses exist on the sources of petroleum hopanes. Naturally, it is the presence of $C_{27}-C_{35}$ homologs that attracts most attention. Such structures

[4] Chirality of C-22 begins with C_{30} hopanes in this instance (R = sec butyl).

as hop-17(21)-ene (LXXXIII) and hop-22(29)-ene (LXXXIV) were found in ferns and mosses:

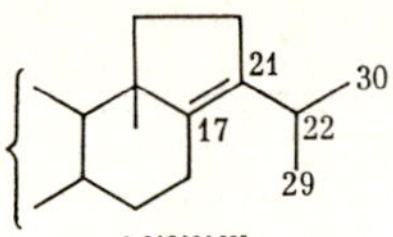

LXXXIII

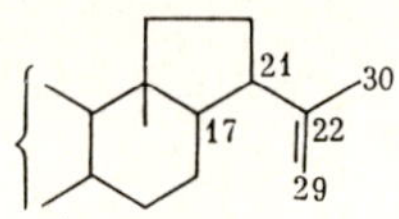

LXXXIV
(Diploptene)

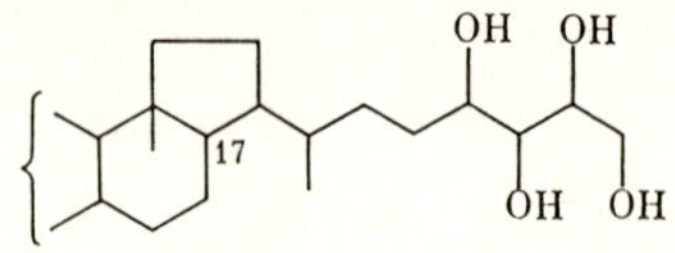

LXXXV

Therefore, a possible method of petroleum hopane synthesis is the alkylation (and degradation) of LXXXIII and LXXXIV with subsequent hydrogenation. However, still another important source of hopanes exists, i.e. "bacteriohopanetetrol", which is a C_{35} tetrahydroxyhopane (LXXXV) (Dididze et al. 1979; Kayukova et al. 1981b; Seifert and Moldowan 1978).

Gradual dehydration, chain degradation and saturation by hydrogenation, i.e. those transformations which are feasible in the presence of clays, may result in the formation of the entire $C_{27}-C_{35}$ regular hopane series. Absence of hopanes higher than C_{35} in most crudes indicates the formation of petroleum hydrocarbons from bacteriohopanetetrol. Indeed, besides regular hopanes, a small amount of $C_{36}-C_{40}$ hopanes, with an unbranched chain (methyl group at C-22 being an obvious exception) (LXXXVI), was reported in a publication devoted to polycyclic hydrocarbons in Thornton bitumen (Rulkötter and Philp 1981):

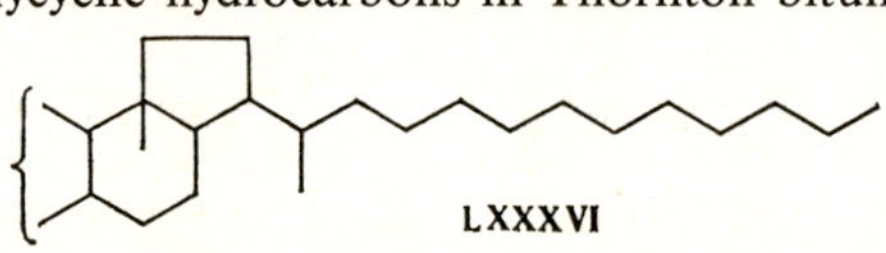

LXXXVI (C_{40})

It is assumed that these hydrocarbons were formed by the alkylation of the bacteriohopane chain.

The majority of crude oils are characterized by the virtual absence of such important natural triterpanes and corresponding hydrocarbons as oleanane, friedelane, gamma-cerane, lupane and others, although their sources are widespread in nature. However, a crude oil, discovered in the Tertiary of the Niger delta, included, besides hopanes, two hydrocarbons of different structure in large quantities (Ekweozor et al. 1979). One of them, 18α-oleanane, is a typical representative of triterpanes with a basic perhydropicene nucleus; the other one is a spirotriterpane. It is assumed that oleanane (a regular component of brown coals) owes its presence to the contribution of continental organic matter, as the residues of higher plants to the sediments in which this crude was generated:[5]

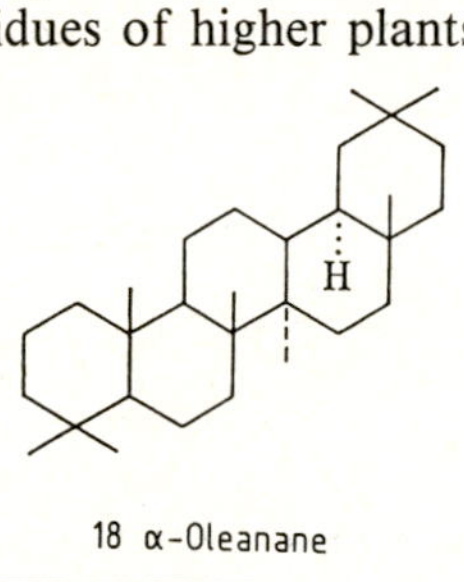

18 α-Oleanane

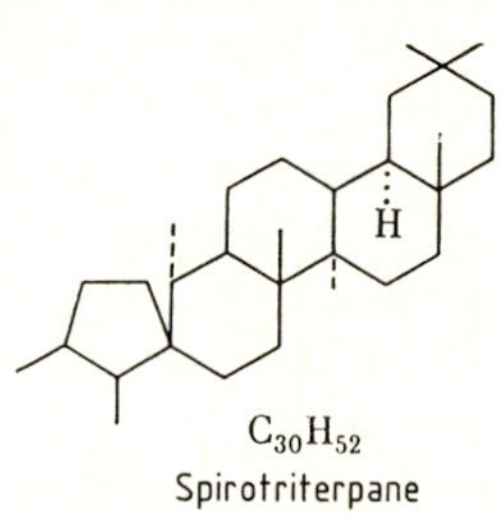

$C_{30}H_{52}$
Spirotriterpane

[5] Recently, oleanane and gamma-cerane were discovered in other crudes of continental origin (Indonesia, China, Australia).

Both hydrocarbons have similar structural and genetic connections:

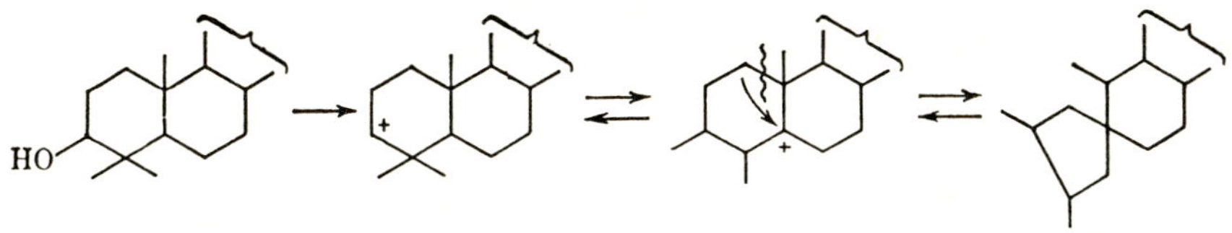

Unusual saturated polycyclic $C_{58}-C_{80}$ isoprenoids consisting of three homologous series: (A) octa-cyclic, (B) deca-cyclic and (C) trideca-cyclic, were found in residues of a crude oil (Ten Fu Yene and Chilingarian 1976; Oelert et al. 1973). These hydrocarbons are evidently based on dimers and codimers (biopolymers) of sterenes and hopenes:

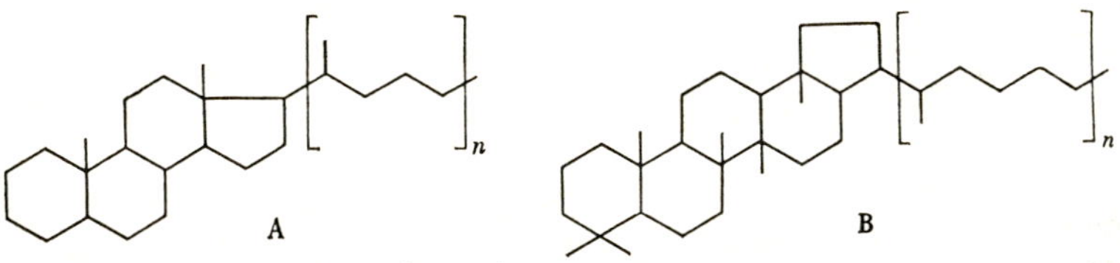

A

B

(n = 2,3)

The homologous series C is assumed to be produced by copolymerization of two sterene and one hopene molecule.

Geochemical Significance of Hopanes

Two aspects in hopane geochemistry should be emphasized. Similar to the steranes discussed above, hopane hydrocarbons may be used in evaluating catagenetic transformations and in various genetic correlations.

Appropriate stereochemical modifications of molecules are of special significance in the study of catagenetic transformation. The important role of newly generated 14*β*,17*β*,20*S*- and 20*R*- and 5*α*,20*S*-steranes has already been demonstrated. Stereochemical maturation of hopanes also plays an important role. The following indicators may be used in evaluating the catagenetic maturity of the organic matter:

1. Ratio of 17*α*,21*β*/17*β*,21*β* hopanes, i.e. the percent of new petroleum hopanes.
2. Ratio of 22*S*/(22*R* +22*S*) hopanes (for C_{31}, C_{32}, etc.).
3. Ratio of $C_{29}+C_{30}$ 17*β*,21*α*-moretanes to $C_{29}+C_{30}$ 17*α*,21*β*-hopanes.

Kinetically, the first reaction proceeds much faster than the latter two. Figure 54 demonstrates changes in the ratios of *ββ*/*αβ*-hopanes and (22*S*/22*R* +*S*)-hopanes in Toarcian shales with burial depth. Table 42 also shows changes in the ratio of the sums of moretanes and hopanes in the same organic material (Seifert et al. 1981). We should recall that practically no 17*β*-hopanes are found in crudes, while the 22*S*/22*R*-isomer ratio is 60:40.

The ratio of moretanes to 17*α*-hopanes in crudes varies from 0.08 to 0.1, which generally corresponds to the estimated thermodynamic equilibrium (0.14).

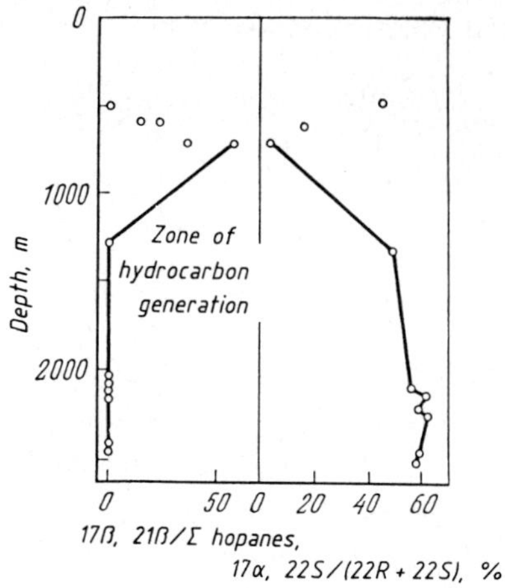

Fig. 54. Changes in 17*β*,21*β* hopane concentrations and 17*α*,22S hopane formation as a function of the depth of burial (Toarcian shales)

Table 42. Changes in the $C_{29}-C_{30}$ moretane to $C_{29}-C_{30}$ 17*α*H-hopane ratio in Toarcian shales with increasing depth

Depth (m)	700[a]	700[a]	1200	2000	2100	2300	2400
Moretanes/Hopanes	0.75	0.84	0.22	0.21	0.19	0.07	0.08

[a] Different samples.

Another aspect of the geochemistry of petroleum hopanes is the genetic application in "oil-oil" and "oil-potential source rocks (sources)" correlations.

Indeed, the distribution pattern of the 17*α*-hopanes is a perfect genetic indicator and, as mentioned previously, a specific "fingerprint" of crudes from a given area. This distribution is as informative as the widely used pristane/phytane ratio. A comparison of the distributions of hopanes in crudes and dispersed organic material (bitumens) helps to determine petroleum sources. These genetic characteristics may be especially significant. For example, the presence of C_{28} hopanes in Sivinski and some other related crudes allows one to establish precise constraints for their formation. Concentrations of adiantane higher than those of hopane in crudes from Tatariya signify their common source. From this viewpoint, identical distributions of hopanes in Tertiary crude oils from Apsheron and Western Turkmenistan are also noteworthy. The ratios between hopane/sterane may serve as an additional criterion. They range from 1.8 to 2.2 for Baku crudes and 3.0–3.2 for Samotlor crudes, while they exceed 10 in the hopane-rich oils from Tatariya and the Krasnodar region (Petrov et al. 1976).

The total concentrations of petroleum hopanes are in the same order of magnitude, if not slightly higher, as those of steranes. Unfortunately, some crudes contain hopanes in small concentrations, which requires their concentration by thermal diffusion or liquid-chromatography on 10X zeolites or aluminium oxide prior to analysis. Hopanes are commonly found in the 450°–520 °C fraction. It should be recalled that the above ratios of hopanes and steranes relate only to the sum of the respective chromatographic peaks of these hydrocarbons. At the same time, according to group-type MS data, the overall concentration of tetracyclic hydrocarbons is always higher, and tetracyclic/pentacyclic hydrocarbon ratios in various crude oils range between 1.6 and 2.8.

The determination of hopane hydrocarbons, as well as of steranes, is also possible in insoluble materials, such as kerogen. Thus, wide use is made of pyrolysis of preliminary kerogen with subsequent analysis of the released biomarkers.

Besides standard geochemical methods of potential source-rock analysis, such as isotope analysis, isolation of organic carbons, etc., biomarker analyses represent a most successful additional, albeit decisive factor in making a final conclusion. These paleoreconstructions are usually aimed at establishing four main factors: (1) sources; (2) maturation; (3) migration and (4) biodegradation (MacKenzie et al. 1980). Detailed analysis of these problems can be found in the original publications (Seifert and Moldowan 1978, 1979b; Seifert 1975, 1978; Seifert et al. 1979; van Dorsselaer et al. 1975).

GC Analysis of Steranes and Hopanes

In conclusion, Table 43 summarizes the sterane and hopane hydrocarbons identified in crudes, as well as their relative retention times. A chromatogram of a mixture of these hydrocarbons in the Samotlor crude oil can be seen in Fig. 55. As can be seen, the determination of hopanes is an easier task, since there is practically no "hump" in the region where the $C_{29}-C_{35}$ hopanes elute, which is characteristic of sterane elution.

Steranes, Hopanes and the Optical Activity of Crudes

Steranes and hopanes are the main sources of the optical activity of crude oils. Figure 56 illustrates changes in the optical activity of petroleum fractions as a function of the boiling temperature (Tissot and Welte 1981). As can be observed, the maximum optical activity occurs in the 420°–550 °C fraction, containing hydrocarbons with molecular weights between 350 and 450, i.e. $C_{27}-C_{35}$ steranes and triterpanes. This is only natural, since these compounds contain a large number of chiral centres (8–9 in steranes and 9–10 in hopanes). Moreover, the optical activity of chiral centres in the cyclic molecular part is characterized by high absolute values. The problem arises elsewhere. How can optical activity be preserved in catagenetic environments under the possible influence of acidic catalysts? Especially since we have demonstrated previously the important role of epimerization reactions in the formation of petroleum hydrocarbons:

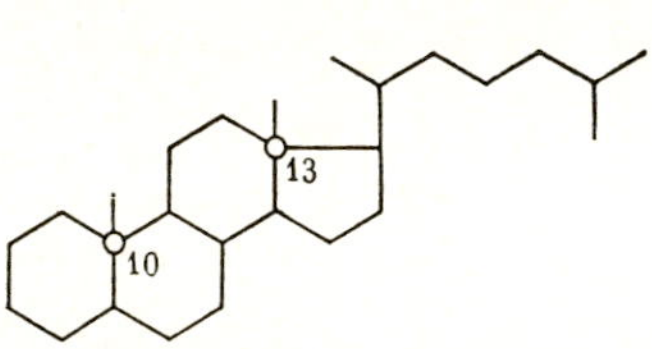

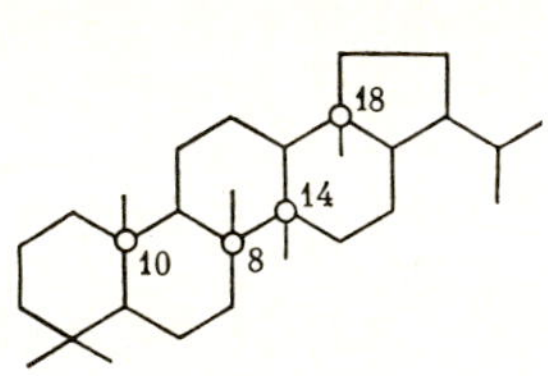

Table 43. Relative retention times of steranes and hopanes identified in crudes

Hydrocarbons	Number of carbon atoms	Sterical configuration[a]	Retention times relative to		Peak number in Fig. 55
			α-Cholestane	17αH, 21βH-Hopane	
β,α-Diacholestane	27	10α,13β,17α, 20*S*	0.56	0.28	$P_{27}I$
β,α-Diacholestane	27	10α,13β,18α, 20*R*	0.60	0.30	$P_{27}II$
24-Methyl-β,α-diacholestane	28	10α,13β,17α, 20*S*	0.69	0.34	$P_{28}I$
24-Methyl-β,α-diacholestane	28	10α,13β,17α, 20*R*	0.74	0.36	$P_{28}II$
24-Ethyl-β,α-diacholestane	29	10α,13β,17α, 20*S*	0.81	0.40	$P_{29}I$
24-Ethyl-β,α-diacholestane	29	10α,13β,17α, 20*R*	0.87	0.43	$P_{29}II$
Isocholestane[b]	27	5α,14β,17β, 20*R*	0.90	0.44	Iso.27I
Isocholestane[b]	27	5α,14β,17β, 20*S*	0.92	0.45	Iso.27II
α-Cholestane	27	5α,14α,17α, 20*R*	1.00	0.49	α-27
24-Methylisocholestane	28	5α,14β,17β, 20*R*	1.12	0.55	Iso.28I
24-Methylisocholestane	28	5α,14β,17β, 20*S*	1.15	0.56	Iso.28II
24-Methyl-α-cholestane (α-ergostane)	28	5α,14α,17α, 20*R*	1.26	0.62	α-28
24-Ethylisocholestane	29	5α,14β,17β, 20*R*	1.33	0.65	Iso.29I
24-Ethylisocholestane	29	5α,14β,17β, 20*S*	1.35	0.66	Iso.29II
24-Ethyl-α-cholestane (α-sitostane)	29	5α,14α,17α, 20*R*	1.52	0.74	α-29
Trisnorhopane ITs	27	18α	1.20	0.59	$H_{27}I$
Trisnorhopane IITm	27	17α	1.30	0.64	$H_{27}II$
Bisnorhopane	28	17α,18α,21β	1.62	0.79	H_{28}
Norhopane (adiantane)	29	17α,21β	1.70	0.83	H_{29}
Normoretane	29	17β,21α	2.01	0.98	M
Hopane	30	17α,21β	2.04	1.00	H_{30}
Moretane	30	17β,21α	2.29	1.12	M
Hopane Homologs:					
$X = C_2H_5$[c] (Homohopane)	31	17α,21β, 22*S*	2.51	1.23	$H_{31}I$
$X = C_2H_5$ (Homohopane)	31	17α,21β, 22*R*	2.59	1.27	$H_{31}II$

$X = C_3H_7$ (Bishomohopane)	32	17*a*,21*β*, 22*S*	2.92	1.43	$H_{32}I$
$X = C_3H_7$ (Bishomohopane)	32	17*a*,21*β*, 22*R*	3.05	1.50	$H_{32}II$
$X = C_4H_9$ (Trishomohopane)	33	17*a*,21*β*, 22*S*	3.53	1.73	$H_{33}I$
$X = C_4H_9$ (Trishomohopane)	33	17*a*,21*β*, 22*R*	3.72	1.82	$H_{33}II$
$X = C_5H_{11}$ (Tetrakishomohopane)	34	17*a*,21*β*, 22*S*	4.34	2.13	$H_{34}I$
$X = C_5H_{11}$ (Tetrakishomopane)	34	17*a*,21*β*, 22*R*	4.64	2.27	$H_{34}II$
$X = C_6H_{13}$ (Pentakishomohopane)	35	17*a*,21*β*, 22*S*	5.41	2.65	$H_{35}I$
$X = C_6H_{13}$ (Pentakishomohopane)	35	17*a*,21*β*, 22*R*	5.84	2.86	$H_{35}II$

[a] The orientation of the hydrogen atom is indicated.

[b] Coelution isomers of ergostane and sitostane in crudes.

[c] X = R; see LXXIX, page 122.

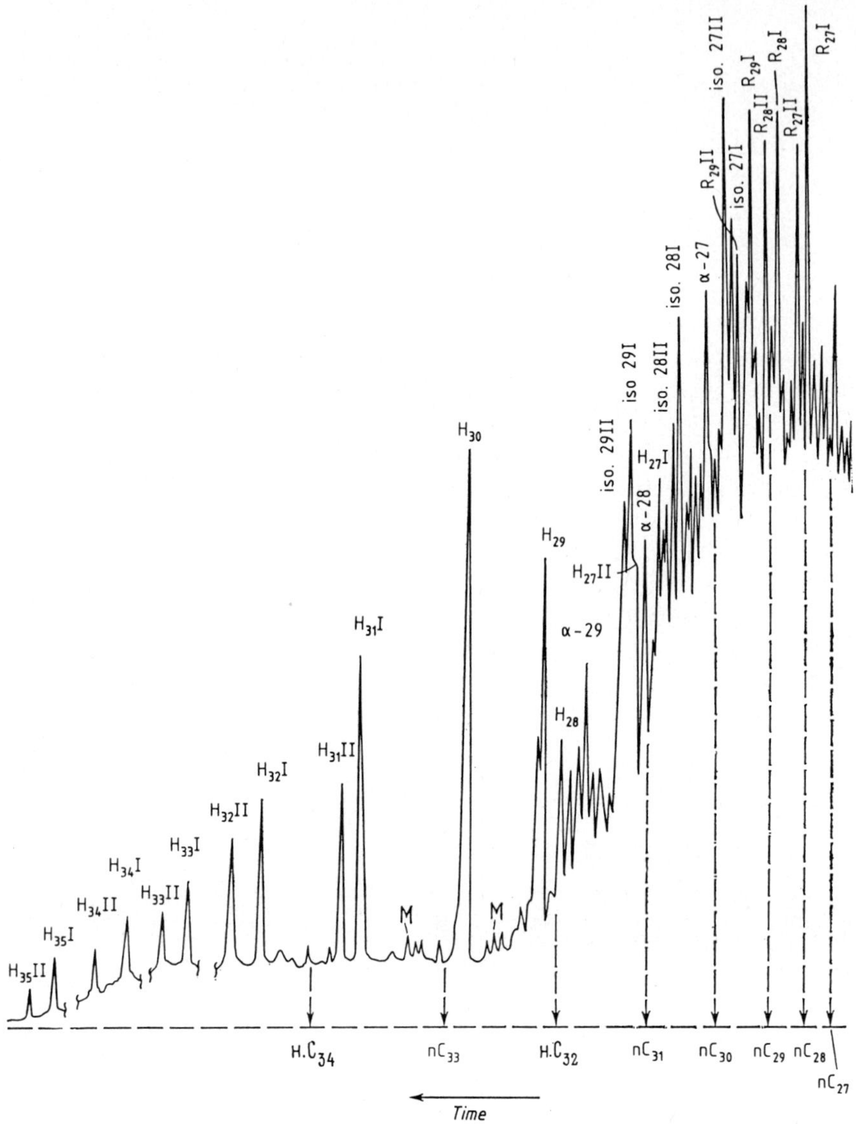

Fig. 55. Chromatogram of a concentrate of steranes and hopanes in the Samotlor crude. Capillary column 80 m, Apiezon; 300 °C. Normal alkane elution positions marked on the abscissa. Peak identifications can be found in Table 43

In our opinion, preservation of optical activity is mostly due to such chiral centres which are not liable to epimerization because of their specific structure. These cites should primarily include the quaternary C-10 and C-13 carbon atoms in steranes, and the C-10, C-8, C-14 and C-18 carbon atoms in hopanes (marked by open circles).

The absence of a hydrogen atom at these chiral centres makes their epimerization unlikely under acid catalysis or in other transformations. Characteristically,

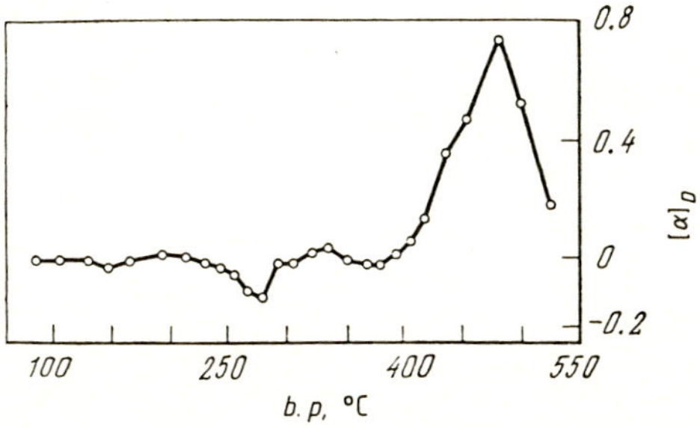

Fig. 56. Changes in the optical activity of saturated petroleum hydrocarbons with an increase in boiling temperature

an equilibrium isomerizate of epimeric cholestanes produced after 50 h heating at 320 °C in the presence of hydrogen and a platinum on charcoal catalyst, retained its high optical activity, being almost equal to that of the original α-cholestane. Optical activity of polycyclane concentrates in Gurgyanskoe crude, consisting of a hopane and sterane mixture (450°–500 °C fraction separation by thermal diffusion) had an optical rotation of $\alpha^D = 19°–20°$.

CHAPTER 4

Aromatic Hydrocarbons (Arenes)

Aromatic hydrocarbons may be subdivided into two main groups.

The first group includes alkylaromatic hydrocarbons with only *aromatic rings* and aliphatic substituents. The most widely occurring homologous series in this instance are those of alkylbenzenes, alkylnaphthalenes, alkylphenanthrenes, alkylchrysenes and alkylpicenes (only aromatic nuclei are illustrated):

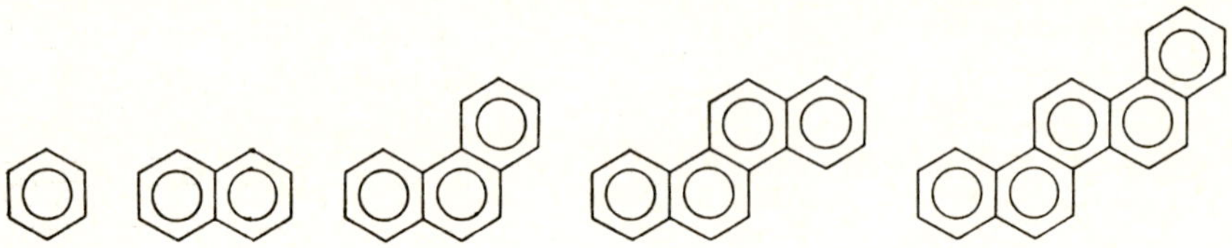

The general formula of these compounds is C_nH_{2n-z}, where factor "z" is correspondingly 6, 12, 18, 24 and 30 (the "z" factor and the distributions of petroleum hydrocarbons depending on it are widely used in MS) (Dooley et al. 1974).

The second, no less important group of aromatic hydrocarbons, which is especially typical of crude oils, consists of hydrocarbons of naphthenoaromatic structure, i.e. those containing both *aromatic* and *naphthenic rings,* as well as, naturally, aliphatic substituents. Homologous series are much more numerous in this case than in the first one, since numerous variations of aromatic and naphthenic rings may occur. The following groups, differing in the number of aromatic rings per molecule, are usually identified (this classification may also be applicable to the first subdivision).

1. *Monoaromatic hydrocarbons:* indanes (tetralins), di-, tri-, and tetranaphthenobenzenes and others (substituents are also omitted, only the cyclic molecular part is represented):

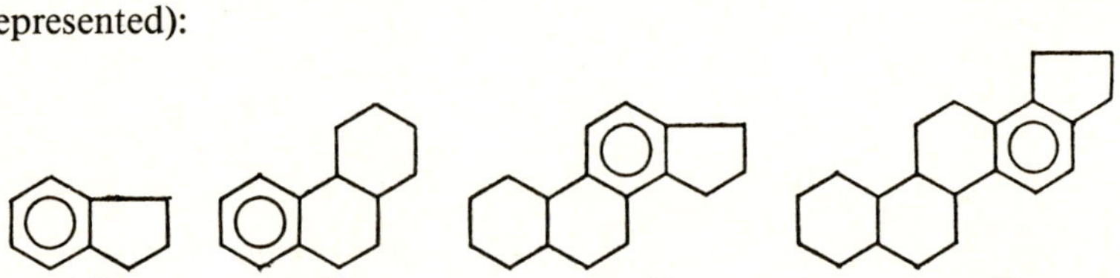

$z=8$ $z=10$ $z=12$ $z=14$

2. *Diaromatic hydrocarbons:* mono- and dinaphthenonaphthalenes, etc. (included are structures most commonly occurring in crudes):

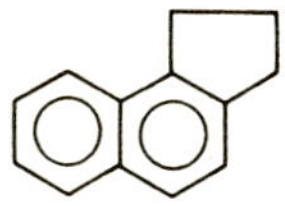
z=14

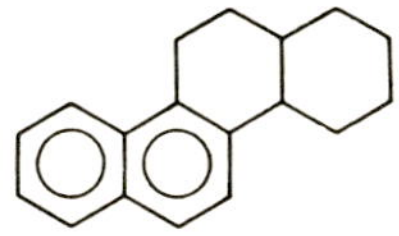
z=16

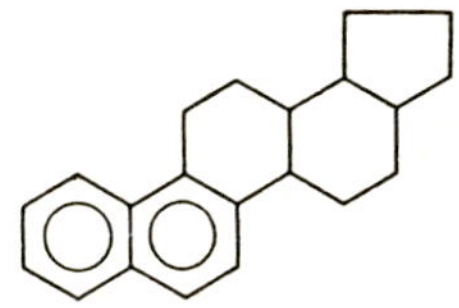
z=18

3. *Hydrocarbons with three and more aromatic rings:* naphthenophenanthrenes, etc. (the most widely spread systems with *cata*-condensation of aromatic nuclei are represented):

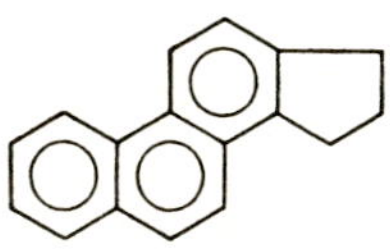
z=20

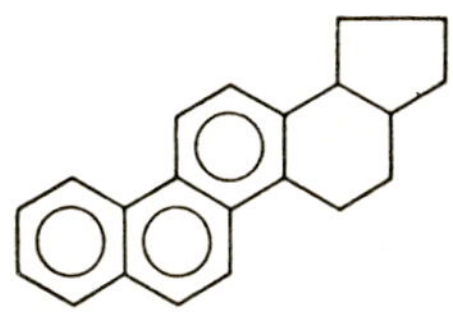
z=22

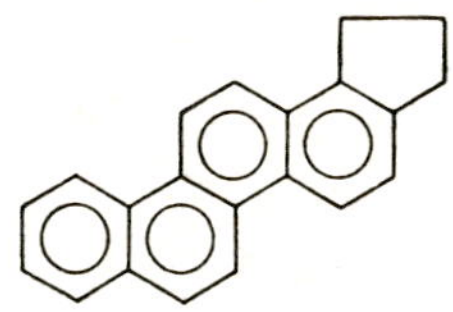
z=26

The division of aromatic hydrocarbons into mono-, di- and polyaromatic ones depends on their properties during liquid chromatographic separation, which occasionally allows the separation of these three groups. However, hydrocarbons of alkylaromatic and naphthenoaromatic structure types often elute together in such experiments.

Additionally, petroleum chemistry and geochemistry pay special attention to such structures as alkylbiphenyls, acenaphthenes, fluorenes and pyrenes (corresponding aromatic nuclei are presented below):

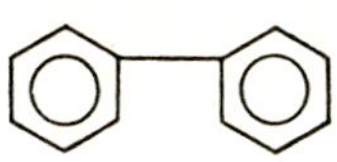
z=14

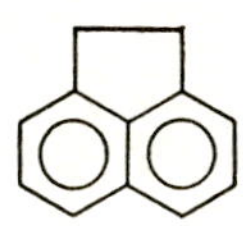
z=14

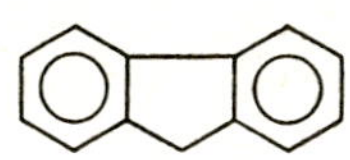
z=16

z=26

As can be observed, the z-factor value may in certain instances be identical for hydrocarbons with different numbers of aromatic nuclei; therefore for purposes of their MS (low resolution) group analysis, it is advisable to make a preliminary separation by groups.

Group-Type Distribution of Aromatic Hydrocarbons in Crude Oils (Aromatic Fingerprint)

A wealth of information of the distribution of aromatic hydrocarbons exists according to structure types. Table 44 presents some data on the composition of aromatic hydrocarbons in various Soviet crudes. These were collected through MS (matrix analysis) and represent a fingerprint of sorts of a particular group of crude oils. In analogy with the naphthenic fingerprint (see Chap. 1), this distribution may be called the aromatic fingerprint. It is true, however, that contrary to

Table 44. Relative distribution of aromatic hydrocarbons by structure types (200° – 430 °C fraction) in crudes (%)

Hydrocarbon	Factor "z"	Deposit, Petroleum type			
		Starogroznenskoe, B^1	Starogroznenskoe, A^1	Samotlor, A^2	Samotlor, A^1
	Monoaromatic:				
Alkylbenzenes	6	14.9	17.9	22.3	21.3
Indanes	8	15.9	17.4	16.8	16.5
Dinaphthenobenzenes	10	11.8	15.2	9.8	9.0
	Diaromatic:				
Naphthalenes	12	20.7	20.9	19.6	18.6
Acenaphthenes	14	16.8	16.2	17.2	17.6
Fluorenes	16	9.1	5.9	4.5	4.5
	Polyaromatic:				
Phenanthrenes	18	5.1	3.1	3.4	4.3
Naphthenophenanthrenes	20	1.7	0.3	1.3	2.7
Chrysenes	24	3.3	2.6	1.9	2.0
Pyrenes	26	0.7	0.4	3.2	3.5
Σ Monoaromatic		42.6	50.5	48.9	46.8
Σ Diaromatic		46.6	43.0	41.3	40.7
Σ Polyaromatic		10.8	6.5	9.8	12.5

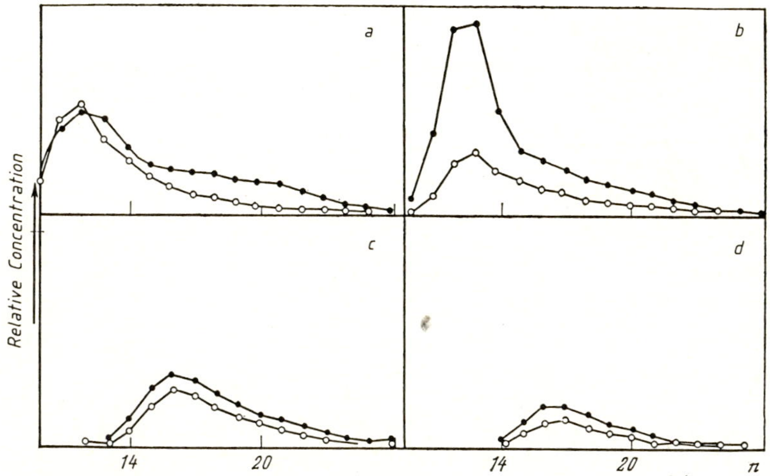

Fig. 57a–d. Relative distribution of various aromatic hydrocarbons depending on the number of carbon atoms per molecule. **a** Indane (tetralin) homologs; **b** naphthalene homologs; **c** fluorene (acenaphthene) homologs; **d** phenanthrene homologs; ● and ○ denote different crudes

the naphthenic fingerprint, the aromatic one of different crudes is more homogeneous.

Table 44 fails to reveal the character or number of aliphatic substituents at the aromatic or naphthenoaromatic nucleus. However, in di- and polyaromatic hydrocarbons the number of carbon atoms in substituents, especially those linked to aromatic nuclei, is small, and as a rule we a have to deal with a few methyl substituents. Aromatic steranes with their aliphatic chains still linked to the naphthenic ring are an exception.

Figure 57 shows the distribution of certain di- and triaromatic petroleum hydrocarbons depending on molecular mass (number of carbon atoms per molecule). In contrast, as will be demonstrated below, monoaromatic hydrocarbons (especially benzene homologs) have long aliphatic substituents.

Data on the distributions of aromatic hydrocarbons by structure types in high-boiling fractions (370°–530 °C) are of particular importance. These are tabulated in Table 45. An appropriate analysis was conducted after preliminary separation of an aromatic mixture into mono-, di- and polyaromatic compounds (Dooley et al. 1974).

Individual Aromatic Petroleum Hydrocarbons

Since most aromatic petroleum hydrocarbons are genetically linked to cyclanes, they will be reviewed similarly to saturated hydrocarbons, i.e. starting with monocyclic hydrocarbons (benzene homologs) and followed by bicyclic (indanes, tetralins) and tricyclic (with different rings in the molecule), etc.

Table 45. Relative distribution of aromatic hydrocarbons by structure types in the high-molecular-weight fraction of crudes

Hydrocarbon	Factor "z"	Deposit (petroleum fraction)	
		Wilmington (370° – 535 °C)	Recluse (370° – 530 °C)
Alkylbenzenes	6	5.71	11.18
Alkylnaphthenobenzenes	8	16.05	18.39
Alkyldinaphthenobenzenes	10	20.64	19.13
Alkyltrinaphthenobenzenes	12	20.95	16.54
Alkyltetranaphthenobenzenes	14	17.18	15.10
Alkylpentanaphthenobenzenes	16	12.02	11.31
Alkylhexanaphthenobenzenes	18	7.45	8.35
∑ Monoaromatic[a]		100 (16.8)	100 (11.7)
Alkylnaphthalenes	12	7.47	5.46
Alkylnaphthenonaphthalenes	14	21.75	15.77
Alkyldinaphthenonaphthalenes	16	29.45	18.50
Alkyltrinaphthenonaphthalenes	18	20.07	19.26
Alkyltetranaphthenonaphthalenes	20	11.37	18.77
Alkylpentanaphthenonaphthalenes	22	5.82	13.76
Alkylhexanaphthenonaphthalenes	24	4.07	8.48
∑ Diaromatic[a]		100 (12.4)	100 (5.06)
Alkylbenzoindanes (Fluorenes)	16	9.75	2.64
Alkylnaphthenobenzoindanes	18	20.96	8.24
Alkyldinaphthenobenzoindanes	20	21.76	17.10
Alkyltrinaphthenobenzoindanes	22	21.19	21.47
Alkyltetranaphthenobenzoindanes	24	16.41	21.05
Alkylpentanaphthenobenzoindanes	26	6.55	16.60
Alkylhexanaphthenobenzoindanes	28	3.38	12.90
∑ Polyaromatic[a]		100 (17.3)	100 (6.12)

[a] In the fraction.

Monoaromatic Hydrocarbons

Monocyclic Hydrocarbons (Benzene Homologs). Benzene and its C_7-C_{10} homologs were thorougly investigated in many crude oils (Kolesnikova 1972; Martin et al. 1964). As always, concentrations of substituted hydrocarbons considerably exceed those of the unsubstituted ones, i.e. benzenes. Table 46 presents some data on the distribution of isomeric C_8-C_{10} arenes in A^1 crudes. These were drawn from the publication of Martin et al. (1964), which contains exhaustive material on benzene homologs in light-boiling petroleum fractions.

Equilibrium data for C_8-C_{10} alkylbenzenes are presented in Table 47 (Stall et al. 1964).

In comparing data in Tables 46 and 47, it becomes evident that petroleum alkylbenzene mixtures are also far from equilibrium. In effect, di- and trisubstituted structures are found in maximal concentrations in crudes, while tri-

Table 46. Relative distribution of C_8-C_{10} alkylbenzenes in crudes (%)

Hydrocarbon	Deposit		
	Alida	Beaver Lodge	Darius
Benzene	12.28	19.27	14.46
Toluene	87.72	80.73	85.54
∑ Benzene + Toluene	100	100	100
C_8 Composition:			
Ethylbenzene	27.94	19.10	14.95
1,4-Dimethylbenzene	12.50	15.17	17.76
1,3-Dimethylbenzene	41.18	46.63	40.19
1,2-Dimethylbenzene	18.38	19.10	27.10
∑ Alkyl-Substituted	100	100	100
C_9 Composition:			
Isopropylbenzene	3.6	2.77	2.13
n-Propylbenzene	10.3	8.46	4.44
1-Methyl-3-ethylbenzene	21.46	19.23	16.65
1-Methyl-4-ethylbenzene	9.44	7.69	10.18
1-Methyl-2-ethylbenzene	8.0	5.69	12.03
1,3,5-Trimethylbenzene	8.50	10.77	10.18
1,2,4-Trimethylbenzene	27.47	33.85	35.15
1,2,3-Trimethylbenzene	11.16	11.54	9.24
∑ Alkyl-Substituted	100	100	100
C_{10} Composition:			
∑ Mono-substituted	11.03	36.72	3.9
∑ Di-substituted	43.19	29.67	40.84
∑ Tri-substituted	28.17	20.49	33.69
∑ Tetra-substituted	17.61	13.12	21.57
∑	100	100	100

and tetra-substituted C_9, C_{10} hydrocarbons should dominate under equilibrium. Relative concentrations of mono-substituted alkylbenzenes are also high in crudes. As was supported by further analysis, mono- and di-substituted hydrocarbons also prevail among higher boiling compounds. Certain C_{10} alkylbenzenes, such as 1,2- and 1,4-methylisopropylbenzenes, are of obvious relict character, hence, their relative concentrations are fairly high. High concentrations of normal alkylbenzenes are based on peculiarities of petroleum hydrocarbon formation from fatty acids (see Chap. 5).

The works of Solli et al. (1979), Ostroukhov et al. (1982, 1983) and Gala et al. (1977) contain valuable data on the compositions of $C_{11}-C_{20}$ alkylbenzenes in medium and higher petroleum fractions.

Monoaromatic hydrocarbons isolated in an aromatic concentrate (200°–430 °C fraction) represent a homologous series of different structures, mostly benzene homologs. In our opinion, monoaromatic hydrocarbons of both regular and mixed types, i.e. compounds which are closest to saturated cyclic petro-

Table 47. Equilibrium concentrations of C_8 and C_9 alkylbenzenes (%)

Hydrocarbon	300 K	500 K	Hydrocarbon	300 K	500 K
C_8 Composition:			C_9 Composition:		
Ethylbenzene	0.5	3.7	Isopropylbenzene	0.01	0.3
1,4-Dimethylbenzene	23.7	23.2	n-Propylbenzene	0.01	0.5
1,3-Dimethylbenzene	54.5	52.8	1-Methyl-4-ethylbenzene	1.1	5.9
1,2-Dimethylbenzene	16.3	20.4	1-Methyl-3-ethylbenzene	1.2	8.3
			1-Methyl-2-ethylbenzene	0.2	2.3
			1,3,5-Trimethylbenzene	36.1	22.1
			1,2,4-Trimethylbenzene	56.1	51.1
			1,2,3-Trimethylbenzene	5.3	9.5

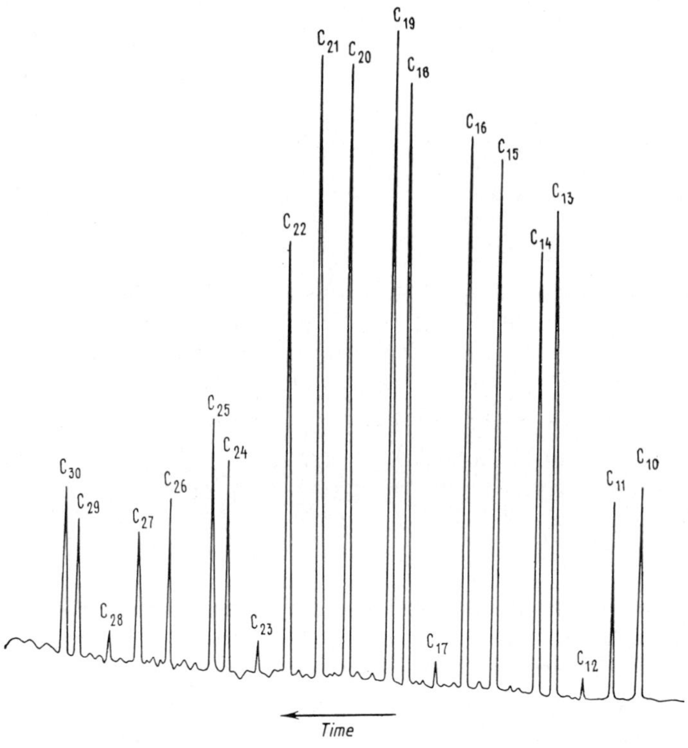

Fig. 58. Chromatogram of a mixture of 1,3,4-trimethyl-2-alkylbenzenes (I), isolated from Shakalyk-Astan crude. The total number of carbon atoms per molecule is identified

leum hydrocarbons, deserve special attention in chemistry, especially in petroleum geochemistry. Among them, relict structures of indubitable origin (such as monoaromatic steranes, etc.) are found. Moreover, monoaromatic hydrocarbons as a group may be easily and clearly isolated in a mixture of aromatic compounds through liquid chromatography on aluminum oxide.

A review of high-boiling monoaromatic hydrocarbons (benzene homologs) will begin with compounds having long isoprenoid chains.

A chromatogram of a monoaromatic mixture of tetra-substituted benzene homologs (isolated from the Shakalyk-Astana crude) is shown in Fig. 58. They possess a long, irregular isoprenoid chain. The structure of the highest molecular weight hydrocarbon (proven by GC-MS and comparative synthesis of a standard is reproduced in the following formula (I)):

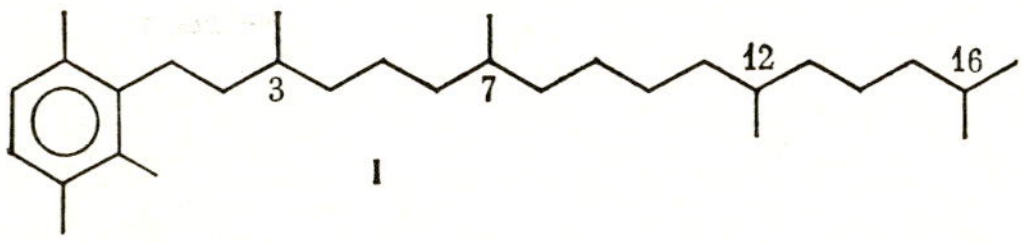

$C_{30}H_{54}$

i.e. this hydrocarbon is a 1,3,4-trimethyl-2-(3′,7′,12′,16′-tetramethylheptadecyl) benzene.

As can be seen in the chromatogram reproduced, the homologous series under analysis, starting with C_{10} and through C_{30}, is present in crude oils.

The long isoprenoid chain is reconfirmed by the absence of C_{12}, C_{17} and C_{23} homologs. Moreover, hydrocarbon pairs eluting immediately after the "concentrational dip", such as C_{13} and C_{14}, C_{18} and C_{19}, etc. have closer boiling temperature than other homologs. This may be explained by the emergence of new methyl substituents in the main chain of the C_{14}, C_{19}, C_{25} hydrocarbons. The homolog distribution in the series clearly indicates carotenoid substances as the appropriate sources.

The aforementioned aromatic hydrocarbons are genetically related to the 1,1,3-trimethyl-2-alkylcyclohexane series described in Chapter 2. The only difference is the migration of geminal methyl groups which results in the formation of aromatic hydrocarbons (Scheme 1). *M/z* 133 ions appearing in *β*-cleavage of C-C-bonds in the aliphatic chain are the key fragments for the aromatic hydrocarbons discussed. Accordingly, mass spectra of tri-substituted dimethylalkylbenzenes will be characterized by the presence of *m/z* 119; di-substituted, 105; mono-substituted, 91.

Scheme 1

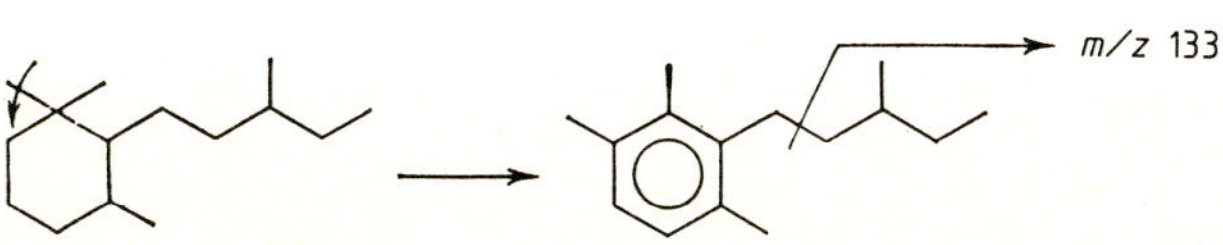

The same Shakalyk-Astana crude contains another series of benzene homologs with an isoprenoid chain, which is again genetically related to carotenoids. These are $C_{10}-C_{30}$ hydrocarbons as well and have the type-II structure:

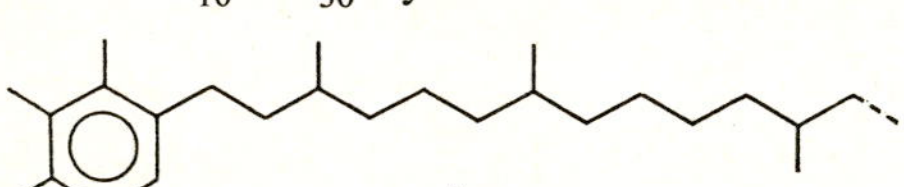

II

i.e. they are 1,2,3-trimethyl-4-alkybenzenes.

A homologous series of other aromatic hydrocarbons, having a $C_{14}-C_{30}$, regular isoprenoid chain, was detected in a Kyurov-Dag crude oil (Tertiary sediments, Azerbaijan). The structure of one of theses hydrocarbon – type-III – is as follows:

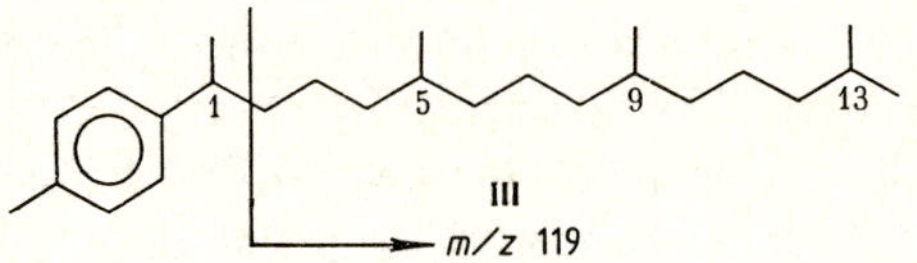

III

$C_{25}H_{44}$

Contrary to the preceding series (trimethylalkylbenzenes), there is a marked predominance of $C_{15}-C_{20}-C_{25}-C_{30}$ hydrocarbons in this one (monomethyl-alkylbenzenes), which signifies a different source. (Most probably the role of the initial compounds in this case was played by isoprenoid acids, such as farnesanic and phytanic acids, etc.) The mass fragmentogram of these hydrocarbons is reproduced in Fig. 59.

High concentrations of individual aromatic hydrocarbons are sometimes found in crudes. The 1-methyl-3-(3′,7′,11′,15′-tetramethylhexadecyl) benzene (IV), discovered in the same Kurov-Dag crude, may serve as an illustration:

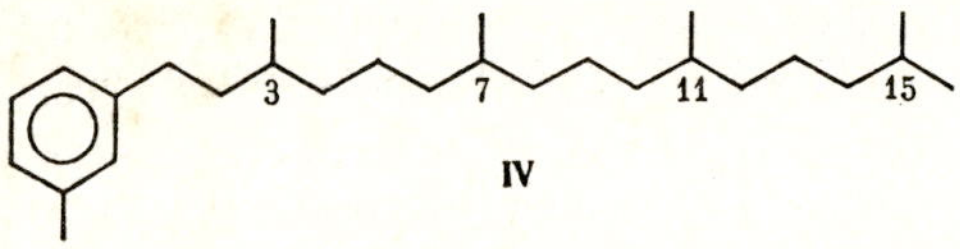

IV

$C_{27}H_{48}$

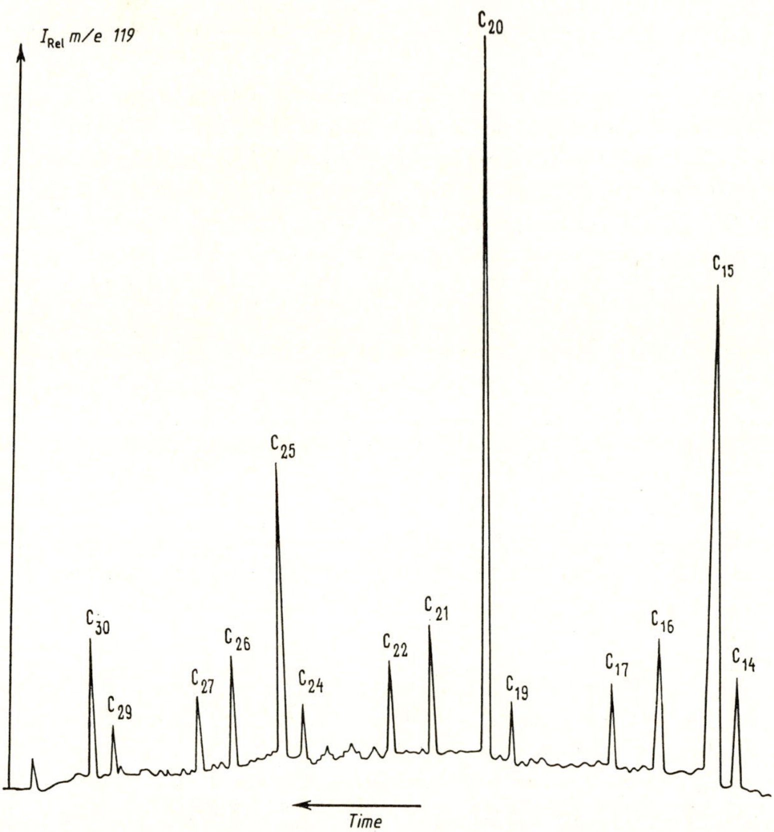

Fig. 59. Mass fragmentogram of a mixture of 1-methyl-4-alkylbenzenes (with isoprenoid chain) (III) isolated from Kyurov-Dag crude

A peculiar phytadiene "dimer" of the structural type V was described in the publication of Simoneit and Burlingame (1973):

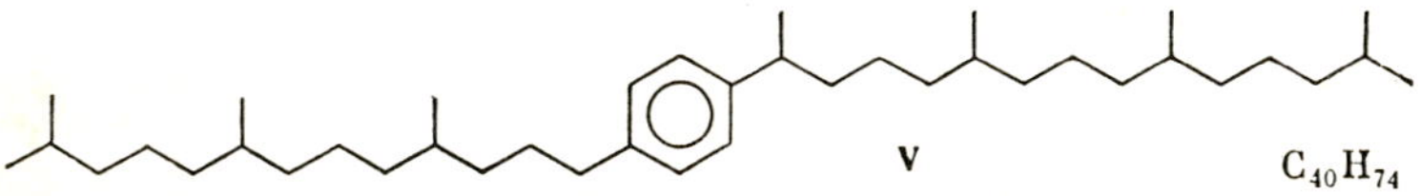

V $C_{40}H_{74}$

Apparently, aliphatic chain degradation products of this "dimer" may also be found in petroleums.

Let us now turn to other homologous series of alkylbenzenes, this time with an unbranched aliphatic chain (Ostroukhov et al. 1983). The $C_{10}-C_{32}$ n-alkylbenzenes, as well as the ortho-, meta- and para-methyl-n-alkylbenzenes isomeric to them, were dicovered in Samotlor crude oils. Appropriate material is reproduced in Fig. 60. The hydrocarbon structure was established by GC-MS, and comparative synthesis of some standard compounds. Key ions are *m/z* 91 (for n-alkylbenzenes) and *m/z* 105 for methylalkylbenzenes (*m/z* 106 is the base peak in meta-substituted hydrocarbons). It is noteworthy that concentrations of para-substituted methylalkylbenzenes are universally, considerably inferior to those of other di-substituted isomers. These are most probably products of dehydration cyclization of unsaturated fatty acids (see Chap. 5).[1] Long-chain alkylbenzenes were also found in kerogen pyrolytic products (Solli et al. 1979).

Bicyclic Hydrocarbons. Let us first review petroleum hydrocarbons of mixed structure, i.e. having both aromatic and naphthenic cycles. Unsubstituted hydrocarbons of this type, belonging to the series of indane (VI) and tetralin (VII), are fairly well known:

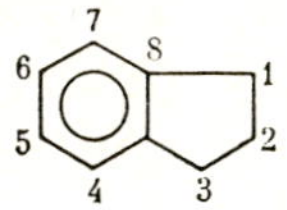

VI

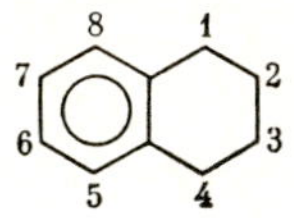

VII

They were found in various crude oils (Ponka City, West Edmon, Romashkino, Arlanskoe, etc.). The presence of 1-, 2-, 4- and 5-methylindanes, tetralin and its mono-, di- and trimethyl homologs were also reported (Sokolov et al. 1972).

An indane homolog with a long aliphatic chain of type VIII was found in relatively large quantities in Shakalyk-Astana crude:

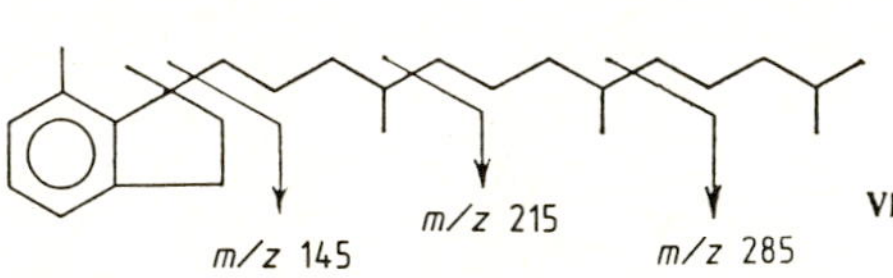

VIII C_{27}

Evidently, this hydrocarbon is produced by cyclization of compound (IX), under acidic conditions, which appears in phytol alkylation of certain aromatic compounds (such as toluene):

[1] Alkylbenzenes with long isoprenoid chains may be regarded as specific analogs of isoprenoid alkanes, while alkylbenzenes with long, normal alkyl chains correspond to normal petroleum alkanes.

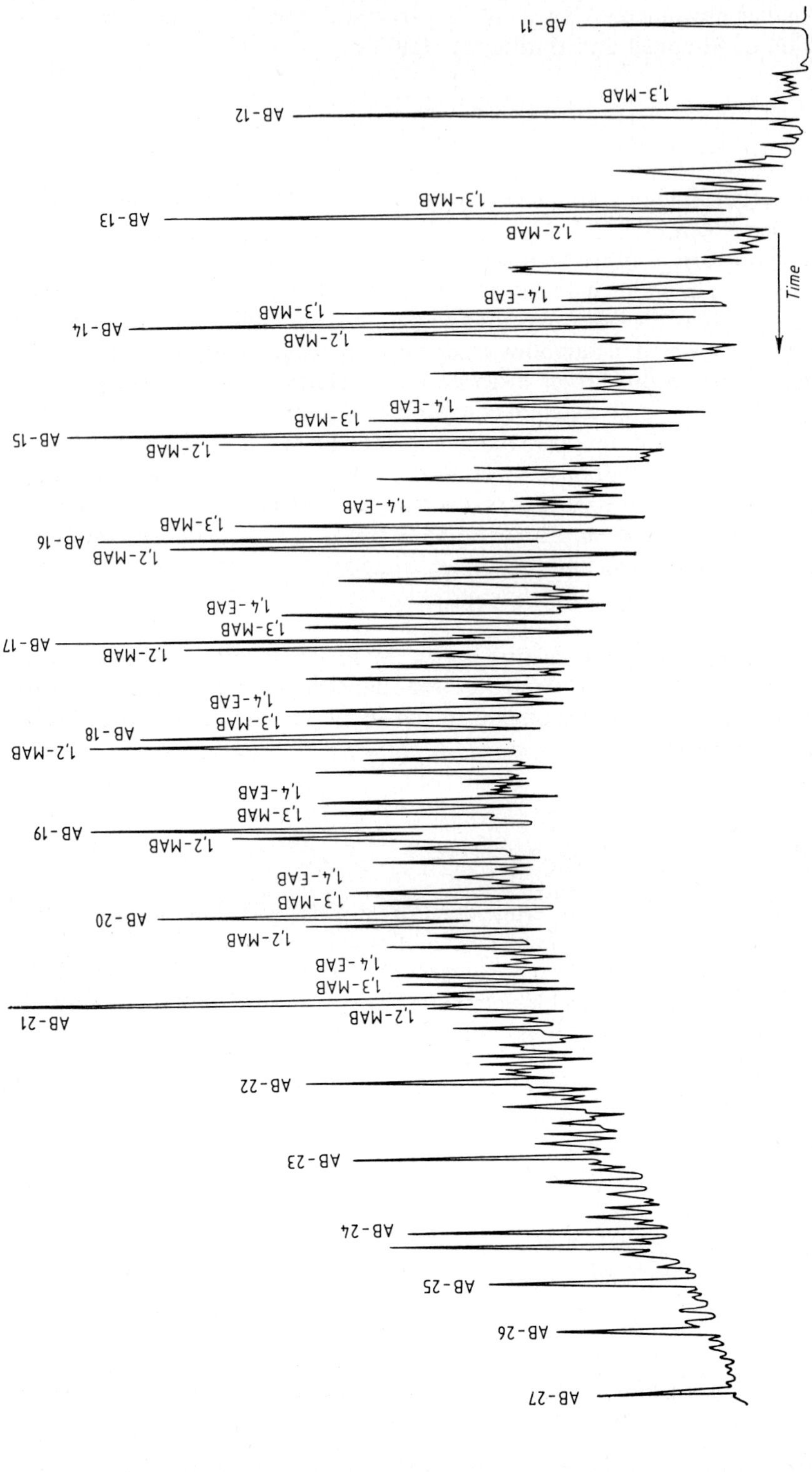

Fig. 60. Chromatogram of a mixture of n-alkyl-, 1,2-, 1,3- and 1,4-methylalkylbenzenes of different molecular mass, isolated from Samotlor crude: AB-12, AB-13, etc.; C_{12}, C_{13}, etc.; n-alkylbenzenes; 1,2-MAB-1-methyl-2-alkylbenzenes, 1,3-MAB-1-methyl-3-alkylbenzenes, etc.; 1,4-EAB-1-ethyl-4-alkylbenzenes. Capillary column, 50 m, Dexil; linear temperature programming, 100°→2°/min

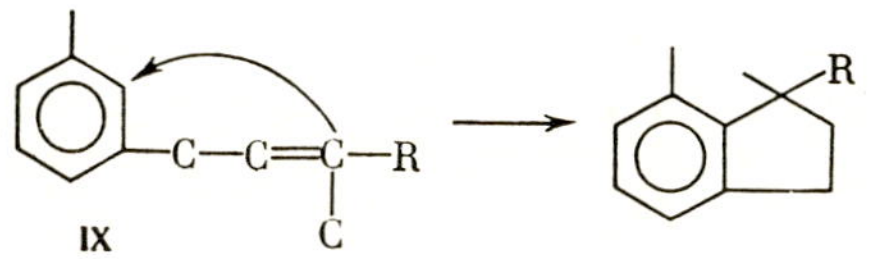

IX

An interesting group of $C_{14}-C_{27}$ phenylcyclohexylalkanes was found in Green River shales (Ten Yu Yene and Chilingarian 1976). These hydrocarbons may also be present in crudes. Their structure may be represented by the following formula (X); normethyl analogs were also discovered:

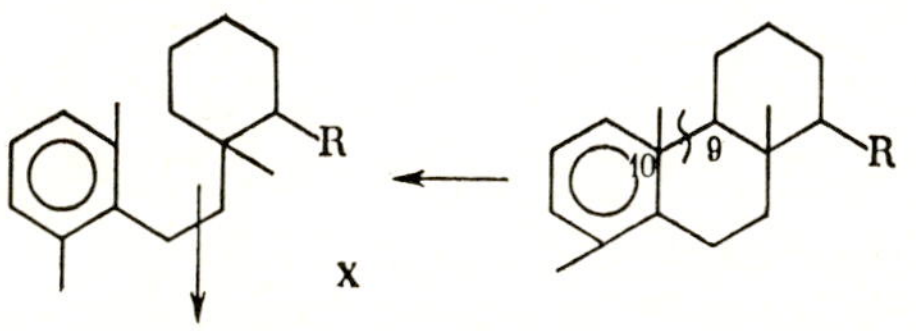

X

m/z 119

Their formation is apparently based on the cleavage of the 9–10 bond in the well-known tricyclic isoprenoid hydrocarbons which are squalene cyclization products, or of other isoprenoid structures (see Chap. 3).

Tricyclic Hydrocarbons (Homologs of Octahydrophenanthrene). Tricyclic aromatic petroleum hydrocarbons are mainly represented by numerous methyl (alkyl) derivatives of phenanthrene. However, let us first analyze the ring junction and stereochemistry of some octahydrophenanthrenes, i.e. compounds that model structural fragments of highly important petroleum hydrocarbons, monoaromatic steroids and hopanoids.

Octahydrophenanthrene, as well as its homolog with an angular methyl group, are the simple model hydrocarbons that successfully illustrate the stereochemical properties of those hydrocarbon molecules which comprise, together with the aromatic ring, six-membered naphthenic rings (Zubenko et al. 1981).

Actually, the connection of the flat aromatic ring with a chair-shaped cyclohexane results in considerable distortions of the saturated cycle similarly to that which occurs in cyclohexenes. Thus, a better steric connection is that of a flatter cyclopentane and a benzene ring, as in indanes.

Distortion of the chair-type conformation of one of the six-membered rings in octahydrophenanthrene is reflected in the relative thermodynamic stability of isomers having this type of junction of saturated rings.

Equilibrium ratios of *cis*-and *trans*-isomers of octahydrophenanthrene (XI), as well as 4a-methyloctahydrophenanthrene (XII), were analyzed in the paper of Zubenko et al. (1981):

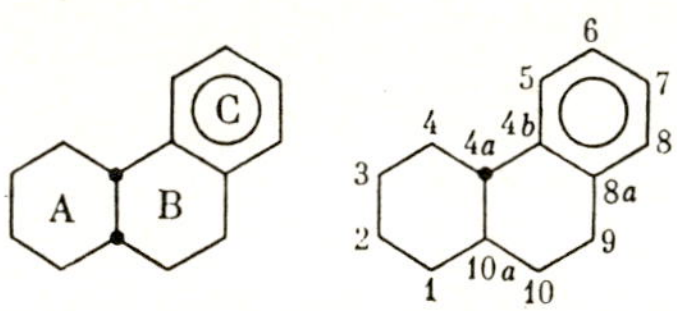

cis- XI *trans-*

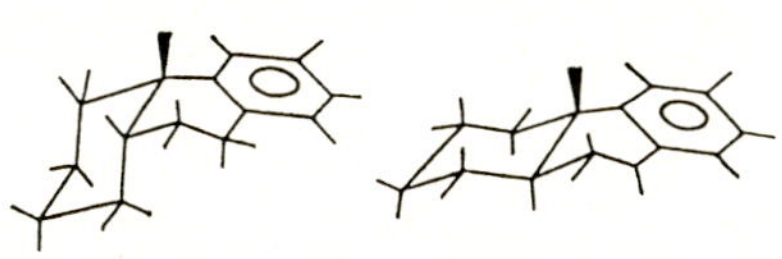

cis- XII *trans-*

Since hydrocarbons XI and XII are regarded by us as phenanthrene derivatives, we use carbon atom numbering accepted for phenanthrene in this instance. However, the numbering used for steranes and hopanes may be preferred. The angular substituent (in compound XII) is bound to atom C – 10 in this case.

A-rings in all of the reviewed monoaromatic structures have chair conformations, while B-rings have semi-chair conformations and C-rings are planar. Their *trans*-isomers are characterized by a rigid ring junction, while *cis*-isomers may undergo conformational changes. The *cis*-isomer has one gauche butane interaction (1-10a-10-9) in the unsubstituted structure XI; such interactions are absent in *trans*-isomers. It is true, however, that due to the distorted, chair-shaped conformation of the B-ring interaction 1-10a-10-9 is only partial.

Under equilibrium conditions (300 K) compound XI was found to contain 23% *cis*- and 77% *trans*-isomer, which closely coincides with the estimated data. It is noteworthy that under similar conditions there is only 2% *cis*- and, correspondingly 98% *trans*-isomers in decalins (or appropriate perhydrophenanthrenes) (Petrov 1981). Thus, an aromatic ring visibly increases the stability of isomers with *cis*-ring junction.

This tendency was clearly reconfirmed in the study of *trans*- and *cis*-isomers of compound XII in equilibrium. Introduction of an angular methyl substituent leads to considerable variations in the energy balance of isomeric octahydrophenanthrenes. As proved experimentally, the *cis*-isomer, amounting to 70% at 500 K (85% at 300 K), becomes more stable in this case. (In an equilibrium situation saturated analogs, i.e. angular substituted perhydrophenanthrenes, will contain 84% *trans*-isomers in equilibrium at 300 K.)

Greater thermodynamic stability of *cis*-4a-methyloctahydrophenanthrene may be explained by only one fully skew butane interaction in this epimer (Me-4a-10a-10). As was pointed out, interaction 1-10a-10-9 is only partial. At the same time, there were three butane interactions with a methyl group participation in the *trans*-isomer. (A more stable *trans*-isomer from among the saturated analogs of compound XII has one skew interaction less.)

Thus, polycyclic structure with angular substituted octahydrophenanthrene moieties may be expected to have greater thermodynamic stability of stereoisomers with *cis*-A/B ring junction, as opposed to regular steranes which were shown to be characterized by a predominance of isomers with *trans*-A/B junction in equilibrium situations (Petrov 1981) (see also Chap 3).

Diaromatic Hydrocarbons

Bicyclic Hydrocarbons (Homologs of Naphthalene and Biphenyl). Naphthalene and its homologs were identified in crudes some time ago. As a rule, the content of naphthalene proper is insignificant, while its mono-, di- and trimethyl homologs are present in much higher concentrations (Savastyanova et al. 1967; Yew and Mair 1966).

Data on methyl- and dimethylnaphthalene distributions in a number of crudes are presented in Table 48.

Table 48. Relative distribution of methyl- and dimethylnaphthalenes in crudes and in equilibrium at 300 K (%)

Hydrocarbon	Deposit, Crude type					Equilibrium concentration (%)
	Arlanskoe, A[1]	Isl. Peschanyi, B[1]	Neftyanye Kamni, B[1]	Gryazevaya Sopka, B[1]	Ponka, A[1]	
2-Methylnaphthalene	58.6	55.8	60.8	74.7	–	65.0
1-Methylnaphthalene	41.4	44.2	39.2	25.3	–	35.0
2,6-Dimethylnaphthalene	17.5	11.6	10.3	18.6	19.4	13.8
2,7-Dimethylnaphthalene	13.5	12.7	12.4	24.0	14.9	13.8
1,7-Dimethylnaphthalene	14.7	15.0	15.7	9.6	21.2	19.9
1,6-Dimethylnaphthalene	27.4	19.8	18.3	11.2	28.2	14.9
1,3-Dimethylnaphthalene	7.2	11.8	11.9	6.3	0.2	19.9
2,3-Dimethylnaphthalene	5.9	5.8	6.6	10.4	7.2	11.7
1,4-Dimethylnaphthalene	4.4	5.8	7.2	8.4	2.8	5.1
1,5-Dimethylnaphthalene	2.7	6.4	8.6	4.3	3.7	7.4
1,2-Dimethylnaphthalene	6.7	11.1	9.0	7.2	2.4	7.3

Note: 1,8-Dimethylnaphthalene was not found in the crudes analyzed.

The maximal concentration of dimethylnaphthalenes usually occur in the case of 1,6- and 1,7-isomers. Ethylnaphthalenes are found in much smaller quantities. 1,2,5-(XIII), 1,2,6- and 1,2,7-trimethyl isomers were identified among trimethyl-substituted isomers, and 1,2,5,6-tetramethylnaphthalenes (XIV) among those tetra-substituted (Nekrasov and Ptitsina 1957):

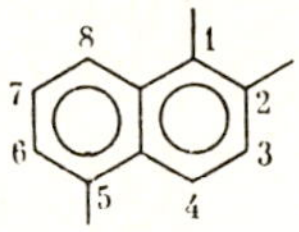
XIII

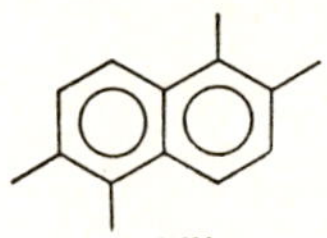
XIV

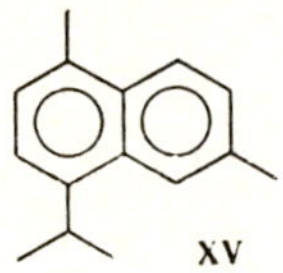
XV

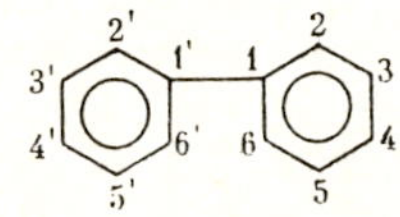
XVI

Their structure is evidently related to those of steroid and triterpenoid degradation (aromatization) products.

A peculiar hydrocarbon with naphthalene structure, [cadalene (XV)] was discovered by Bendoraitis in Loma Novia crude (Bendoraitis 1974). The concentration of cadalene (1,6-dimethyl-4-isopropylnaphthalene) reaches 0.4% in petroleum. Its special structure leaves no doubts as to its relict nature and genetic links with bicyclic sesquiterpenes.

Bicyclic hydrocarbons also include biphenyl (XVI) and its first homologs (2-, 3- and 4-methyl-), found in Ponka crude (Yew and Mair 1966). However, concentrations of biphenyl hydrocarbons in crudes are much smaller than the concentrations of naphthalene hydrocarbons.

Tricyclic Hydrocarbons. Tricyclic hydrocarbons with two aromatic and one naphthenic ring – alkylacenaphthenes and alkylfluorenes – are widely represented in crudes. However, the first homologs in these homologous series are present in smaller quantities in crudes, and may chiefly be found in coals and Recent sediments.

Triaromatic Hydrocarbons

Triaromatic hydrocarbons with three aromatic rings, mainly phenanthrene and its homologs (Mair and Barnewall 1964; Mair and Martinez-Pico 1962; Carruthers 1957), are also present in crudes, Thus, 1-methylphenanthrene (XVII); 2-methylphenanthrene (XVIII), 3-methylphenanthrene (XIX), 1,8-dimethylphenanthrene (XX); 1,2,8-trimethylphenanthrene (XXI) were isolated:

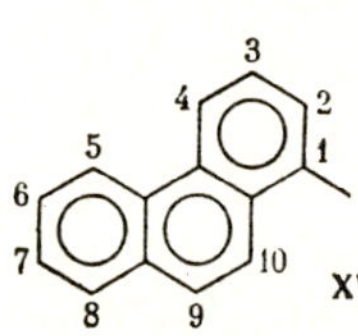
XVII

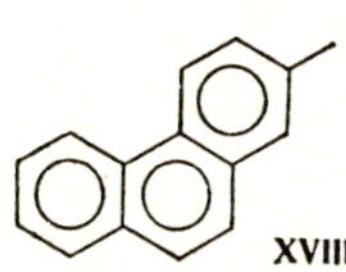
XVIII

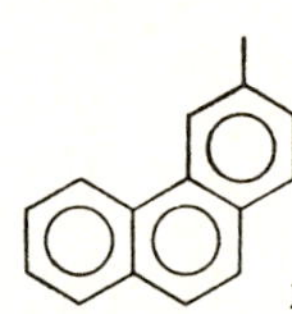
XIX

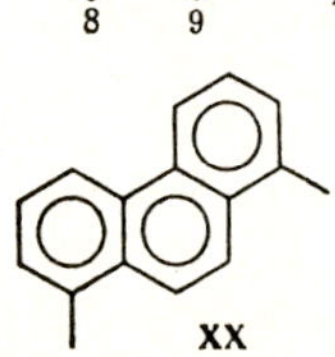
XX

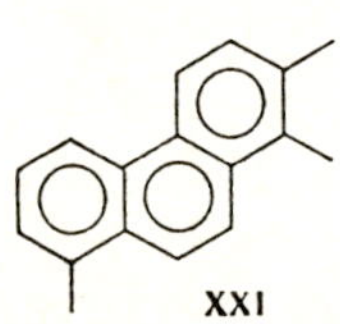
XXI

It is noteworthy that not a single methylphenanthrene so far identified had methyl groups in positions 4 and 5. This is in full agreement with the fact that steroids (apparent sources of phenanthrenes) are also missing methyl substituents at corresponding carbon atoms (Mair 1964). Evidently, 4-methyl-sterols are precursors of 1,8-dimethyl- and 1,2,8-trimethylphenanthrenes.

1-Methyl-7-isopropylphenanthrene (XXII, retene) – as a degradation product of diterpenoids – also deserves special attention:

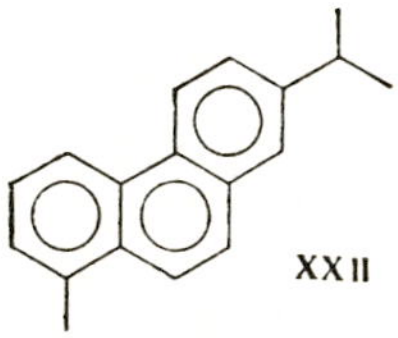

XXII

Comprehensive and profound analysis of crystal aromatic hydrocarbons isolated in a Tertiary Norio crude (Georgia) was undertaken under the guidance of Melikadze (Melikadze and Lekveishvili 1977; Lekveishvili et al. 1979; Polyakova et al. 1982).

The majority of these hydrocarbons were based on a phenanthrene nucleus. They were successfully isolated in an adduct-formation reaction with maleic anhyride, followed by adduct decomposition. Consequently, numerous $C_{15}-C_{19}$ alkylphenanthrenes, as well as naphtheno- and dinaphthenophenanthrenes were identified in the crude. The 2-, 3-, 9- and 1-methylphenanthrenes as well as 9-ethyl, 9-isopropyl- and 9-butylphenanthrenes were identified among the former hydrocarbons. A group of dimethyl (2,5-, 2,3- and 2,7-) and trimethylphenanthrenes was also established. Of special interest are hydrocarbons of the naphthenophenanthrene ($C_{19}H_{18}$) and dinaphthenophenanthrene ($C_{24}H_{26}$) series which may structurally be connected to triaromatic steroid and hopanoid hydrocarbons.

The ratio of α- and β-methylphenanthrenes, i.e. the methylphenanthrene index, acquires special significance on its own merit in helping establish the maturity of organic matter (Radke et al. 1982). The ratio of $\Sigma 2$ and 3-methyl-/$\Sigma 1$- and 9-methylphenanthrenes is in common use. The concentrations of α-substituted structures (1- and 9-methylphenanthrenes are higher in the initial organic material. The concentrations of more stable β-substituted methylphenanthrenes increase with greater catagenetic transformations. Similar regularities evidently apply to β- and α-methylnaphthalenes.

Aromatic Steroid Hydrocarbons

Monoaromatic steroid hydrocarbons, widely represented in various crude oils and shales (Tissot and Welte 1981; Seifert and Moldowan 1978; Schaefle et al. 1978), are of utmost importance for petroleum chemistry and geochemistry. It was established in the earliest research that the C-ring of steroids is the easiest to be aromatized, resulting in the loss of methyl groups and key fragments at *m/z* 253 (also *m/z* 239, 267 for normethyl- and methylsteroids, respectively) (see Scheme 2).

Scheme 2

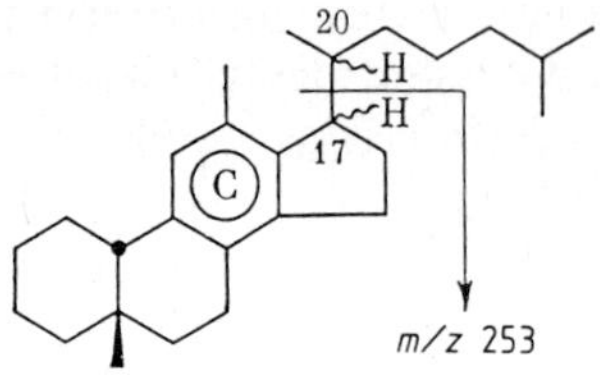

The synthesis of four diastereomeric aromatic C_{27} steroid hydrocarbons with the structure presented below, was reported in the publication of Schaefle et al. (1978). Later, one of the isomers was identified in crudes.

These isomers have a characteristic configuration of C-17 and C-20 chiral centres (carbon atom numbering used for monoaromatic steroids and hopanoids is that of their saturated analogs).

M/z 253 fragmentograms are used in various geochemical correlations, particularly in evaluating crude oil maturity (Tissot and Welte 1981; Seifert and Moldowan 1978).

Numerous sterols, which are present in various plant residues, are the source of monoaromatic steroids. The formation of such steroids occurs concurrently with that of regular steranes in contact between sterols and clays at moderate temperatures. These reactions are extensively analyzed in subsequent chapters.

Besides monoaromatic C_{27} steroid hydrocarbons, the structures of which were discussed above, crudes contain monoaromatic C_{28} and C_{29} steroids (i.e. respectively, monoaromatic 24-methyl- and 24-ethyl-substituted homologs). Each of them is represented by four diastereomers with identical mass spectra.

Figure 61 presents chromatograms of a monoaromatic steroid hydrocarbon concentrate isolated in the Anastasievskoe crude, as well as identical compounds obtained through dehydrogenation of cholestane, ergostane and sitostane. Publications on monoaromatic steroid hydrocarbon stereochemistry (Zubenko et al. 1980, 1981; Meney et al. 1973; Seifert et al. 1981; MacKenzie et al. 1981) helped clarify their structure. There are reasons to believe that methyl substituents in such petroleum steroids[2] (XXIII) occupy positions C-10 and C-17:

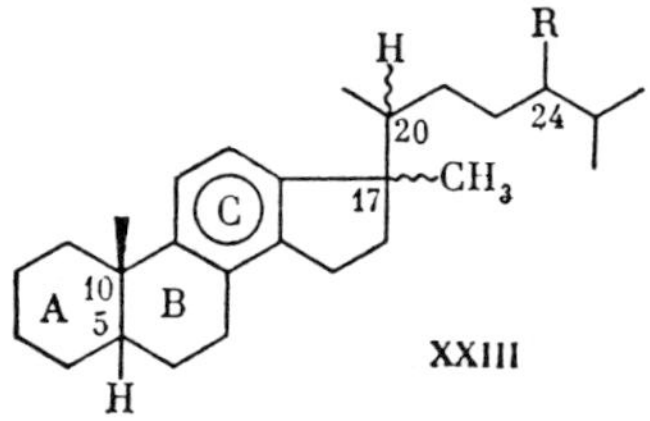

XXIII

($R = H, CH_3, C_2H_5$)

The four peaks, corresponding to each homolog in the chromatograms (see Fig. 61,b–d), belong to *5β,20S, 5β20R, 5α20S* and *5α20R* epimers, respectively.

An important role in establishing the structure of monoaromatic steroids was played by prior research of various physical and chemical properties of specially

[2] Naturally, we cannot completely exclude the presence of rearranged monoaromatic steroids, i.e. with a methyl group at C-5.

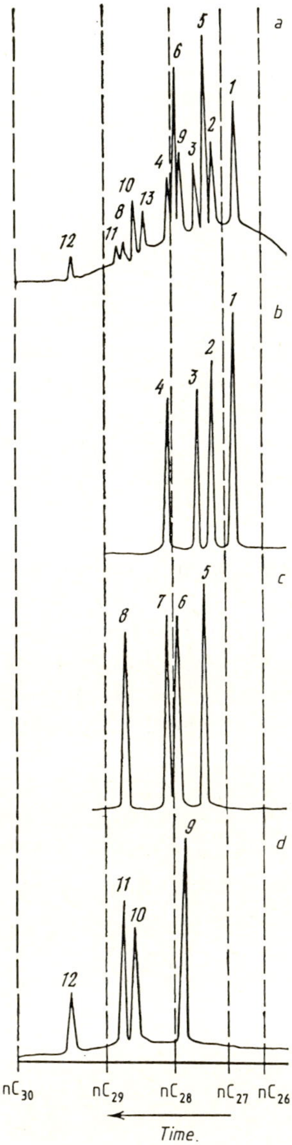

Fig. 61a–d. Chromatograms of monoaromatic steroid hydrocarbons. **a** Isolated from Anastasievskoe crude (concentrate); generated from: **b** cholestane, **c** ergostane, **d** sitostane; *1–4, 5–8, 9–12* different epimers of $C_{27}-C_{29}$ steroids. Normal alkane elution points indicated on the abscissa. Capillary column 50 m, Apiezon

synthesized hydrocarbons (XXIV–XXVIII), which are partial structures of the molecule under discussion:

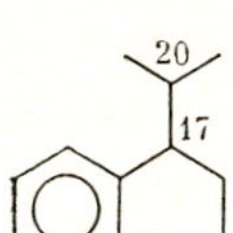

XXIV

(One epimer)

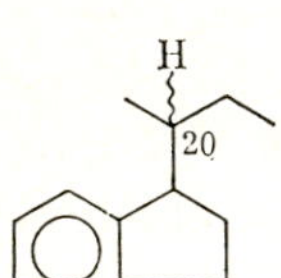

XXV

(Two epimers)

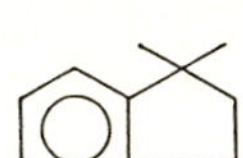

XXVI

(One epimer)

(continued on page 156)

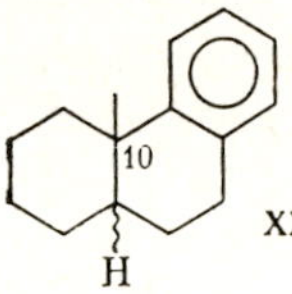

XXVII

(Two epimers)

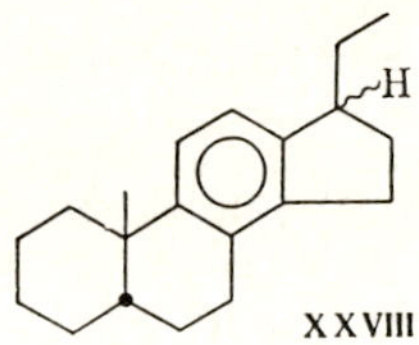

XXVIII

Monoaromatized pregnane
(Two epimers)

Figure 62 shows histograms of $C_{27}-C_{29}$ monoaromatic steroid mixtures, isolated from the Samotlor (A^1) and Gyurgyanskoe (type B^2) crude oils. Table 49 contains information on the relative concentations of these compounds in some crudes (Zubenko et al. 1980). Data on the distribution of monoaromatic and saturated steroid hydrocarbons are tabulated for comparison. A certain resemblance in the distributions of monoaromatic and saturated steroids may be observed. Small quantities of monoaromatic 4-methylsteroids were also identified in a computer-reconstructed mass chromotogram of the characteristic fragment ion *m/z* 267.

Besides C-ring monoaromatic steroids, those with an aromatic A-ring were identified in crudes. Ring aromatization is accompanied by methyl substituent migration from C-10 to C-1 and C-4. The mass spectra of these hydrocarbons are characterized by the high intensity of the molecular ion (40–80%). *M/z* 211 is the base peak corresponding to *m/z* 217 in regular steranes.

Besides monoaromatic steroids, appreciable quantities of those (XXIX) with three aromatic rings (A, B, C) were found in crude oils (Schaefle et al. 1978; MacKenzie et al. 1981; Laflamme and Hites 1979):

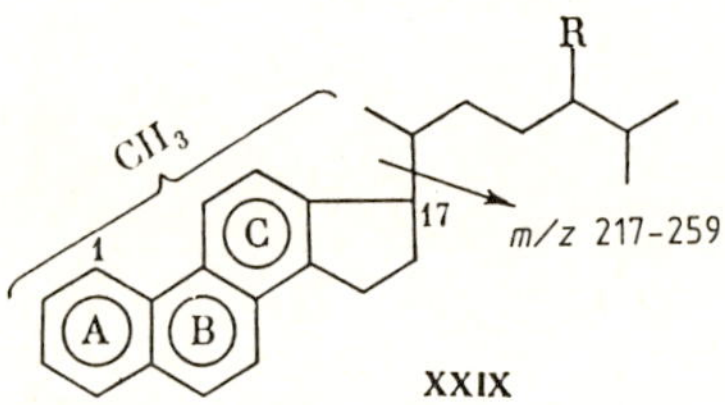

XXIX

m/z	217,	231,	245,	259
CH_3	0	1	2	3

($R = H, CH_3, C_2H_5$)

$C_{20}-C_{21}$ and $C_{26}-C_{30}$ hydrocarbons were identified. A substituent at C-17 in triaromatic C_{20} and C_{21} steroids is ethyl and isopropyl, respectively. The base peak in triaromatic steroids, without a methyl group in the nucleus, is at *m/z* 217; in triaromatic steroids with a single methyl group, *m/z* 231; and *m/z* 245 in those with two such groups. The fragment ion *m/z* 259 is the key fragment for aromatic steroids derived from 4-methylsteroids. The methyl group positions in triaromatic $C_{27}-C_{29}$ steranes were not established. It can only be assumed that one of these substituents is positioned at C-17, the other at C-1.

As a rule, the methyl substituent at C-10 is eliminated in the formation of triaromatic steranes: therefore, $C_{26}-C_{28}$ triaromatic steroids correspondingly derived from cholestane, 24-methyl- and 24-ethylcholestane predominate in crude oils. Each hydrocarbon is usually represented by two epimers with different con-

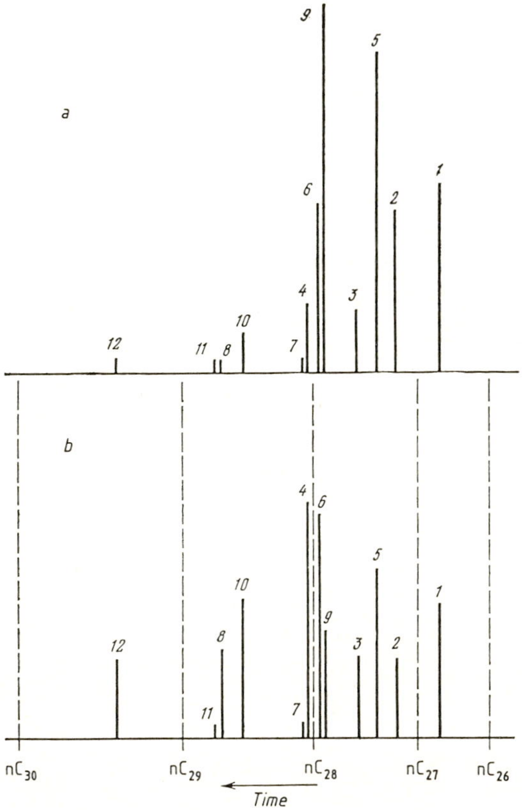

Fig. 62a, b. Bar diagrams of monoaromatic steroid hydrocarbon mixtures. **a** Isolated from Samotlor crude (A^1); **b** isolated from Gyurgyan crude (B^2). Peak identifications are the same as in Fig. 61

Table 49. Relative distribution of saturated and monoaromatic steroid hydrocarbons in crudes (%)

Crude		Saturated-type steranes			Monoaromatic steroids		
		C_{27}	C_{28}	C_{29}	C_{27}	C_{28}	C_{29}
Samotlor	A^1	23.2	33.3	43.5	33.3	32.9	33.8
Gyurgyanskoe	A^2	33.5	26.1	40.4	32.9	29.6	37.5
Anastasievsko-Troitskoe	B^1		Trace		25.9	42.6	31.5
Sivinskoe	B^2		Trace	100	17.2	22.5	60.3

figurations of the C-20 atom. Figure 62A, as described by MacKenzie (1984) presents a chromatogram of a triaromatic steroid hydrocarbon mixture.

The 1,2-cyclopentanophenanthrene (XXX) and its 3-methyl-substituted homolog (XXXI), the so-called Diels hydrocarbon (Mair 1964), were also identified in crudes:

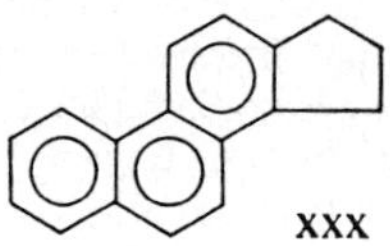

XXX

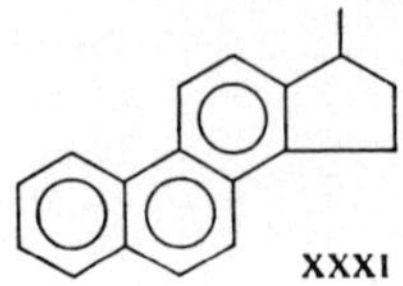

XXXI

Mono- and triaromatic steroids reveal their geochemical significance in establishing the degree of maturation (aging) of crudes and organic matter. In the process of catagenesis, monoaromatic steroids undergo degradation of the aliphatic side chains and are transformed into triaromatic hydrocarbons. It should be noted that once the B-ring aromatization (slow stage) in monoaromatic steroids (aromatic C-ring) is completed, subsequent A-ring aromatization proceeds with considerable speed. This results in low concentrations of diaromatic steroids (B- and C-rings).

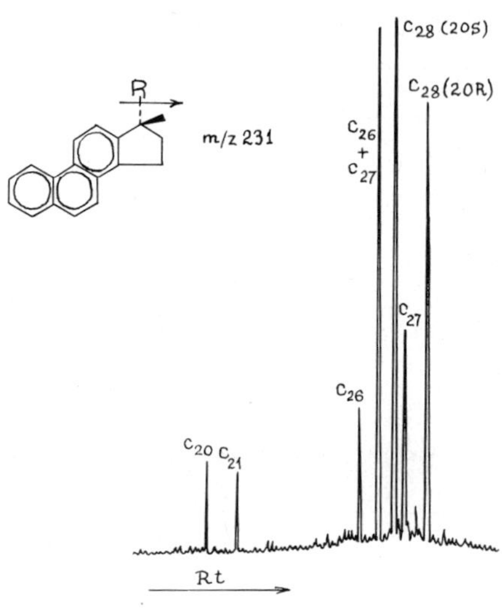

Fig. 62A. Mass fragmentogram of a mixture of petroleum triaromatic steroid hydrocarbons

Aromatic Triterpenoids and Hopanes

Bendoraitis was the first to announce the presence of aromatic hydrocarbons of triterpenoid origin in crude oils (Bendoraitis 1974). He discovered two series of monoaromatic hydrocarbons, which could be produced with the help of triterpenoids in Loma Novia crudes.

Hydrocarbons of the former series included three naphthenic rings, and one aromatic ring; those of the latter, four naphthenic and one aromatic ring. Judging by their mass spectra, it was suggested that tetracyclic hydrocarbons may have structures XXXII and XXXIII, since it was assumed that their genesis was due to a rupture of the 11 – 12 bond in dihydro-β-amyrin:

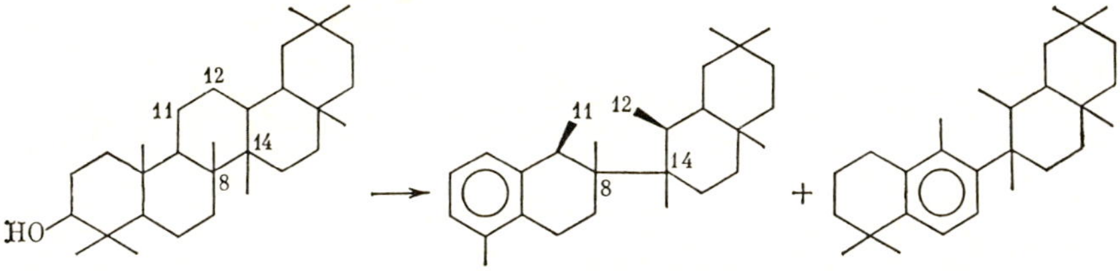

Dihydro-β-amyrin XXXII C_{28} XXXIII C_{28}

Hydrocarbons belonging to the second (pentacyclic) series had two hydrogen atoms less. No data on their structure were given.

Extensive research of pentacyclic aromatic hydrocarbons was undertaken in connection with the study of the Eocenic Messel oil shale (Greiner et al. 1976; Spyckerelle et al. 1977a, b, c, d; Greiner et al. 1977a, b; Wakeham et al. 1980b). Aromatic hopanes isolated from this shale differed in structure and comprised one to four aromatic rings. Structures of many hydrocarbons discovered were reconfirmed by the synthesis of standards. Thus, a stepwise scheme (Scheme 3) of the transformation of original triterpenoids (hopene or adiantanone) was established.

Scheme 3

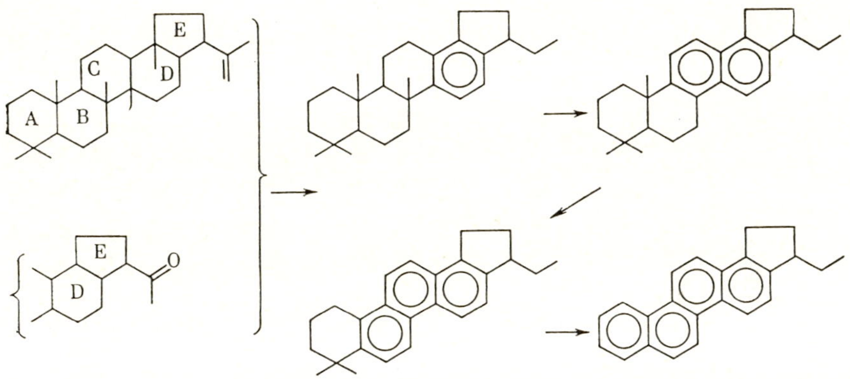

A gradual ring aromatization occurs in petroleum formation in the following sequence: D→C→B→A, with a concurrent loss of methyl substituents.

A vast group of 1,2-cyclopentanochrysenes was submitted to a specially thorough analysis. The hydrocarbons identified are presented in Scheme 4 (also given are molecular and main fragment ions).

The analysis of synthetic adiantane monoaromatization products demonstrated that they are represented by two stereoisomers (21αH and 21βH) inseparable by GC on regular non-polar phases (Scheme 5).

Mass spectra of the two stereoisomers, separated by thin-layer chromatography, proved to be identical.

The publication of Zubenko et al. (1979) is devoted to the study of monoaromatic hopanoids in Soviet crude oils. A chromatogram of these hydrocarbons isolated from the Anastasievsko-Troitskoe crude (horizon IV) is shown in Fig. 63. GC-MS showed that monoaromatic hopanoids are represented by two

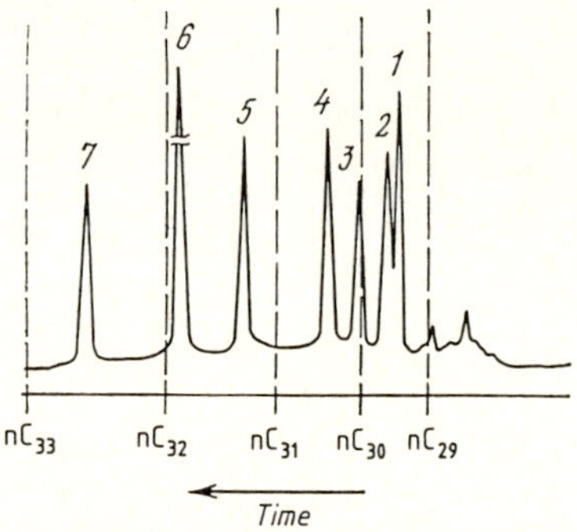

Fig. 63. Chromatogram of a mixture of monoaromatic hopanoids isolated from Anastasievskoe crude. Hydrocarbons *1–4*: monoaromatic 8,14-secohopanoids; hydrocarbons *5–7*: pentacyclic monoaromatic hopanoids. Normal alkane elution points indicated on the abscissa

Scheme 4

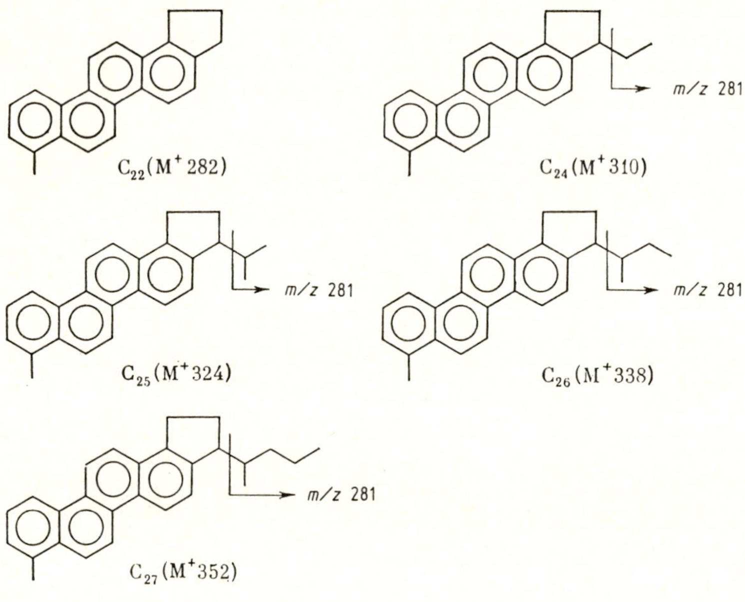

Scheme 5

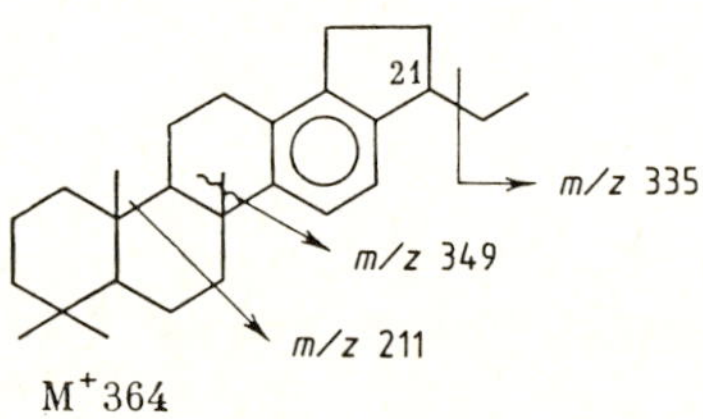

groups of compounds (Scheme 6). Those of the former belong to hopane derivatives with three naphthenic rings only. Hydrocarbons of the latter group have four naphthenic rings per molecule.

Scheme 6

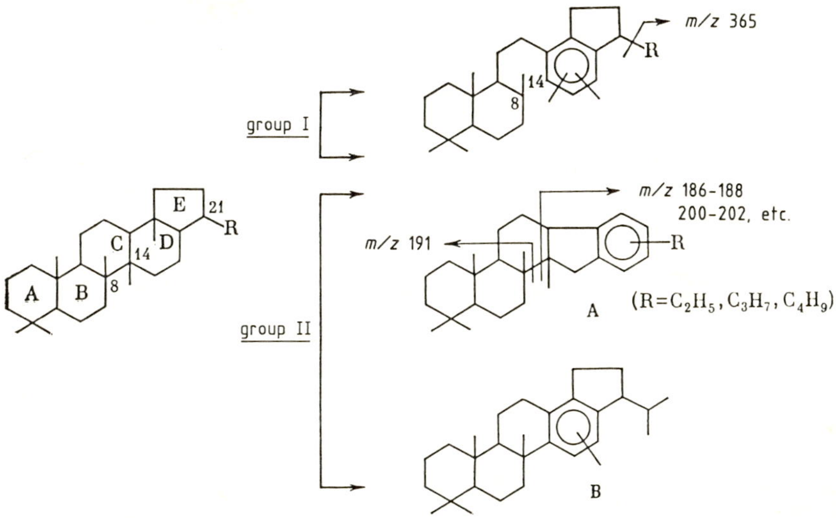

The structures and mass spectra of the former hydrocarbon group, which are monoaromatic 8,14-secohopanoids resemble those of monoaromatic hopanoids isolated by Bendoraitis in the crude of Loma Novia (Bendoraitis 1974). There are reasons to believe that cleavage occurs of the least stable 8–14 bond. Characteristically, dehydrogenation of petroleum hopanes, as well as their heating in the presence of aluminosilicates (clays), unalterably result in the formation of monoaromatic 8,14-secohopanoids (Kayukova et al. 1981). Hydrocarbons 1–4 (see Fig. 63) have similar mass spectra with the base peak always at *m/z* 365, due to the loss of an alkyl group and a fragment at *m/z* 159, suggesting the presence of an indane moiety. The hydrocarbons differed only in their molecular masses (C_{29} and C_{30}) and, apparently, in the relative position of the methyl substituents in the aromatic ring. In effect, they belong to tetracyclic compounds, their links with hopanes being mostly of genetic nature.

It is much more complicated to establish the structures of hydrocarbons belonging to the second group (5–7), which are real monoaromatic hopanoids. The first two hydrocarbons have a molecular mass of 392, which corresponds to monoaromatic C_{29} hopanes, with the third one having 30 carbon atoms in the molecule.

The substituents R in hydrocarbons 6 and 7 are C_3 and C_4 respectively, which, for this molecular mass, indicates a loss of one methyl group in the polycyclic system. Hydrocarbon 5 is a monoaromatic adiantane. However, the mass spectra of these compounds are very complex, which prevents the establishment of the aromatic ring position in a molecule. The possibility should not be excluded that we have to deal here with an E-ring, modified by a hydrindane rearrangement (which results in the E-ring becoming six-membered, while the D-ring is then five-membered). Besides fragment ions *m/z* 363 and 349 (loss of group R), the mass spectra of hydrocarbons 5–7 reveal sets of fragment ions characteristic of both substituted indanes (tetralins): *m/z* 200–202, 186–188, 159, 145, as well as saturated hopanes: *m/z* 191, 205. At this point, it is difficult to choose between structures A and B. Some available information refers to them as structure B, however, data published on monoaromatic adiantane mass spectra fail to support this conclusion (Greiner et al. 1977b). Hence, the need for further study of monoaromatic hopane structure is apparent. Hydrocarbons of the first and second groups analyzed are present in approximately equal amounts.

The publications of Spyckerelle et al. (1977a, b) reported the identification of other possible degradation products of triterpenoids in crudes: 3,3,7-trimethyl-1,2,3,4-tetrahydrochrysene (XXXIV) and 3,3,7,12a-tetramethyl-1,2,3,4,4a,11,12,12a-octahydrochrysene (XXXV), while identification of 2,2,9-trimethyl-1,2,3,4-tetrahydropicene (XXXVI) is mentioned in the work of Carruthers and Walkins (1963):

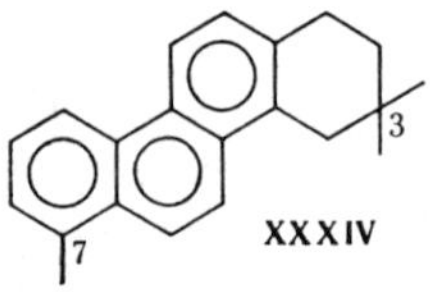

XXXIV

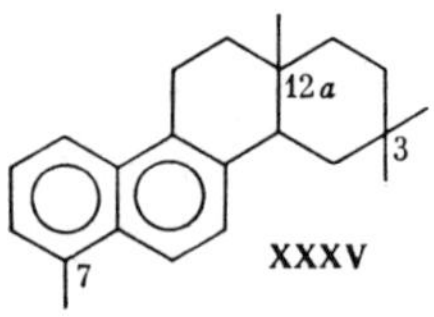

XXXV

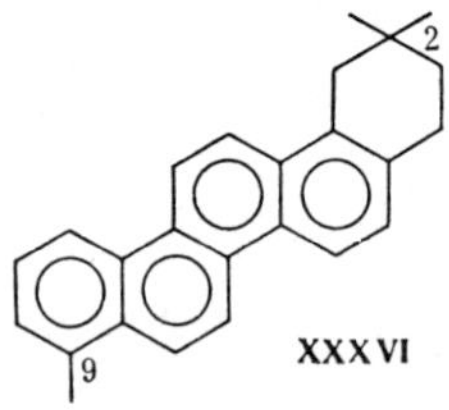

XXXVI

Recently, a new, unique homologous series of monoaromatic $C_{32}-C_{35}$ hydrocarbons of the hopane type, however, with five saturated cycles per molecule (Ostroukhov et al. 1983), was discovered in a number of Soviet crudes (Scheme 7).

Scheme 7

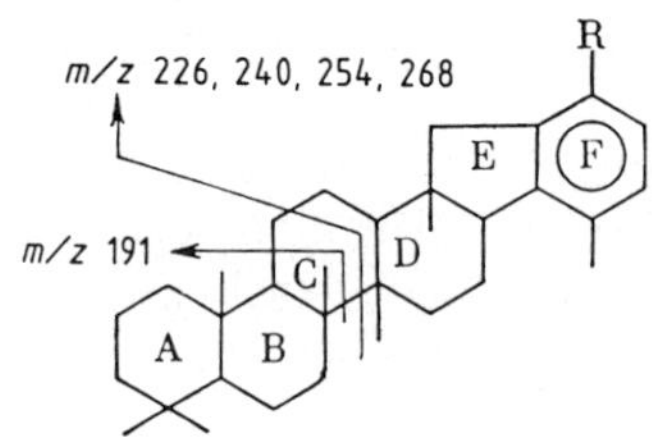

($R=H, CH_3, C_2H_5, C_3H_7$)

Triplets with *m/z* 156–158, 170–174 etc. are characteristic fragment ions, indicating the presence of indane groups. The structure of these hydrocarbons is related to that of their precursors, which is, in our opinion, represented by bacteriohopane, mentioned in the previous chapter. A possible scheme of the formation of these hexacyclic hydrocarbons is as follows (Scheme 8):

Scheme 8

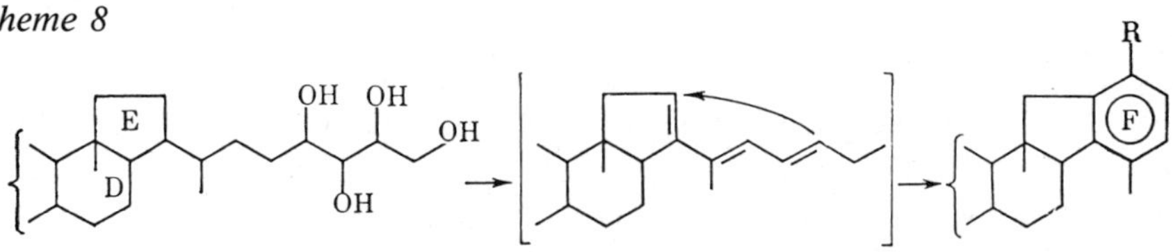

Dehydration of secondary alcohols, which occurs easily in clays, results in a series of trienes capable of intramolecular cyclization with the formation of an

E-ring. This is accompanied by the partial loss of aliphatic chain fragments. Characteristically, only $C_{32}-C_{35}$ hydrocarbons of this structure type were found in crudes, i.e. only those which may be formed starting with bacteriohopane.

Other Polyaromatic Hydrocarbons

Polyaromatic hydrocarbons unconnected in their origin with triterpenoids are often found in crude oils. These include the well-known perylene (XXXVII) and 1,12-benzperylene (XXXVIII) (Maksimov 1981; Carruthers and Cook 1954), as well as pyrene (XXXIX) and its methyl homologs and 3,4-benzpyrene (XL):

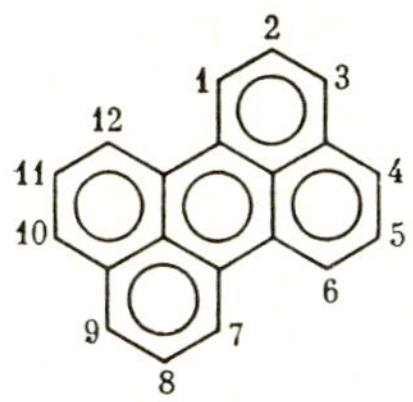

XXXVII

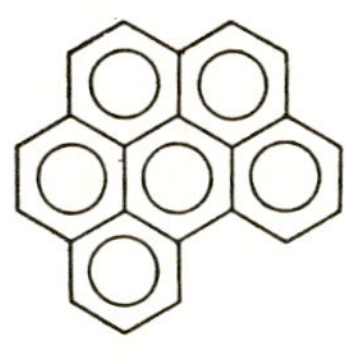

XXXVIII

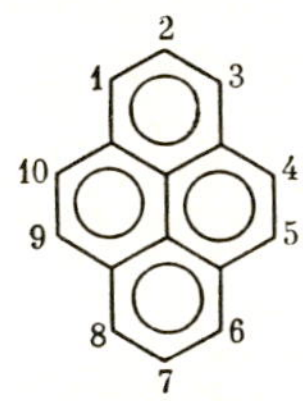

XXXIX

XL

Hydrocarbons XXXVII–XL found wide application in various geochemical investigations (Maksimov 1981; Aizenshtat 1973).

Chrysene and its methyl homologs, 1,2-benzfluorene, as well as triphenylene were identified in crudes and shales (Anders et al. 1973):

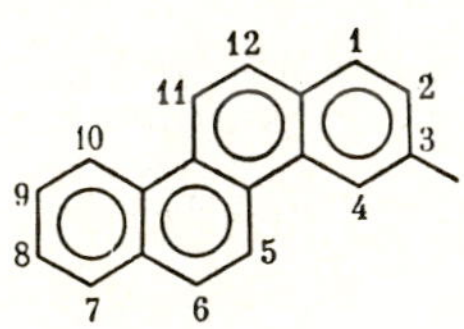

3-Methylchrysene

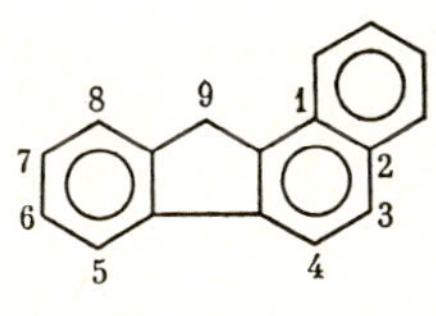

1,2-Benzfluorene

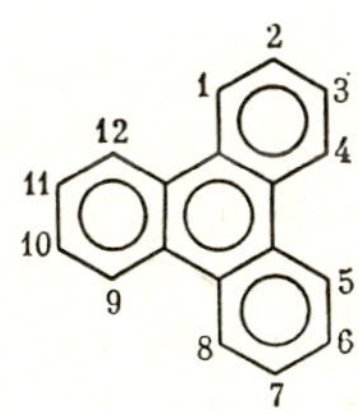

Triphenylene

Polyaromatic hydrocarbons in petroleum and dispersed organic matter are presently subjected to thorough analysis, especially in view of environmental problems. GC with high resolution capillary columns is the best method of analysis, which has been extensively illustrated (Aizenshtat 1973; Wakeham et al. 1980a; Giger and Schaffner 1978; Laflamme and Hites 1978; Lee et al. 1979).

Valuable data on the mean statistical distribution of aromatic hydrocarbons in crudes, depending on the number of aromatic rings per molecule, have been presented in the monograph of Tissot and Welte (1981). This information was obtained in the analysis of 121 crude oils belonging to different deposits:

1. Monoaromatic hydrocarbons: 43.0%
2. Diaromatic hydrocarbons: 30.5%
3. Triaromatic hydrocarbons: 16.5%
4. Tetra- and polyaromatic hydrocarbons: 10.0%

This research reconfirms the important role of monoaromatic hydrocarbons (both benzene homologs and monoaromatic steroids and hopanoids) in the overall balance of aromatic petroleum components. It is worthwhile to reiterate that these hydrocarbons are closest to saturated hydrocarbons in their structure, and usually preserve the latter's relict character.

To conclude the section devoted to individual petroleum hydrocarbons (see Chaps. 2–4), a short perusal of data related to their quantitative content will be presented.

Evaluation of the Quantitative Content of Various Hydrocarbons in Crudes

The concentrations of various hydrocarbons in crudes vary considerably and range within 1×10^{-3} to 5%. All the hydrocarbons analyzed above may be subdivided schematically into three groups:

1. hydrocarbons in concentrations exceeding 1% (in petroleum);
2. hydrocarbons in concentrations ranging between 0.1–0.99%, and
3. hydrocarbons in concentrations <0.1%. Understandably, most petroleum hydrocarbons belong to the last group.

Among the relatively rare representatives of the first group, mention should be made of C_6-C_{26} n-alkanes; concentrations of their homologs attain 2–3%. Concentrations of pristane and phytane are also high, attaining similar levels. Light petroleum fractions are characterized by high concentrations of methylcyclohexane (up to 2–5% in gas condensates), *cis*-1,3-dimethylcyclohexane and 2-methylheptane.

The second group of hydrocarbons may include: 2-methylpentane, cyclohexane, C_7-C_{18} 2- and 3-methylalkanes, methylcyclopentane, toluene, *trans*-1,2-dimethylcyclopentane, *trans*-1,2-dimethylcyclohexane, 1,1,3-trimethylcyclopentane, 1,1,3-trimethylcyclohexane, ethylcyclohexane, C_9-C_{18} isoprenoid alkanes, 1,2,4-trimethylcyclopentane – *trans, cis,* 1,2,3-trimethylcyclopentane – *trans, trans,* 12- and 13-methylalkanes, 7-methylheptadecane, 1,2,4,-trimethylcyclohexane-*trans, cis, m-* and *o*-xylen, ethylbenzene, n-propyclclohexane, 2-methyl-3-ethylheptane, 2-methylnaphthalene and some other hydrocarbons. More elaborate data on the relative distribution of C_5-C_{12} hydrocarbons are presented in the monograph of Kolesnikova (1972). However, the majority of hydrocarbons reviewed have concentrations <0.1%, sometimes <0.01% in petroleum. At the same time, the geochemical importance of these compounds cannot be overestimated.

CHAPTER 5

Sources and Reactions of Petroleum Hydrocarbon Formation

The present chapter is devoted to biogenic compounds serving as starting material for petroleum hydrocarbons. Presented are results of laboratory experiments in the simulation of natural petroleum formation processes, also discussed are possible mechanisms of reactions thus involved. Moreover, it reviews and analyzes contemporary perceptions related to transformations of organic molecules during diagenesis and catagenesis, together with reactions and important intermediates in these transformations.

Sources of Petroleum Hydrocarbons

Petroleum hydrocarbon sources are represented by biogenic substances of various compound classes, mostly their lipid (fatty) components.

These sources may be approached from different angles and reviewed in a purely biological sense, as algae, phyto- and zooplankton, bacteria, lipide portions of higher plants, etc. They may also be analyzed depending on the gross composition of the organic mass: sapropel matter, humic matter (its lipid component is important for crude oils), etc. Original materials may also be investigated according to the types of organic molecules they contain, such as acids, alcohols, ethers, etc., which may serve as sources of petroleum hydrocarbons. It is this latter aspect that will be the centre of our attention. Those wishing to learn more about the conditions of formation and the accumulation and composition of organic matter are referred to the important monograph of Tissot and Welte (1981).

It has been clearly established that the most important role in petroleum hydrocarbon formation, among the main natural organic substances, is played by algae, bacteria (especially lipid of cell membranes), phyto- and zooplankton, as well as higher plants (Tissot and Welte 1981). As previously stated, the lipid components perform the main function in petroleum formation. Though, in general, the lipid part of the plant kingdom is compositionally, fairly homogeneous, i.e. represented by sets of similarly structured molecules, certain variations exist, which permit the establishment of the participation of a given source material in the formation of a particular crude oil.

In effect, the entire lipid part in the plant kingdom is limited to two main classes: (1) compounds consisting of molecules based on an unbranched (or slightly branched) chain, and (2) compounds based on aliphatic and acyclic isoprenoid units. Other compounds are also possible, consisting of various parts belonging to different classes, such as waxes, which consist of complex ether molecules of higher fatty acids, aliphatic alcohols and sterols, which are polycyclic isoprenoid alcohols.

As demonstrated above, petroleum hydrocarbons may be regarded as derivatives of these two most important classes of organic compounds.

Let us now review certain typical representatives of the lipid part of the plant kingdom. $C_{12}-C_{26}$ (sometimes higher) fatty acids may be included among the most important unbranched molecules. Both saturated and unsaturated acids and hydroxy acids exist. Fatty acids (usually present as triglycerides) are chiefly characterized by even numbers of carbon atoms in molecules, since they are synthesized from C_2-acetate units in nature.

Natural waxes differ from fats in that instead of glycerin they comprise sterols and higher fatty alcohols, again with an even number of $C_{16}-C_{36}$ carbon atoms. Fatty acids in waxes have even numbers of carbon atoms and similar range of molecular mass.

Slightly branched acids, such as iso- and anteisoacids (or their esters have similar structures, with methyl substituents at the end of the aliphatic chain opposite to the carboxyl group. Suberin and cutin (compounds in various plant elements) are special initial materials. These compounds are polymerized and cross-connected fatty acids and alcohols. Among the components of suberin, $C_{12}-C_{26}$ dicarboxylic acids and hydroxy acids play an important role. Cutin is based on hydroxy acids. These compounds are particularly resistant to microbial and fermentation influence, which evidently protects aliphatic chains from biological oxidation at early stages of diagenesis. Schematically, elements of the cutin structure (I) are given below (Tissot and Welte 1981).

Molecules based on isoprenoid units have a much more variable composition. The following types may be distinguished: (1) monoterpenes (C_{10}), both aliphatic and monocyclic; (2) sesquiterpenes (C_{15}), aliphatic, mono- and bicyclic; (3) diterpenes (C_{20}), aliphatic (phytol), bi-, tri and, possibly, tetracyclic; cyclic diterpenes are usually present in higher plants; (4) triterpenes (C_{30}), aliphatic as well as tri-, tetra- and pentacyclic. Among tetra- and pentacyclic compounds we should mention those important for petroleum chemistry: $C_{27}-C_{30}$ sterols and various triterpenols, such as hopane derivatives with functional groups in particular (see page 167):

The aliphatic lycopene, as well as mono- and bicyclic carotenoids (α- and β-carotenes), are found among the most important tetraterpenes (C_{40}). An important role in petroleum formation is also played by higher isoprenoids (polyprenols), such as the aliphatic nonaprenoid solanesol (C_{45}), as well as compounds with long aliphatic chains: ubi- and plastoquinones.

Let us now analyze the most important types of molecules found in the main petroleum formation sources (Tissot and Welte 1981).

Bacteria. Various $C_{27}-C_{30}$ sterols (and possible 4-methylsterols) and hopanes were discovered in the bacterial lipid fraction. Fatty acids are usually represented by $C_{10}-C_{20}$ compounds. $C_{14}-C_{18}$ iso- and anteisoacids as well as certain higher aliphatic isoprenoids are the most typical bacteria.

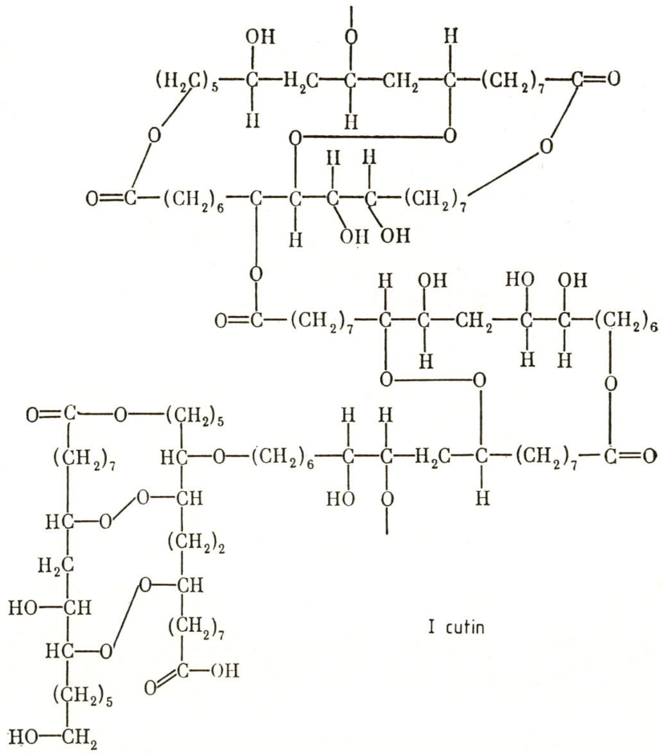

I cutin

An interesting confirmation of the extensive role played by bacteria in the formation of petroleum hydrocarbons may be found in the publication of Ourisson et al. 1984).

Phytoplankton and Seaweeds comprise a large amount of free fatty acids, both saturated and unsaturated, mostly $C_{12}-C_{20}$. Palmitic (C_{16}) and stearic (C_{18}) acids are found in especially high concentrations. Polyunsaturated acids, usually absent in higher plants, were also identified.

The lipid fraction of algae also contains up to 5% aliphatic hydrocarbons, mostly normal $C_{14}-C_{32}$ alkanes. N-heptadecane may sometimes dominate. Present are also iso- and anteisoalkanes, as well as isoprenoid alkanes, pristane and botryococcene. Sterols and carotenoids have variable distributions in the lipid fraction. Ergosterol dominates in some instances, while, on the other hand, C_{29} sterols are sometimes found in higher concentrations. Total sterol contents reach 0.3% of the dry weight. Lipids may amount to 10–20% in algae and 30–50% in bacteria (bacterial membranes).

Lipids of Higher Plants are also one of the main sources of petroleum hydrocarbons. These lipids possess many specific features which allow the differentiation between hydrocarbons produced from higher plant remains (organic matter of continental origin), and those generated from marine lipids and bacteria. Lipids are concentrated in certain, specific parts of plants: e.g. spores, cortex, pollen, fruits, etc.

Normal $C_{10}-C_{40}$ alkanes, with a marked predominance of odd-numbered hydrocarbons in the n-C_{23}−n-C_{35} range, were identified in higher plants (concentrations of n-C_{27}, n-C_{29}, and n-C_{31} hydrocarbons are especially high).

Aliphatic alcohols with an even number of carbon atoms ($C_{24}-C_{36}$) are found in great variety in waxes. $C_{12}-C_{26}$ saturated fatty acids with an even number of carbon atoms also have wide occurrence. Present are also monounsaturated acids ($C_{14}-C_{20}$) and hydroxy acids ($C_{12}-C_{26}$). Resins are characterized by tricyclic diterpenes: abietic acid and its derivatives are characterized by retene and fichtelite. Triterpenoids are chiefly represented by pentacyclic terpanes with a six-membered E-ring. The most characteristic sterols are C_{29} compounds: β-sitosterol (II), stigmasterol (III), campesterol (IV) and cholesterol (V):

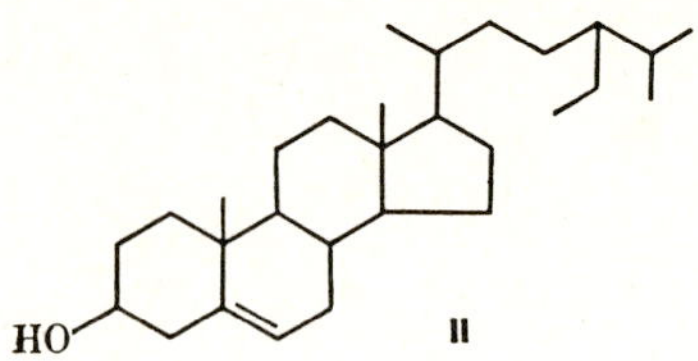
II

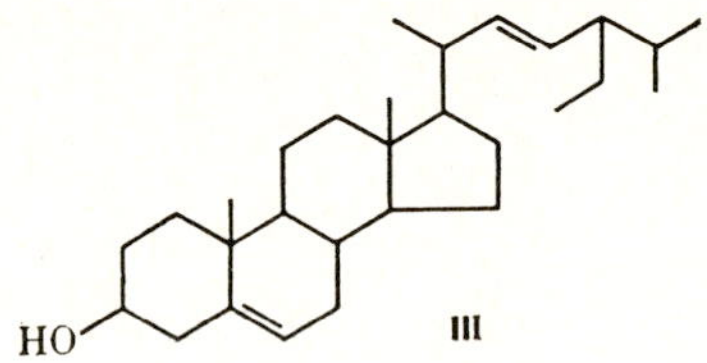
III

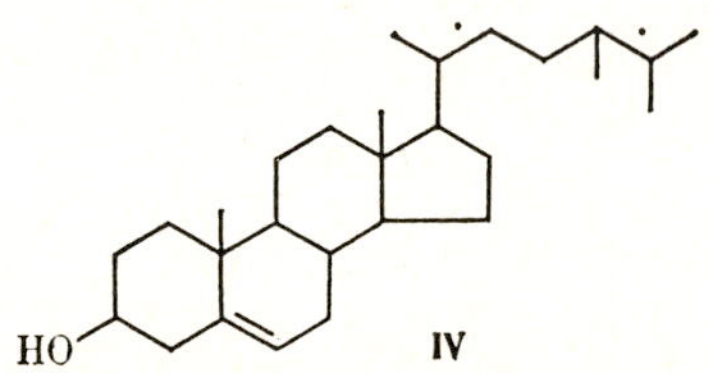
IV

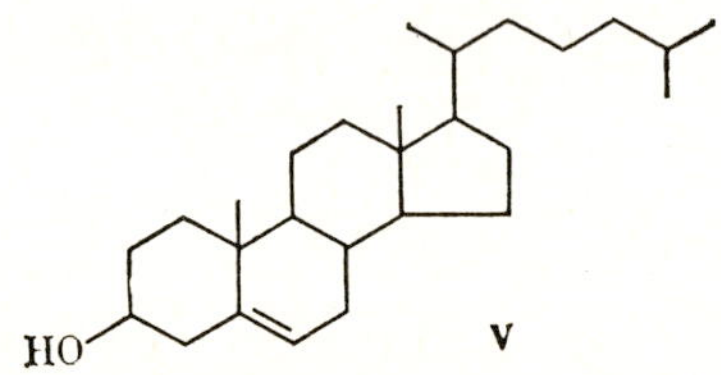
V

Transformations of Organic Molecules in Sediments (Dia- and Catagenesis)

Let us now review present-day concepts on the ways and mechanisms of transformations of organic lipids into petroleum hydrocarbons. These reactions are complex and occur in multiple steps. Only a small portion of the original molecules finds its way into petroleum in an unaltered or moderately altered form. The main alteration of organic matter in sedimentary rocks is the formation of an insoluble product (geopolymer), commonly called kerogen. Besides remnants of initial organic molecules, kerogen includes a non-organic component, usually represented by argillaceous minerals.[1] Detailed descriptions of kerogen's composition, properties, and structure are found in the monographs of Tissot and Welte (1981) and Yen and Chilingarian (1980).

In order to understand the mechanism of the transformation of organic matter, it should be borne in mind that at a certain stage, the organic molecules are chemically bound to their non-organic matrix. Along with the deeper burial of kerogen in the sedimentary cover of the earth's crust, i.e. with an increase in temperature (which is especially important) and pressure, various microbiological and

[1] Strictly speaking, only the insoluble organic matter of rocks is called kerogen. But the presence of kerogen in rocks in an intimate mixture with a non-organic constituent is of cardinal importance for petroleum formation reactions. Clayey materials play a special role in this case.

chemical transformations occur in kerogen. Usually two main stages in kerogen formation and transformation are identified: (1) diagenesis or sedimentogenesis (Tissot and Welte 1981; Vassoevitch et al. 1976) and (2) catagenesis.

At the diagenetic stage the formation of kerogen proper occurs, followed by its condensation by the loss of water and heteroelements. Of particular importance is also an intense microbiological activity, which results primarily in selective saturation of multiple bonds (Vleet and Quinn 1979). However, microbial alterations of organic matter in diagenesis are as yet poorly investigated. At the same time, only these alterations can explain the extensive saturation of such labile compounds as carotenoids, which nevertheless retain their specific structural features. Diagenetic processes proceed at relatively moderate temperatures, and are not accompanied by extensive destruction of organic matter.

Later, along with the accumulation of sedimentary material in a given basin, a gradual subsidence of the products of diagenetic condensation occurs, i.e. kerogen, to considerable depths and under extreme temperature and pressure conditions.

It is at this point that the most important stage in petroleum formation begins, i.e. catagenesis. In the process of catagenesis the maturation and rearrangement of organic molecules, their stereochemical modification and, most importantly, a separation between the mineral part and organic molecules occur. Moreover, a statistical rupture of C–C bonds in long aliphatic chains, resulting in the petroleum hydrocarbon homology occurs. Petroleum is constantly produced from kerogen. Reserves of kerogen in the earth's crust are immense, and according to the information of Philp et al. (1978), they are 500–1000 times greater than those of coal, with kerogen's overall amounts nearing 1.3×10^{18} tons. It is noteworthy that according to IR spectra, kerogen resembles petroleum asphaltenes and some grades of coals. Different authors give varying evaluations of catagenetic temperatures, with the most commonly held estimates ranging between 80°–150°C (Tissot and Welte 1981). Higher upper limit temperatures relate to gas condensates.

Numerous publications (Tissot and Welte 1981; Yen and Chilingarian 1980) were devoted to the study of kerogen. This geopolymer is insoluble in regular solvents, but can be separated from the non-organic constituent by the latter's dissolution in hydrofluoric acid.

As was already mentioned, kerogen transformations involve the destruction of C–C bonds, as well as of K–O–R and $K-O-\underset{\underset{O}{\|}}{C}-R$ bonds (where K is the kerogen matrix), i.e. the main types of organic compounds such as alcohols and acids, which later generate petroleum hydrocarbons (a cleavage at the bond between carbon and heteroatoms: N, S, O, also occurs). It has been suggested that ruptures with the non-organic part of the kerogen matrix are made easier through an alcohol group, which allows preferential formation of steranes, hopanes, and isoprenoids, i.e. the most important biological markers (Tissot and Welte 1981).

The diagram in Fig. 64 reflects the main stages in kerogen evolution (Tissot and Welte 1981). Various stages of the maturation of organic matter are also given in accordance to the degree of coal metamorphism, vitrinite reflectance (R_0) and

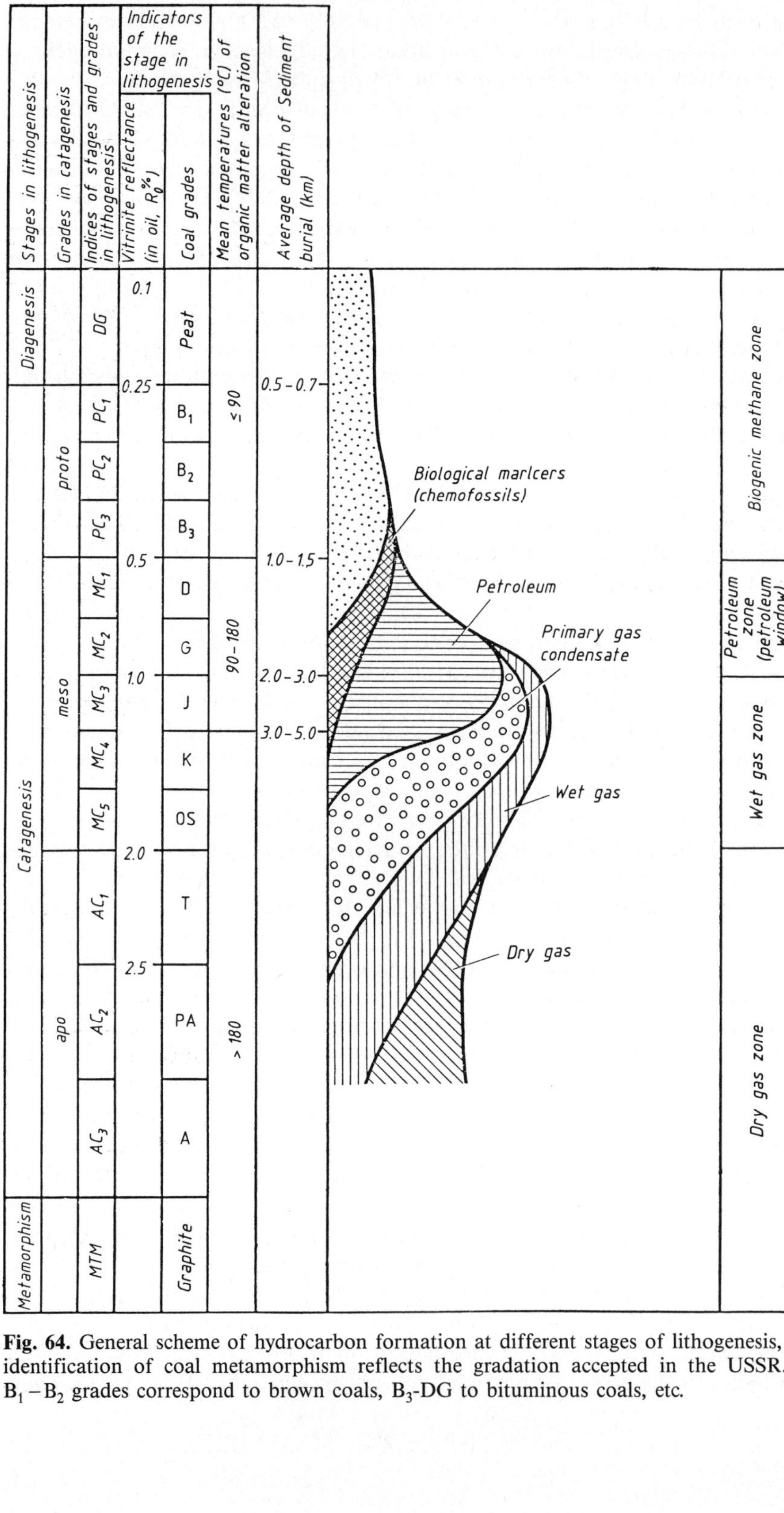

Fig. 64. General scheme of hydrocarbon formation at different stages of lithogenesis, i.e. the identification of coal metamorphism reflects the gradation accepted in the USSR. Thus, B_1-B_2 grades correspond to brown coals, B_3-DG to bituminous coals, etc.

the catagenetic scale proposed by Vassoevitch et al. (1969). A fairly narrow interval in the maturation of the organic matter at the D–G stage, or at $R_0 = 0.5-1.0\%$ is called "the petroleum window" or "the main phase in petroleum formation" (Vassoevitch et al. 1969; Perregaard and Schiener 1979). The latter is followed by zones of gas condensate of wet and later dry gas formation.

The most widely accepted current scheme of the conversion of organic matter into petroleum hydrocarbons can be seen in Fig. 64. Required temperatures, and especially burial depths, may naturally vary. Rates of transformation may also vary between marine and continental organic matter. A number of problems remain unclear. One of them is the possibility of a large number of crudes appearing at early stages of catagenesis or late stages of diagenesis (stages PC_3; $R_0 \leqslant 0.5\%$). Crudes of this type are called immature.

However, according to their genetic properties, the bulk of reserved petroleum reserves corresponds to fairly mature organic matter (stages MC_1-MC_2, or $R_0 = 0.5-1.0\%$). Consequently, catagenetically generated crude oils represent mature fossil fuels. Moreover, their formation (or expulsion into a separate liquid phase) proceeds within the restricted limits of organic matter maturation (the petroleum window is fairly narrow). Therefore, hopanes and isoprenoid alkanes are usually represented in crudes by stereochemically mature compounds, thus there is no clear-cut predominance of odd- or even-numbered normal alkanes.

Numerous attempts of artificial maturation of kerogen were undertaken. However, artificial aging (maturation) of kerogen and simulation of the main phase of petroleum formation under laboratory conditions differ considerably in their results from processes occurring in sediments. In fact, the decomposition of the kerogen matrix and the formation of hydrocarbons under laboratory conditions require much higher temperatures, hence, artificial pyrolysis of kerogen has to be conducted at temperatures of up to 400 °C and above, while petroleum formation in sediments proceeds at 130 °– 150 °C, but evidently lasts for millions of years. Hence, it becomes obvious that the composition of kerogen pyrolytic products varies from that of petroleum hydrocarbons, for example because pyrolytic products usually contain large amounts of olefins, such as alk-1-enes, which is never observed in crude oils.[2] Generally, the composition of high-temperature kerogen pyrolytic products resembles that of some shale oils. At the same time, hydrocarbons are generated only gradually at long-term, low-temperature heating of kerogen. In this case there is time for hydrogen rearrangement (see below), which is characteristic of transformations of organic compounds adsorbed on aluminosilicates (clays). As a result, saturated and aromatic hydrocarbons, i.e. hydrocarbon mixtures compositionally similar to those in petroleum, emerge.

The publication of MacKenzie et al. (1981) is important in this sense. The authors subjected immature Toarcian shales to a relatively low-temperature

[2] A paraffinic crude oil from the Bradford deposit (Pennsylvania) is the only one with a large amount of unsaturated hydrocarbons (Hoering 1977). Unsaturated hydrocarbons include alk-1-enes (40%), branched alkenes (30%) and monocyclenes with double bonds in the aliphatic chain (30%). Similarity in the distribution of these structural groups in saturated and unsaturated portions of the Bradford crude suggests that unsaturated hydrocarbons occurred as a result of thermal cracking, which is supported by extensive tectonic activity in the region, as well as the presence of magmatic intrusions.

heating (201 °–257 °–285 °C) for 20–80 days. Transformation products, consisting mostly of newly formed hydrocarbons, underwent all stereochemical and structural modifications registered earlier in the increasing burial of those same shales. Epimerization of pristane (formation of the *dl*-form), epimerization of chiral centres C-14, C-17 and C-20 in steranes and of C-22 in hopanes were observed. A decrease of the mono-/triaromatic hydrocarbon ratio was registered in aromatic steroids, as well as side-chain cracking leading to the formation of aromatic C_{20} steroids. The relative rate of epimerization decreases in the sequence: C-6 atom in pristane, C-22 in hopanes, C-20 in 20*R*, 5*α*, 14*α*, 17*α*-steranes, C-14 in steranes, i.e. we are dealing with relative rates similar to those under natural catagenesis, i.e. during burial of Toarcian shales. Characteristically, no unsaturated hydrocarbons are generated by extended low-temperature heating of shales (as opposed to pyrolysis).

Valuable results were also obtained by heating petroleum asphaltenes. Details of these transformations are treated in the next chapter.

The degree of catagenetic maturity of the starting material is of importance in the artificial transformation of kerogen, as well as the type of organic material it contains. Larger amounts of petroleum hydrocarbons are produced in pyrolysis of mature kerogen. More comprehensive information on the processes involved in laboratory simulation of natural hydrocarbon formation is presented in the recently published paper of Mouin et al. (1979). An exhaustive list of publications devoted to this subject may be found in this work as well as in Seifert (1978).

In our opinion, the pyrolytic reactions under review may find practical applications, e.g. the determination of the composition and distribution of relict structures, especially steranes and hipanes in kerogen. Extensive research in this direction was undertaken by Seifert (1978); Seifert and Moldowan (1979).

While comparing molecular-mass distribution of steranes and hopanes in crude oils and the pyrolytic products of kerogen, which were assumed to be the source of these crudes, we can reconfirm this assumption. Naturally, possible variations in the composition and stereochemical properties of the pyrolytic products should be taken into account.

The publication of Perregaard and Schiener (1979) contains an interesting analysis of the thermal maturation of organic compounds, which combines elements of artificial and natural aging (prolonged heating at moderate temperatures). Thus, Fig. 65 presents alterations in the composition of organic matter and the degree of its maturity depending on the distance to the screen, bordering the rocks heated by a magmatic intrusion, while Fig. 66 reproduces chromatograms of crudes occurring in different sections of the rocks. As can be observed, the "petroleum window" is fairly narrow, according to catagenetic limits, and corresponds to the limits of R_0 (0.5–1.2%), or the D-G stages in coal metamorphism.

The high catagenetic maturity of crude oils should be stressed (see Fig. 64). However, some aspects of the generally recognized scheme of petroleum formation, on the basis of conclusions drawn from this material, deserve additional consideration. Firstly, compositional modifications of hydrocarbons generated at different stages of kerogen transformation must still be clarified. As was already mentioned, we can only assume that mixtures, richer in biomarkers, are the first

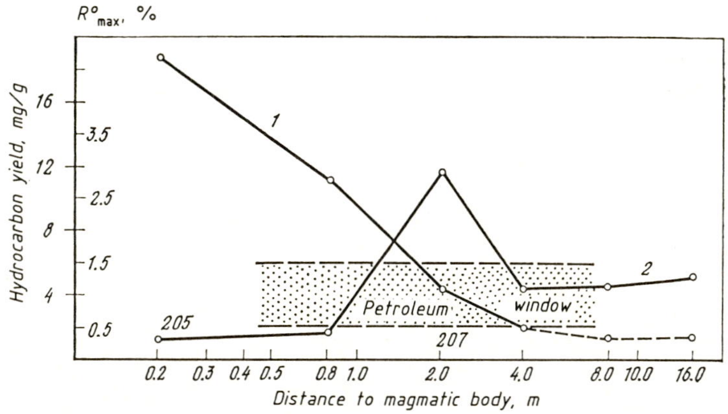

Fig. 65. Changes of kerogen properties with the distance from the magmatic heating source. Identified are changes in the vitrinite reflectance (*1*) and yield of hydrocarbons generated (*2*). Hydrocarbon compositions (samples Nos. 205 and 207) are given in Fig. 66

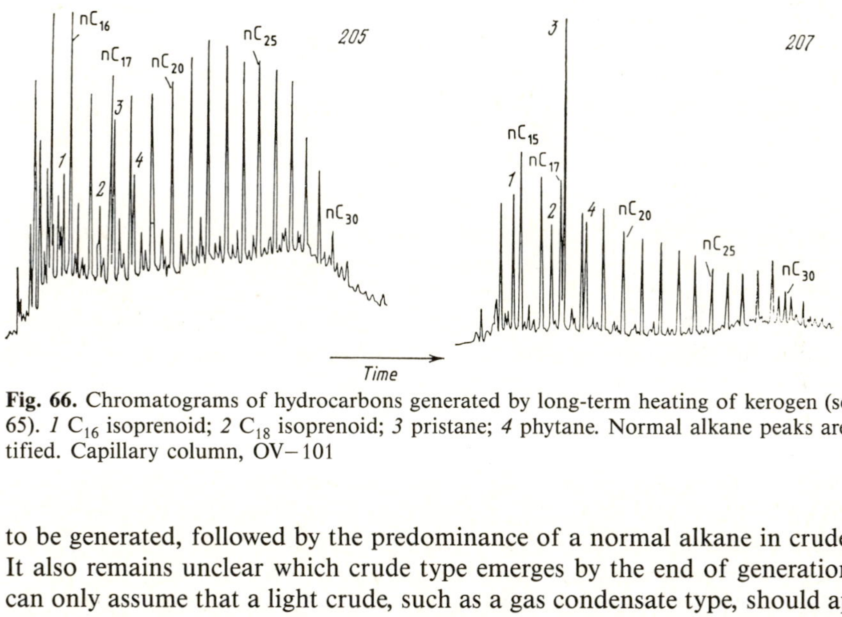

Fig. 66. Chromatograms of hydrocarbons generated by long-term heating of kerogen (see Fig. 65). *1* C_{16} isoprenoid; *2* C_{18} isoprenoid; *3* pristane; *4* phytane. Normal alkane peaks are identified. Capillary column, OV−101

to be generated, followed by the predominance of a normal alkane in crude oils. It also remains unclear which crude type emerges by the end of generation. We can only assume that a light crude, such as a gas condensate type, should appear at this point. It is surprising that the formation of large amounts of methane is usually postulated below the main phase of petroleum formation, i.e. in the deep catagenetic zone. Thermal decomposition of organic compounds cannot lead to the formation of methane alone. Its higher homologs, as well as unsaturated hydrocarbons, should also be present in such a situation.

Unfortunately, researchers are deprived of the possibility of studying crude oil compositions (at the molecular level), occurring at different stages of organic matter maturation. (The narrowness of the "petroleum window" has already been often mentioned.) Hence, the study of fossil fuels, i.e. extractable coal hydrocarbons, acquires special importance, since they represent all the stages in catagenesis: from immature peat and brown coals, to anthracites. Such an analysis provides a better understanding of the modifications in the composition and stereochemistry of the most important biological markers, which one typical of various stages in catagenesis.

Compositional Transformations of Coal Hydrocarbons at Various Stages of Catagenesis

An analysis of this type is important for elucidating the chemistry of organic compound catagenetic transformations. Studies of the composition of extractable coal hydrocarbons are of special interest, since, firstly, they allow objective evaluations of the degree of catagenesis on the basis of vitrinite reflectance (R_0), and secondly, for choosing groups of coals of similar genesis and compatible petrographic composition. Moreover, the range of catagenetic variations in coals is much wider than in crude oils, and contrary of crudes, coals and shales are not subjected to migration. Therefore, deposits of these caustobioliths are true deposits, as opposed to crude oil deposits. This implies an unequivocal determination of their geological age and the stratigraphic attribution of their source materials. We will concentrate mostly on coals, since shale hydrocarbons were comprehensively analyzed in the monograph of Yen and Chilingarian (1980). Incidentally, all biological marker structures present in crudes were to a large extent also found in bituminous (oil) shales and coals.

A number of publications (Elington and Murphy 1974; Kontorovitch and Danilova 1973; Rodionova et al. 1973) are devoted to the relative distribution of biological markers in coals at various stages of metamorphism. It was established that relative concentrations and molecular-mass distributions of these hydrocarbons depend entirely on the level of catagenesis, and generally, greater catagenetic transformations are accompanied by a gradual decrease in the concentration of odd-numbered $C_{25}-C_{31}$ normal alkane concentrations, while concentrations of $C_{15}-C_{24}$ alkanes increased.

A more exhaustive analysis of variations in the relative concentrations of numerous biological markers in humic coals at various catagenetic stages was undertaken in the works of Gulyaeva et al. (1976); Gulyaeva and Arefyev (1978) and Gulyaeva et al. (1978) These works not only helped reconfirm previously observed regularities, but also enhanced our knowledge of the composition of biological markers in the original organic matter of coals of various categenetic levels (at different stages of coalification).

Figure 67 presents the relative distribution of normal alkanes at different degrees of metamorphism, isolated from humic coals. (Data on normal alkanes in peat are given for comparison.)

It can easily be observed that the increase in catagenetic transformations is accompanied by a concentrational shift towards lighter, normal alkanes and a simultaneous decrease in odd carbon-number predominance ($C_{odd.}$), especially in the n-$C_{25}-C_{31}$ series. A concentrational increase of lighter, normal alkanes is exemplified by the coefficient $\Sigma C_{19}-C_{24}$ n-alkanes/$\Sigma C_{19}-C_{31}$ n-alkanes. The latter may vary within a wide range: from 0.06 for early brown coal stages to 0.57 for

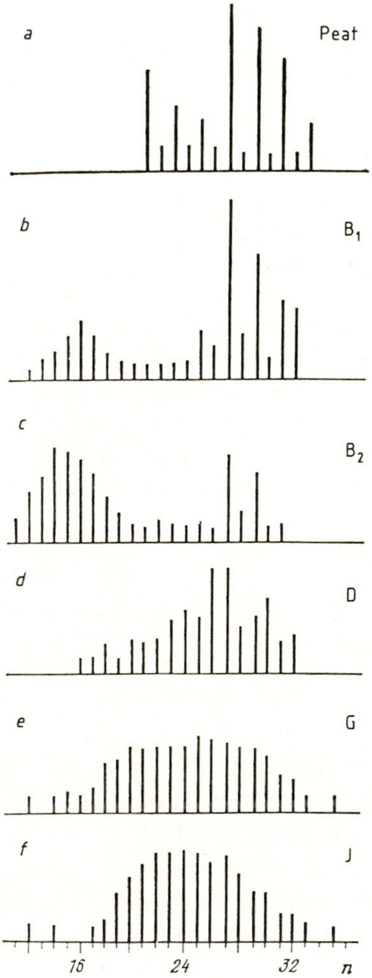

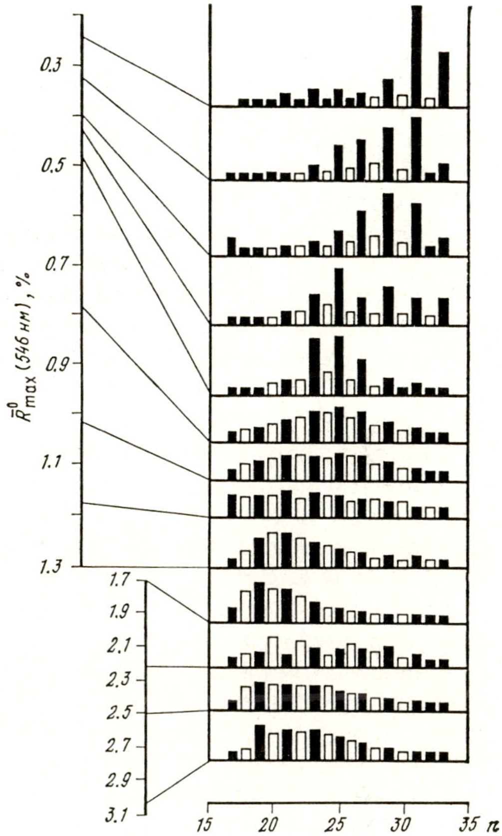

Fig. 68

▲

Fig. 67 a–f. Diagram of normal alkane distributions in peat and coals at various stages of coalification (metamorphism). **a** Peat; **b, c** brown coals; **d–f** bituminous coals. Number of carbon atoms indicated on the abscissa

Fig. 68. Relative distribution of normal alkanes in coals at different stages of coalification, and connections between alkane composition and vitrinite reflectance values (R_0). *1* Odd alkanes; *2* even alkanes

MC_4 coals. Brown coal stages are characterized by a "dip" in $C_{19}-C_{24}$ normal alkanes. (In crudes this coefficient ranges on the average between 0.25–0.35.)

$C_{odd.}$ also varies sharply. The highest levels are reached by normal alkanes isolated in peat, where $C_{odd.} = 12.0$; $C_{odd.} = 1.6-3.4$ for brown coal stages. As in crudes, beginning with stage D and especially G, $C_{odd.}$ approaches 1.

Consequently, as was repeatedly stated, the "petroleum stage in catagenesis" corresponds to stages MC_1-MC_2 of coal catagenesis. Variations in the above coefficients are not only caused by transformations of normal alkanes in the original organic matter. An important role is played by the formation of new, normal alkanes through the cracking of geopolymers, i.e. kerogen, which also results in increased relative concentrations of normal alkanes in coals of advanced metamorphism (stages D-G). Compatible results were registered in the publication of Tissot and Welte (1981) and Leythaeuser and Welte (1968), which presented data on normal alkane distributions in a number of samples at different stages of metamorphism (starting wth peat). The information is presented as a function of the parameter. Corresponding material is reproduced in Fig. 68 (Tissot and Welte 1981).

The distribution of isoprenoid alkanes in various coals is also quite revealing. Only at relatively advanced stages (D-G), isoprenoid distribution (prevalence of phytane and especially pristane) begins to resemble that of petroleum hydrocarbons. Appropriate material is given in Fig. 69.

It should be noted that the high values for the pristane/phytane ratio (2.3 – 5.6) are usually registered only in mature (bituminous) coals. At the brown coal stage this ratio may reach 1.0 – 1.6.

Increase in metamorphism brings about unusual changes in the ratio Σ isoprenoid alkanes/Σ normal alkanes. Respective data are presented in Fig. 70. It can be easily observed that maximal isoprenoid concentrations are reached at D-G stages, i.e. again at the petroleum window stage.

The peculiar curve in Fig. 70 may be explained by a very small content of isoprenoids at brown coal stages (they appear only at B_3 stages), while normal alkanes derived from the initial organic matter are found. Later, at higher levels of catagenesis, isoprenoid and normal alkanes are generated, especially up to the D-G stages when isoprenoid formation proceeds at relatively high rates. Apparently, normal alkane formation becomes more intense at deeper stages (K, J).

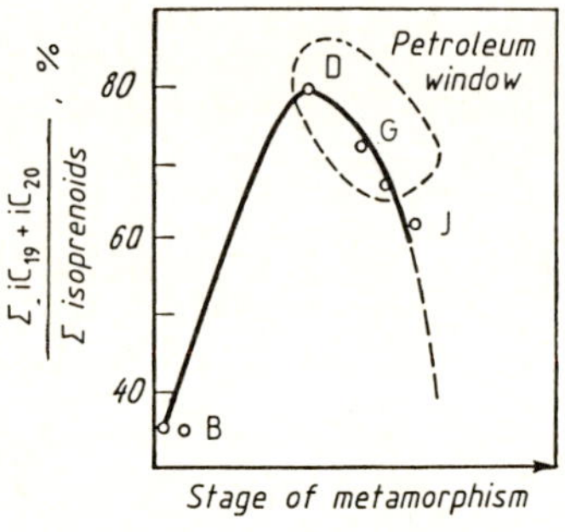

Fig. 69. Relative concentration of pristane and phytane as a function of coal metamorphism

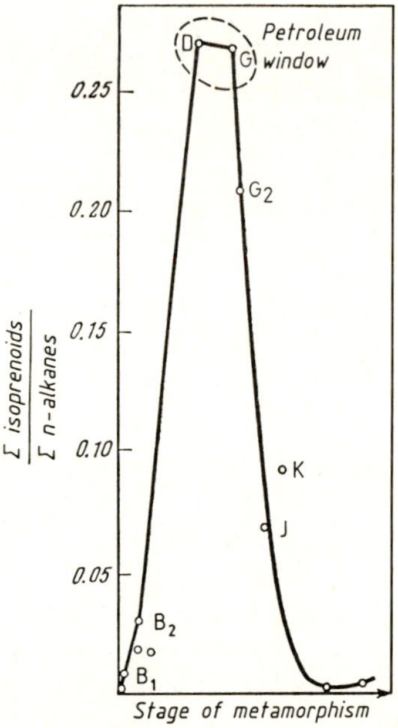

Fig. 70. Correlation between total concentrations of isoprenoids and normal alkanes as a function of coal metamorphism ►

We should also mention similar results reported by Connan (Connan 1973), who studied various aspects of the maturation of organic matter having a higher plant origin (coals and lignites).

Additional, important information related to stereochemical transformations of polycyclic molecules was obtained in the study of triterpenoid hydrocarbons isolated from coals at various stages of metamorphism.

Figure 71 presents normalized chromatograms (obtained by GS-MC) of $C_{27}-C_{32}$ pentacyclic hydrocarbon mixtures, isolated from coals at various stages of metamorphism. As proved by analysis, these compounds may be divided into three groups: (1) hydrocarbons with double bonds; (2) hopane hydrocarbons with a biological (unstable) 17β-configuration; and (3) hopane hydrocarbons with a 17α-configuration characteristic of petroleum hopanes. As is amply demonstrated in Fig. 71, hydrocarbons of all three types are usually present at brown coal stages, while bituminous coals (at stage D and especially G) are characterized by the ex-

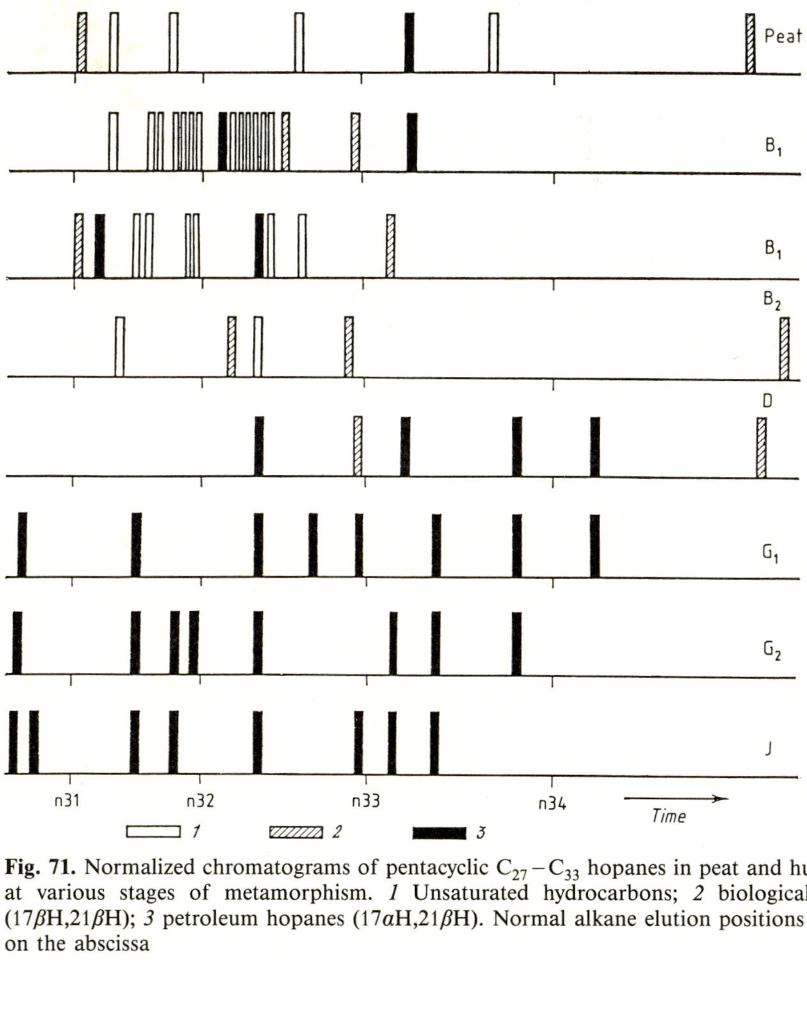

Fig. 71. Normalized chromatograms of pentacyclic $C_{27}-C_{33}$ hopanes in peat and humic coals at various stages of metamorphism. *1* Unsaturated hydrocarbons; *2* biological hopanes (17βH,21βH); *3* petroleum hopanes (17αH,21βH). Normal alkane elution positions identified on the abscissa

clusive presence of 17α-hopanes, i.e. those hydrocarbons which may also be found in crudes.

Since the source material of coals is of purely continental origin, their triterpenoid composition is more varied as compared to crudes. Thus, besides $C_{27}-C_{32}$ hopanes (and hopenes), which still are the main triterpenoids, $\Delta^{18(19)}$-oleanene, $\Delta^{13(18)}$-oleanene (VI), isomultiflorene (VII), as well as their saturated analogs, were also identified.

These hydrocarbons have the following structures:

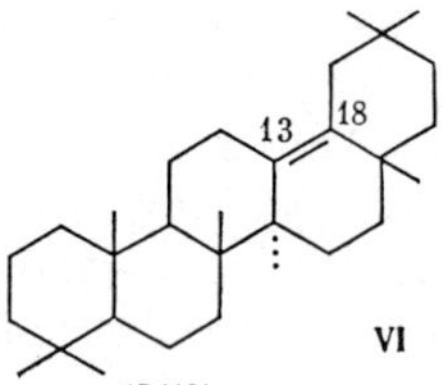

VI

$\Delta^{13(18)}$-Oleanene

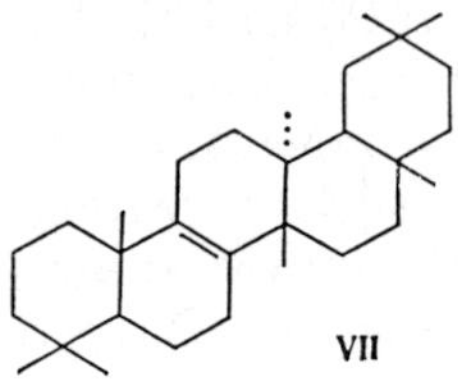

VII

Isomultifloren

Therefore, both in crude oils and in coals, processes of catagenetic maturation of organic molecules are similar. However, compositionally, petroleum hydrocarbons correspond to a fairly deep and narrow stage in the metamorphism of original organic molecules.

Let us now review concrete results of experiments on the laboratory simulation of petroleum formation processes. All these reactions were performed in the presence of clays which are the most probable catalyst in the synthesis of petroleum hydrocarbons in nature.

Laboratory Simulation of Petroleum Formation Reactions (Aluminosilicates as Catalysts)

A famous Soviet physicochemist, A. V. Frost, was the first to mention the special role of clays as catalysts in this process (Frost 1946). As early as 1946 he produced exact equations of reactions involving organic compounds (as possible precursors of petroleum hydrocarbons), which may be catalyzed by aluminosilicates. Since then, extensive research on the simulation of natural petroleum formation under laboratory conditions has been undertaken. Results obtained are comprehensively discussed in well-known monographs and articles (Frost 1945; Petrov 1960; Andreev et al. 1958; Petrov 1971; Solokov et al. 1972). Hence, we will highlight only the more important elements in this research, while focussing most attention on recent advances in this field and on the mechanism of petroleum formation reactions.

What are these reactions of petroleum hydrocarbon formation? Their most important elements are: loss of functional groups (acid, alcohol, ketone, etc.) in the original biological molecules; redistribution of hydrogen, resulting in the generation of saturated hydrocarbons of aliphatic and alicyclic series; cracking reactions and formation of aromatic compounds. It should be added that all these reactions must proceed at temperatures within a 100°–200 °C range. It is perfectly clear that all these processes may basically be catalytic in character and the only real natural substance which is capable of catalyzing these reactions is aluminosilicates (Frost 1945).

Aluminosilicates play an especially important and original role in the redistribution of hydrogen and simultaneous, low-temperature formation of saturated and aromatic hydrocarbons. Mechanisms of these reactions were extensively treated in our monograph (Petrov 1960).

Due to their isomerization potential, aluminosilicate catalysts are capable of generating a large number of structurally and stereochemically modified hydrocarbons having wide occurrence in crude oils. In certain cases these catalysts may preserve relict structures, since isomerization and cracking mainly proceed in compounds with double bonds and, hence, with much greater reactivity in transformations based on the carbonium-ion mechanism. At the same time, the saturated hydrocarbons that emerge have poor reactivity and preserve their structure.

Transformations of Fatty Acids

Fatty acids are the most important class of organic compounds serving as sources of petroleum hydrocarbons. An appropriate comprehensive analysis was undertaken in the USSR under the direction of A. I. Bogomolov. Generally, fatty acids and fatty alcohols are the main suppliers of alkanes and unbranched carbon chains of cyclanes widely represented in crudes. The formation of aliphatic hydrocarbons will be treated further on, while at this point, we will analyze the unusual direction in the transformation of unsaturated fatty acids, namely the mechanism of dehydrational cyclization. These reactions have been extensively discussed (Andreev et al. 1958; Petrov 1971; Sokolov et al. 1972).

Transformations of unsaturated fatty acids result in the production of a wide variety of alkanes, cyclanes and arenes. Thus, for example, a mixture (1:1) of saturated and aromatic hydrocarbons was formed from oleic acid at 250 °C; its MS analysis is represented in Table 50. As can be seen, cyclization of unsaturated acids results in the formation of not only monocyclanes and monoarenes, but also

Table 50. Composition of hydrocarbons derived from oleic acid (200° – 325 °C fraction) (%)

Types of hydrocarbons	Content (%)	Types of hydrocarbons	Content
Saturated hydrocarbons		Aromatic hydrocarbons	
Alkanes	41.6	Alkylbenzenes	32.1
Monocyclic naphthenes	29.5	Indanes (tetralins)	37.9
Bicyclic naphthenes	17.1	Dinaphthenobenzenes	15.0
Tricyclic naphthenes	7.6	Naphthalenes	3.5
Tetracyclic naphthenes	4.2	Acenaphthenes	0.8
		Phenanthrenes	4.5
		Naphthenophenanthrenes	2.1
		Pyrenes	4.1
Σ	100	Σ	100

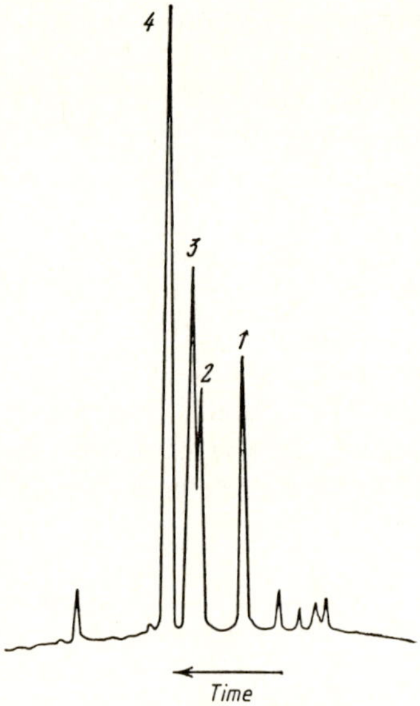

Fig. 72. Chromatogram of a mixture of C_{12} alkyl- and methylalkylbenzenes produced in thermocatalysis of oleic acid. *1* Methyl-3-undecylbenzene; *2* 1-methyl-4-undecylbenzene; *3* dodecylbenzene; *4* 1-methyl-2-undecylbenzene

bi- and tricyclic hydrocarbons. The analysis of reaction products, as well as of intermediate compounds, demonstrates that dehydrational cyclization is the basic process involved, and it proceeds according to the scheme: acid→lactone →ketone→hydrocarbons:

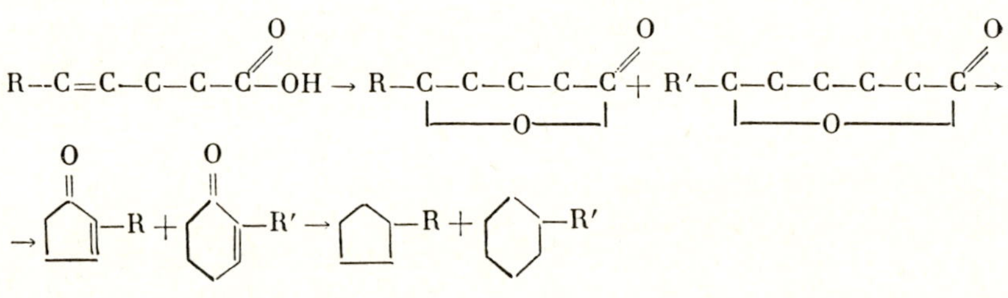

The formation of alkylcyclohexanes as well as methylalkylcyclohexanes from oleic and stearic acids has also been reported (Rubinstein and Strausz 1979).

Figure 72 features a chromatogram of a mixture of C_{18} aromatic hydrocarbons produced from oleic acid over aluminosilicates. According to the chromatogram, methylalkylbenzenes prevail, with an especially high concentration of a ortho-disubstituted isomer. This may be easily correlated to relatively high concentrations of ortho-methylalkylbenzenes in crudes, as mentioned in Chapter 4. Apparently, the formation of methlyalkylcyclohexanes and methylalkylbenzenes is related to an intermediate generation of lactones with a larger cycle, upon the isomerization of which (or isomerization of the hydrocarbons produced), the following compounds emerge:

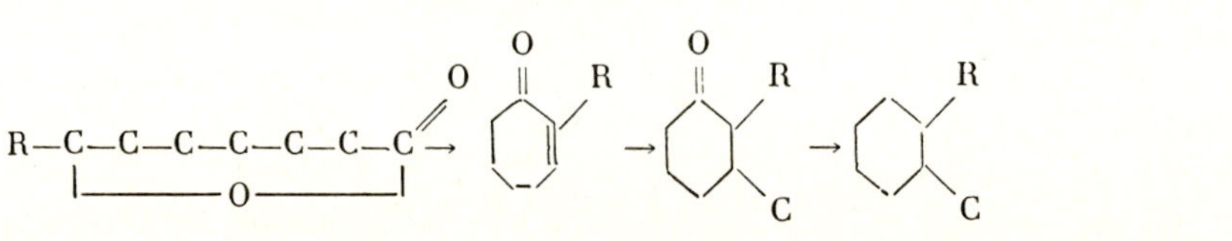

Evidently, the formation of indane series hydrocarbons proceeds through a phase of alkylbenzene formation, according to the scheme:

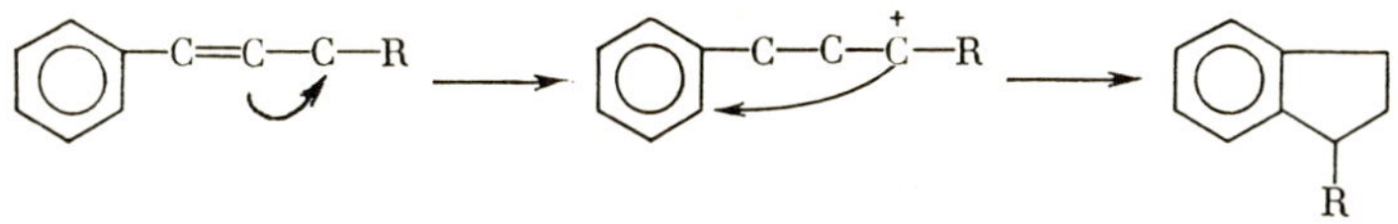

At the same time, the mechanism of alkylcyclohexane formation from saturated acids still remains unclear. As was mentioned, aromatic hydrocarbons are generated in reactions of hydrogen redistribution, in accordance with the following scheme:

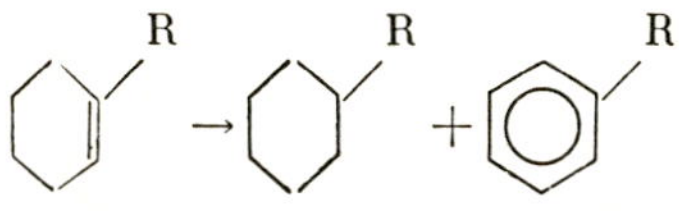

However, successful completion of these reactions necessitates a fairly large molecular mass of the initial cyclane. Cyclohexene transformation under these conditions may be represented in the following way:

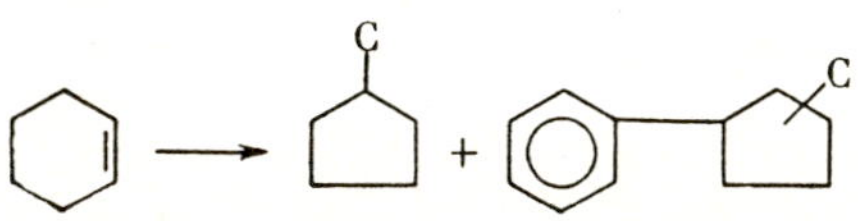

and reaction products contain practically no benzene, nor cyclohexane. Consequently, sometimes aromatization proceeds only after polymerization of a part of the initial cyclene. It is of interest that saturation through hydrogen transfer (ionic hydration) occurs only in methylcyclopentene, i.e. a hydrocarbon capable of generating a stable tertiary carbonium atom. Aromatization of higher-molecular-weight compounds, such as sterenes, proceeds in a "monomeric" form (see below).

Parallel reactions of cracking and polymerization create the entire range of aliphatic, cyclic and aromatic C_6-C_{35} hydrocarbons, which usually emerge during the contact between fatty acids and aluminosilicates.

The scheme of bi- and tricyclic hydrocarbon formation is of special interest. A separate study of alkylcyclopentenones and alkylcyclohexenones revealed that upon aluminosilicate catalysis, these compounds undergo condensation of a crotonic type, with the formation of hydrocarbons belonging to the dicyclohexyl (or dicyclopentyl) series:

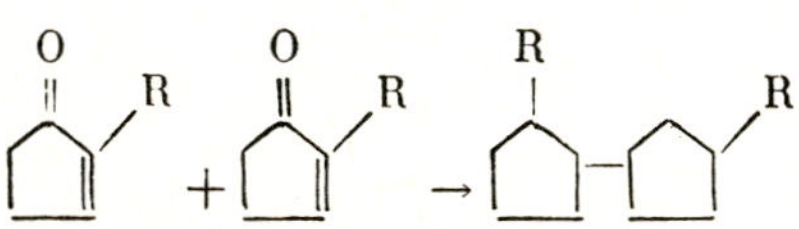

As is known, conjugated hydrocarbons may easily be isomerized into decalin homologs (Petrov 1971).

Besides bicyclic hydrocarbons, reaction products may contain up to 20% tricyclanes. These hydrocarbons may evidently be produced as the result of cyclization in accordance to the Diels-Alder reaction. An analysis of a reaction mixture of cyclopentenone and cyclohexenone demonstrated that besides bicyclic hydrocarbons, reaction products contain a mixture of tricyclo[5.2.2.$0^{2,6}$]undecane (VIII) and tricyclo[6.2.2.$0^{2,7}$]dodecane (IX), which emerge in accordance to the following scheme:

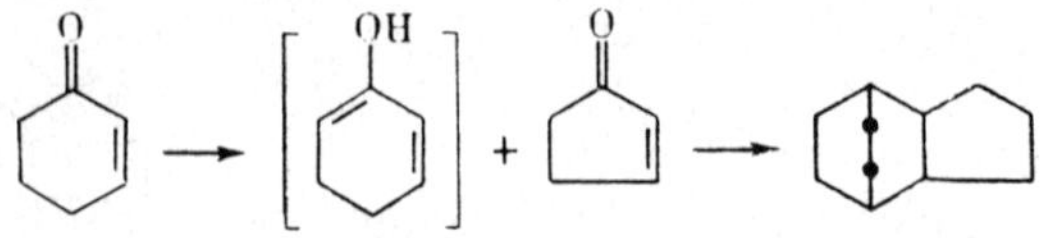

VIII

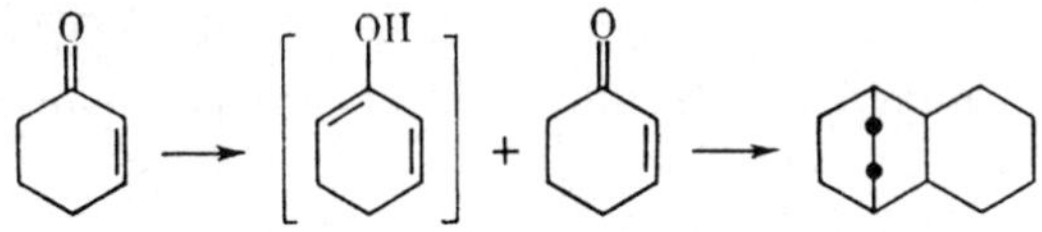

IX

Hydrocarbons VIII and IX may be regarded as reactive protoadamantane structures (see Chapt. 3). (It is noteworthy that no tricyclic hydrocarbons corresponding to the enol form of cyclopentenone could be identified.)

Formation of Normal Alkanes

The process of normal alkane formation in crude oils is fairly complex. It is generally assumed that the main reaction in this instance is that of decarboxylation of fatty acids. Beyond any doubt, this reaction occurs during the contact of saturated acids with aluminosilicates. This was proven by experiments with behenic and stearic acids (Tissot and Welte 1981; Sokolov et al. 1972; Petrov 1974; Vassoevitch et al. 1975). However, even in these simple experiments, besides regular decarboxylation, other reactions occur which result not only in the formation of a respective normal alkane, but a whole series of normal alkanes of both larger and smaller molecular mass. A chromatogram of a mixture of normal alkanes isolated in the products of stearic acid transformations over aluminosilicates can be seen in Fig. 73. Besides n-heptadecane (70%), alkanes of different molecular mass, without predominance of odd- (or even -numbered) hydrocarbons, were produced. Of special interest in this case is the formation of C_{19} and higher hydrocarbons by the ketonization of a part of the acid with the formation of stearone. Later, stearone is subject to cracking and reduction, whereby new hydrocarbons emerge:

$$C_{17}-COOH \rightarrow C_{17}-\underset{\underset{O}{\|}}{C}-C-C-C_{15} \rightarrow C_{17}-\underset{\underset{O}{\|}}{C}-C \rightarrow n\,C_{19}\,, \quad \text{etc.}$$

Therefore, besides direct decarboxylation, other reactions occur, which result in a fairly complex composition of the normal alkanes generated. Free-radical addition of α-olefins to acids with further (and easier) decarboxylation of α-alkyl-substituted acids may also occur (Petrov et al. 1960):

$$C=C-R+R'-CH_2-COOH \rightarrow R'-\underset{\underset{C-C-R}{|}}{C}\wr COOH \rightarrow R'-C-C-C-R$$

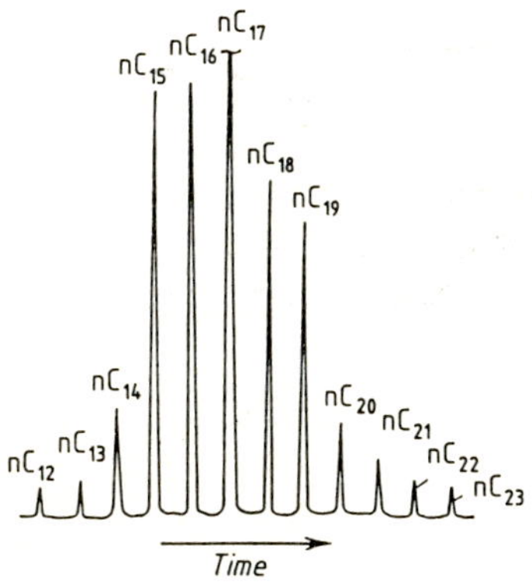

Fig. 73. Chromatogram of a mixture of normal alkanes produced by thermocatalysis of stearic acid

Since olefins in the reaction mixture are represented by a series of homologs; this process leads to the formation of statistical mixture of odd- and event-numbered normal alkanes.

There are also indications (Tissot and Welte 1981) that at fatty acids contact with calcium carbonate, a reaction of β-C–C-bond destruction also occurs leading to the formation of even-numbered, normal alkanes from even-numbered, fatty acids.

Thus, a whole set of transformations exists which leads to the leveling of even- and odd-numbered, normal alkane concentrations. Moreover, petroleum maturation is accompanied by various reactions of cracking of long aliphatic chains (see Chap. 6), which again results in the formation of equal amounts of even- and odd-numbered alkanes. Therefore, the absence of any marked predominance of either even- or odd-numbered normal alkanes in such a catagenetically mature product as petroleum is not surprising. Details on the formation of normal alkanes from fatty acids and mathematical processing of appropriate reactions are presented in the survey of Kvenvolden and Weiser (1967).

Formation of Branched Alkanes

The formation of branched alkanes (of non-isoprenoid type) is a more complicated chemical process than that of normal alkanes, although both reactions are closely related. Two main sources of branched alkanes may be identified. The first one (less important) is represented by iso- and anteisoacids which, in their transformations proceeding according to schemes similar to those of unbranched saturated acids, produce 2-methyl- and 3-methylalkanes. The second one, a more important source of branched alkanes, consists in the homologous destruction of fatty acid aliphatic chains, which leads to the emergence of alk-1-enes, and their further transformation.

Initially, this reaction consists in a simultaneous formation of both saturated and unsaturated hydrocarbons resulting in a statistical mixture of normal alkanes and alk-1-enes of various molecular masses:

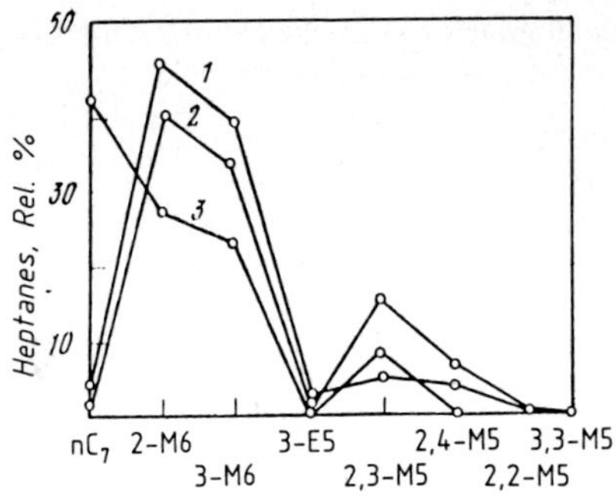

Fig. 74. Distribution of heptanes in a mixture of thermocatalytic products. *1* Oleic acid; *2* heptene-1; *3* stearic acid; 2-M6: 2-methylhexane; 2,4-M5: 2,4-dimethylpentane; 3-E5: 3-ethylpentane, etc.

Table 51. Relative distribution of octanes in a mixture of thermocatalytic products and in crudes (%)

Hydrocarbon	Thermocatalytic products		Crude	
	Oleic acid	Stearic acid	Starogroznenskoe, A^2	Groznenskoe, paraffinic, A^1
n-Octane	4	32	4.3	43.3
2-Methylheptane	30	21	31.3	23.8
3-Methylheptane	29	27	22.6	14.0
4-Methylheptane	6.5	7	13.7	5.8
Σ Monomethyl-substituted	65.5	55	67.6	43.6
2,5-Dimethylhexane	12.5	4	8.6	4.0
2,4-Dimethylhexane	13.5	3	8.7	4.0
2,3-Dimethylhexane	4.0	4	3.1	2.0
3,4-Dimethylhexane	0.5	2	–	1.1
Σ Dimethyl-substituted	30.5	13.0	20.4	11.1
Σ *gem*-substituted	Trace	Trace	7.9	2.0

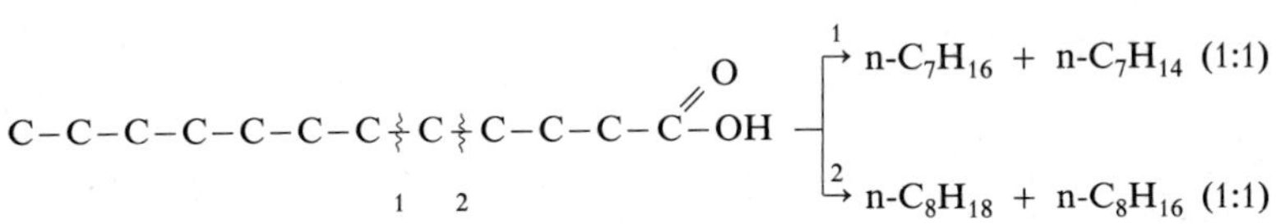

$\xrightarrow{1}$ $n\text{-}C_7H_{16} + n\text{-}C_7H_{14}$ (1:1)

$\xrightarrow{2}$ $n\text{-}C_8H_{18} + n\text{-}C_8H_{16}$ (1:1)

Further evolution of normal alkanes and alk-1-enes is different. While alkanes are directly included into petroleum, alkenes, which are compounds of considerable reactivity, undergo a series of transformations under the catalytic influence of aluminosilicates, finally resulting in a mixture of branched alkanes (Petrov 1960; Bogomolov and Shimanskii 1966) (see transformations of dodec-1-ene, below).

Results of experiments on the simulation of light petroleum, hydrocarbon formation processes on the basis of stearic and oleic acids are presented in Figs. 74 and 75 and Tables 51 and 52. Reactions were performed in closed vessels at 250 °C. Gasolines produced from acids also contained large amounts of cyclanes

Table 52. Relative distribution of nonanes in a mixture of thermocatalytic products and in crudes (%)

Hydrocarbon	Thermocatalytic products		Crudes	
	Oleic acid	Stearic acid	Starogroznenskoe, A^2	Groznenskoe, paraffinic, A^1
n-Nonane	3	28	5.2	34.4
2-Methyloctane	27	22	11.3	10.3
3-Methyloctane	19	20	15.3	9.7
4-Methyloctane	8	5	8.3	6.7
Σ Mono-substituted	54	47	34.9	26.7
2,2-Dimethylheptane	0.5	Trace	3.4	1.6
3,3-Dimethylheptane	1.5	Trace	1.6	2.4
2,3-Dimethylheptane	6.0	4	20.3	12.0
2,4-Dimethylheptane	7.0	3	6.1	3.0
2,5-Dimethylheptane	16.0	11	6.7	3.7
2,6-Dimethylheptane	8.0	7	14.6	10.0
3,5-Dimethylheptane	2	Trace	5.0	1.7
3,4-Dimethylheptane	2	Trace	Trace	2.5
Σ Di-substituted	43	25	57.7	36.9

(cyclane/alkane ratio = 1 : 1). The composition of the cyclanes produced according to the above scheme in the cyclization of unsaturated acid is reproduced in Fig. 75. The formation of those same C_6-C_8 cyclanes, which are usually present in crudes, should be underlined.

Compositionally, gasolines produced from fatty acids resembled those of category A crude. Moreover, stearic acid usually generated mixtures which were similar to gasolines of type A^1 crude, while oleic acid yielded light hydrocarbons similar to those of type A^2. Corresponding comparisons are shown in Tables 51 and 52.

Data in Fig. 74 demonstrate that compositionally, there is an obvious correlation between heptanes produced from hept-1-ene and from oleic acid. A rather consistent compositional correlation exists between gasolines derived from different acids, and those in A^1 and A^2 crudes (see Tables 51 and 52). However, it should be borne in mind that for obvious reasons biomarkers, *gem*-substituted and isoprenoid hydrocarbons in particular, are absent from mixtures of acid transformation products.

Thermodynamically and Kinetically Controlled Mechanism of Hydrocarbon Formation

Figure 76 contains a compositional comparison of dodecane mixtures isolated from crudes, and produced from dodec-1-ene. As can be observed, the composi-

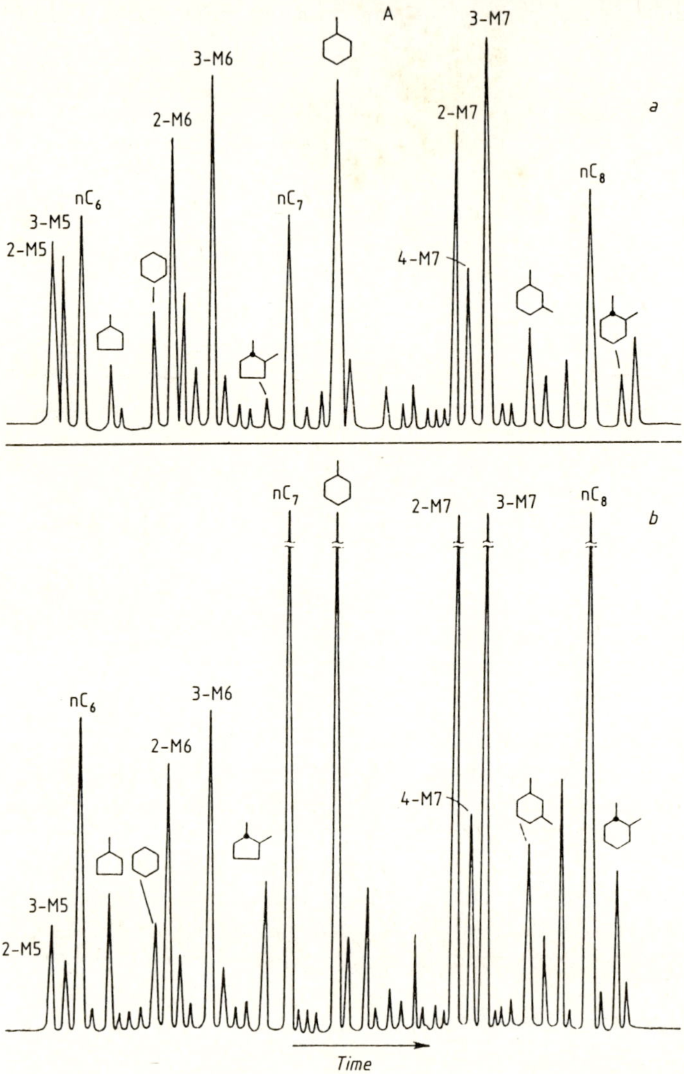

Fig. 75 A, B. Chromatograms of C_6-C_8 saturated hydrocarbon mixtures, produced by thermocatalysis of fatty acids and isolated in crudes. **A:** *a* Produced from stearic acid; *b* isolated from Surgut crude (A^1). **B:** *a* Produced from oleic acid; *b* isolated from Starogroznenskoe crude (A^2). Capillary column 100 m, squalane; linear temperature programming, 50°→1°/min

tion of dodecanes isolated from petroleum is fairly similar to that of a hydrocarbon mixture produced from dodec-1-ene by thermocatalysis over aluminosilicates. Additionally, Fig. 76 shows a chromatogram of an equilibrium mixture of dodecanes (separately for mono- and di-substituted structures). It may appear at a first glance that petroleum hydrocarbons of the same degree of substitution are in a state of equilibrium. However, it is only an apparent equilibrium, since the isomeric composition of dodecane was formed at the stage of dodec-1-ene

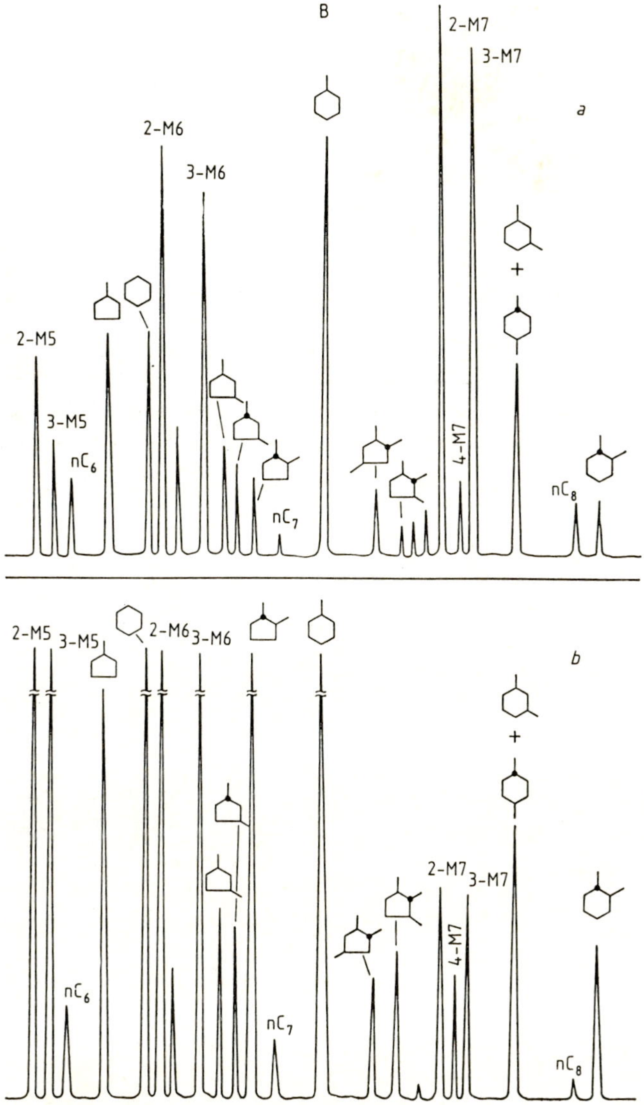

Fig. 75A

transformation in accordance to a thermodynamically controlled mechanism. Other biological markers found in crudes, such as isoprenoid alkanes, 2,3-dimethylalkanes, are present in concentrations exceeding equilibrium, although they were capable of being isomerized into other structures. On the other hand, high concentrations of biological markers appear in crudes due to reactions proceeding on the basis of a kinetically controlled mechanism, i.e. with the preservation of the original molecules' structural properties. Hence, as was already often mentioned, there is no complete equilibrium in crude oils, while certain groups of isomers exist emerging due to a thermodynamically controlled mechanism, that

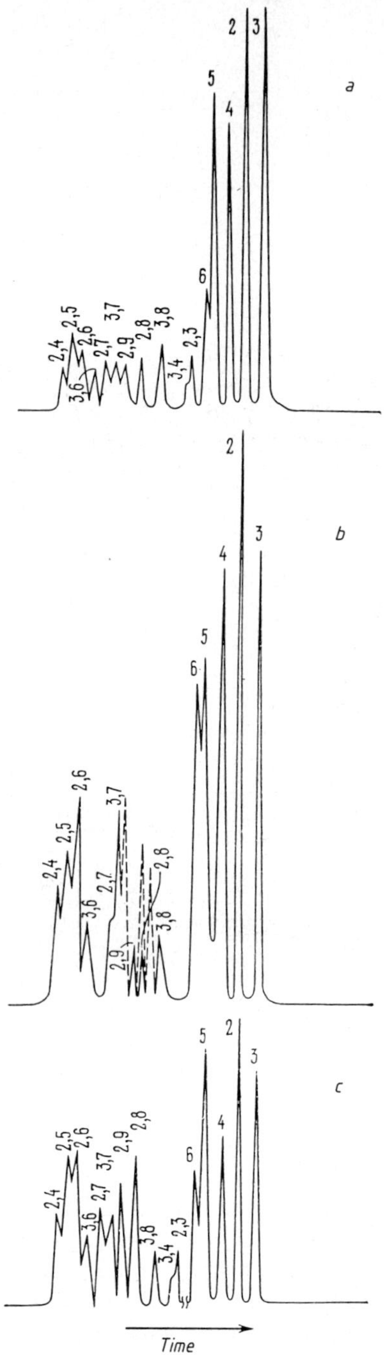

Fig. 76 a – c. Chromatograms of branched dodecane mixtures. **a** Generated by thermocatalysis of dodec-1-ene; **b** isolated from the Surgut crude; **c** equilibrium mixture of methylundecanes and dimethyldecanes. The positions of the methyl substituents are *numbered.* Cyclane peaks are denoted by *dotted lines.* Capillary column 50 m, squalane; 90 °C

coexist with relict structures of different origin. Equilibrium among monomethylalkanes may be obtained through rapid rearrangement of cations (dodecyl ions in this case) following a correlated scheme or through fast migration of a multiple bond in the carbon chain of the original olefin. The mechanism of simultaneous formation of a mixture (close to equilibrium) of methyl-substituted alkanes (using dodec-1-ene as an example) may be represented in the following way (Scheme 1):

Scheme 1

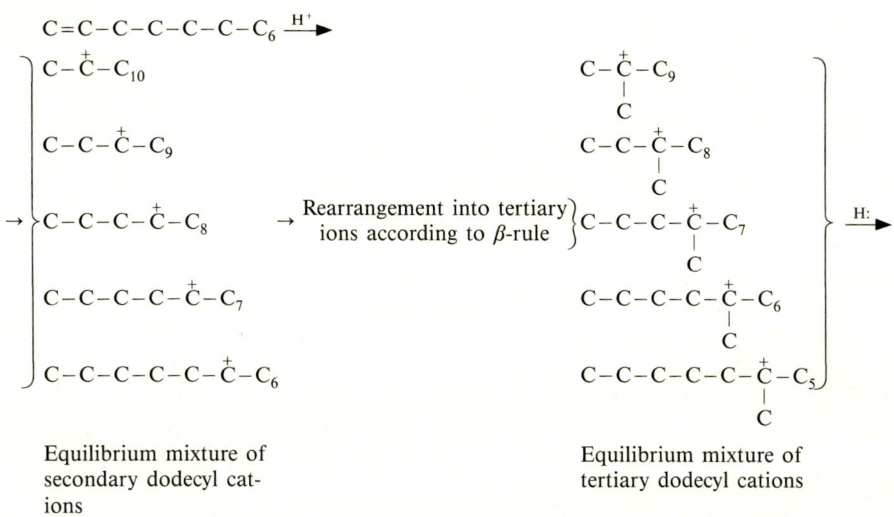

→ stabilization of cations → 2-methyl-, 3-methyl-, 4-methyl-, 5-methyl-, 6-methylundecanes

It is of interest that no concurrent shift of methyl radicals could be observed in saturation (due to hydrogen redistribution) of branched alkanes:

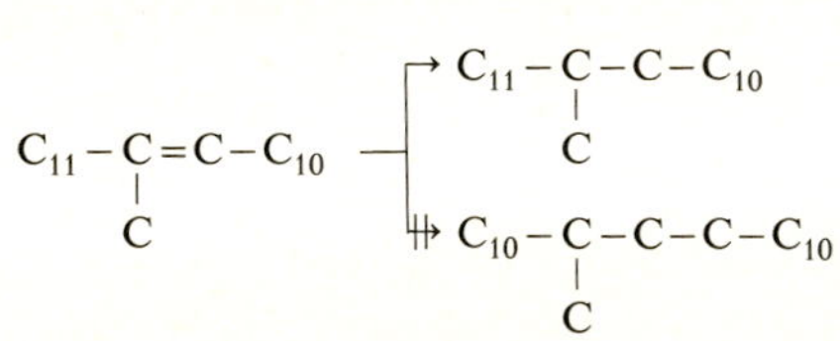

In view of the stability of tertiary carbonations, unsaturated hydrocarbons of this structure are directly transformed into respective alkanes. Apparently, this fact helped preserve 12- and 13-methylalkanes in certain crudes (see Chap. 2).

The above examples reconfirm that the judgement of the equilibrium state of isomers is not enough to determine their equilibrium concentrations. Sources and ways of petroleum hydrocarbon formation should also be taken into account.

A number of conditions should be met in order to evaluate the temperature of petroleum formation, which may, in principle, be determined by analyzing the composition of isomers, generated by thermodynamically controlled reactions (equilibrium mixtures). Firstly, there should be enough data on the ratios of these isomers (under equilibrium conditions) within a broad temperature interval.

Secondly, it is necessary that the K_{eq} between the hydrocarbons analyzed alters considerably with temperature. Thirdly, it is necessary that a quantitative analysis of these isomer concentrations be made, preferably GC methods. And last but not least, the most difficult condition, isomeric equilibrium mixtures formed by thermodynamically controlled reactions should not be enriched by any isomers from reactions of thermal cracking of high-molecular-weight hydrocarbons. Thus, low-molecular-weight hydrocarbons should be considered unsuitable objects, since their composition undergoes constant changes due to higher petroleum hydrocarbon destruction products (see Chap. 6). High-boiling hydrocarbons are more reliable in this sense, however, their research presents serious problems of purely analytical nature.

Compositions of mono-substituted alkane mixtures, formed at 200 °C over aluminosilicates from dodec-1-ene and hexadec-1-ene, are reproduced in Table 53.

Thus, α-olefins, which emerge during the cracking of long aliphatic chains of fatty acid or alkylcyclane molecules, may be regarded as sources of branched petroleum alkanes.

Moreover, most branched alkanes are represented by relict structures originating from various isoprenoids. Genetically, a part of the branched alkanes is evidently connected with branched fatty acids.

So far, the origin of a large number of *gem*-substituted alkanes in category B crudes (usually accompanied by *gem*-substituted cyclanes) remains unclear. It can only be assumed that these structures represent destruction products of steranes and hopanes which contain numerous *gem*-substituted groupings (characteristically, high concentrations of *gem*-substituted alkanes and cyclanes in gasolines are, observed exclusively for C_6-C_8 hydrocarbons). It may also be that the high concentration of these compounds is the result of their accumulation in the biological degradation of less branched hydrocarbons (see Chap. 6).

It is natural for alkenes to be transformed into aliphatic hydrocarbons in the presence of aluminosilicates, however diene transformations under similar conditions was rather unexpected. Over aluminosilicates, nonane-1,9-diol (250 °C) pro-

Table 53. Relative distribution of C_{12} and C_{16} methyl-substituted alkanes in mixtures of thermocatalytic products and in crudes (%)

Hydrocarbon	Thermocatalytic products		Petroleum hydrocarbons	
	Dodec-1-ene	Hexadec-1-ene	C_{12}	C_{16}
2-Methylalkane	27	23	30.0	33
3-Methylalkane	28	25	22.2	26
4-Methylalkane	19	15	19.1	10
5-Methylalkane	19	16	15.2	16
6-Methylalkane	7	21	13.5	15
7-Methylalkane	–	21	–	15
8-Methylalkane	–	21	–	15

Note: Ratios of the total C_{12} mono- and dimethyl-substituted alkanes proved to be similar in both mixtures, i.e. 68 : 32 in thermocatalytic products and 77 : 23 in crudes.

duced a saturated hydrocarbon mixture consisting of 60% branched nonanes and 40% C_9 cyclanes. Monomethyl-substituted hydrocarbons (80%) dominated among nonanes, with only 20% having two methyl substituents (mostly 2,5-dimethylheptane). C_9 cyclanes were chiefly represented by n-propylcyclohexane ($\approx$ 50%) and a methylethylcyclohexane mixture. Since a,ω-diols were identified in the kerogen matrix, the above analysis points at still another possible source of cyclic petroleum hydrocarbon.

Formation of Isoprenoid Alkanes

As was often mentioned, phytol – an unsaturated alcohol – is considered to be the main source of isoprenoid petroleum hydrocarbons. A possible scheme of catalytic phytol transformations has been suggested (Pustilnikova et al. 1973; Maxwell et al. 1971; Brooks and Maxwell 1973; De Leeuw et al. 1973; Eglinton 1971).

As reported in the work of Pustilnikova et al. (1973) a mixture of phytol and an aluminosilicate catalyst (1:1) was heated to 200 °C for 50 h. An appreciable amount of phytane emerged within the first few hours of the experiment (by this time, phytol had disappeared in the mixture of reaction products). After 10 h of heating a C_{14} isoprenoid (2,6,10-trimethylundecane) was identified together with phytane. They were followed by C_{15}, C_{16}, C_{18} isoprenoid alkanes and pristane. The final mixture of hydrocarbons with a wide range of boiling points contained a large quantity of cyclic compounds as well. According to MS analysis, this mixture consisted of 63.6% alkanes, 26.4% mono-, 4.7% bi- and 5.3% tri- and tetracyclic naphthenes. Evidently, naphthenes were formed by the Diels-Alder cyclodimerization reaction, involving the initially generated phytadiene. The concentration of individual isoprenoid alkanes reached only 6%. Higher isoprenoid alkane yields (11%) were obtained when phytol was diluted (1:10) by oleic acid. Dilution of phytol brings simulation of isoprenoid alkane formation closer to natural processes and decreases the process of cyclodimerization. It is noteworthy that a different pristane/phytane ratio was obtained in the mixture of the reaction products in this case. Table 54 presents data on the distribution of isoprenoid hydrocarbons in the above experiments.

It should be underlined that, as a rule, pristane appears with a certain delay, i.e. at later stages. Apparently, a similar process occurs in natural petroleum formation, since the pristane predominance in genetically related crudes is usually observed in catagenetically more mature samples (or, to be more precise, in crudes emerging at the stage of advanced kerogen catagenesis). Since the pristane/phytane ratio is an important genetic indicator, this issue should also be touched upon. According to contemporary views (Tissot and Welte 1981), the most important role in phytol transformations is played by the reductive or oxidative conditions during the initial stages of diagenesis (Welte and Waples 1973). In the former case phytol is reduced dihydrophytol, which then turns into phytane; in the latter case, phytol is oxidized to phytenic acid with further formation of pristene and pristane. In general terms, these transformations are represented in Scheme 2.

Table 54. Relative distribution of isoprenoid hydrocarbons in a mixture of thermocatalytic products (%)

Hydrocarbon	Number of carbon atoms	Thermocatalytic products	
		Phytol	Mixture of phytol and oleic acid (1 : 10)
2,6,10-Trimethylundecane	14	17	1.2
2,6,10-Trimethyldodecane (Farnesane)	15	8	5.8
2,6,10-Trimethyltridecane	16	4	4.0
2,6,10-Trimethyltetradecane	17	–	–
2,6,10-Trimethylpentadecane	18	13	2.8
2,6,10,14-Tetramethylpentadecane (Pristane)	19	11	41.4
2,6,10,14-Tetramethylhexadecane (Phytane)	20	47	44.8

Scheme 2

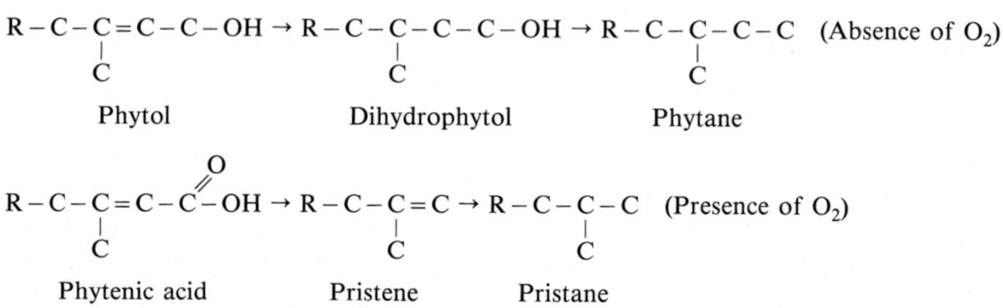

The mechanism of low-temperature phytol decomposition in the presence of montmorillonite is comprehensively analyzed in the paper of De Leeuw et al. (1973). A whole range of compounds, including phytenic and pristenic aldehydes, C_{13} and C_{18} ketones, isporenoid acids, phytadiene, as well as its dimers and trimers were identified in a mixture of reaction products. The principles of catalytic unsaturated hydrocarbon cracking, in accordance to a carbonium-ion mechanism and the β-rule, play a certain role in the formation of low-molecular-weight isoprenoid alkanes:

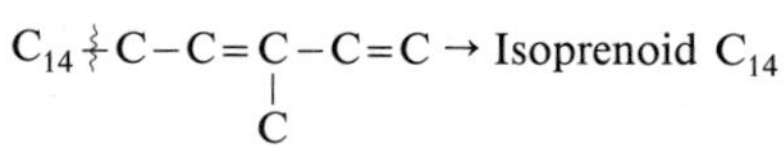

The cracking of aliphatic chains in the saturated phytadiene's dimer X (see Chap. 3) with the formation of C_{14}, C_{15}, C_{16}, C_{18} isoprenoid hydrocarbons may also be feasible:

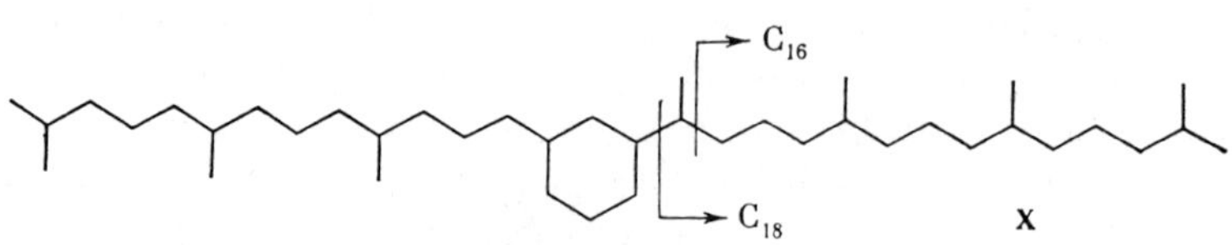

X

These isoprenoids usually appear (in respective ratios of 32:36:17:15) in the thermal decomposition high-molecular-weight phytadiene cyclodimers. At the same time, as expected, almost no pristane and phytane formation occurred in this way. (Generation of isoprenoid alkanes by the cracking of higher petroleum hydrocarbons will be analyzed in Chap. 6.)

Isoprenoid acids found in crudes may serve as additional sources of isoprenoid alkanes. In particular, ketonization of phytanic (3,7,11,15-tetramethylhexadecane) acid may result in the formation of regular isoprenoids with more than 20 carbon atoms (up to C_{39}). Characteristically, if stearic acid can generate n-nonadecane in large concentrations (without counting n-C_{17}), then, in accordance to the analogous, ketone β-decomposition scheme, phytanic acid should be expected to produce a C_{21} isoprenoid, which is actually present in crudes (among those higher than C_{20} isoprenoids) in the highest relative concentration, as was demonstrated in Chapter 2. The significance of high-molecular-weight, irregular isoprenoids in the formation of isoprenoid petroleum alkanes was examined above (see Chap. 2).

Formation of Saturated and Monoaromatic Steroids

A large number of recent publications (see De Leeuw et al. 1973; Rubinstein et al. 1975 for example) was devoted to the study of sterol and stanol transformations at early stages of diagenesis. Different views exist as to the mechanism of steroid transformations. Mainly two factors involved in the modification of biological molecules in sediments are discussed in the literature, i.e. the impact of microorganisms and that of aluminosilicates (clays). Obviously, the former factor is vital for understanding diagenetic processes, while the influence of aluminosilicates at high temperatures relates to the catagenesis of the organic matter. Without going into details of the publications devoted to the role of microorganisms, we should, however, mention that the main reaction in diagenesis is that of the reduction of sterol into stanols (Nishimura 1978; Gagosian and Heinzer 1979).

The paper of Pustilnikova et al. (1982) was devoted to the thorough analysis of sterols (XI) and stanol (XII) transformation reactions over aluminosilicates at temperatures between 180°–250°C. The structure and stereochemistry of emerging hydrocarbon molecules were the centre of attention. Cholesterol, cholestanol, ergosterol and sitosterol were used as initial compounds. Results of experiments with cholesterol under various conditions are presented in Table 55.

The main hydrocarbons to appear consistently in all experiments were: rearranged steradienes (XVII) (key ion, *m/z* 255), rearranged sterenes (XVI) (key ion, *m/z* 257), regular sterenes (XIII) (key ion, *m/z* 215), regular steradienes (XIV) (key ion, *m/z* 213) (Rubinstein et al. 1975; Gagosian and Farrington 1978), regular steratrienes (XV) and A-ring monoaromatic steranes ($m/z = 211$), and monoaromatic steroids (XVIII) (key ion, *m/z* 253). (The positions of multiple bonds in dienes and trienes are given on conjecture.) Among the saturated hydrocarbons, which appeared in amounts of 15–20% under more severe condi-

Table 55. Composition of hydrocarbons produced from cholesterol under various conditions

Cholesterol/ Aluminosilicate	Time (h)	Relative composition (%)		Rearranged hydrocarbons			Monoaromatic cholestanes	Isocholestanes
		Cholestadienes and cholestatrienes[a]	Cholestenes	Cholestadienes	Cholestenes	α Cholestane		
Temperature 180 °C								
1:2	2	16	24	28	21	–	10	1
1:2	20	22	16	19	29	–	12	2
1:2	50	–	12	16	41	3	18	10
1:8	50	–	16	12	17	8	20	27
Temperature 250 °C								
1:2	2	33	19	3	6	–	37	2
1:8	2	29	31	2	4	–	32	2

[a] Cholestatrienes, possibly in a mixture with monoaromatic cholestanes (with an aromatic A ring) predominated in all experiments. In both cases, the key ion was m/z 211.

tions, α- and β-steranes (5α and 5β, 14α, 17α, 20*R*), rearranged steranes (13β, 17α, 20*R*, 20*S*) as well as isosteranes (14β, 17β, 20*R*, 20*S*) were identified:

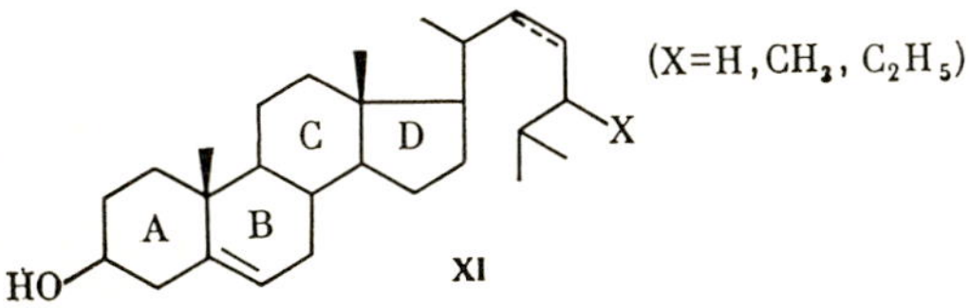

XI

Sterols

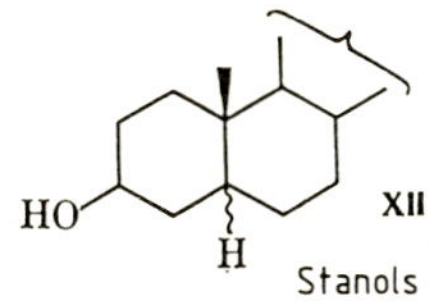

XII

Stanols

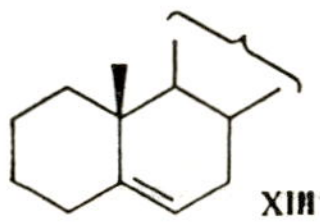

XIII

Regular sterenes

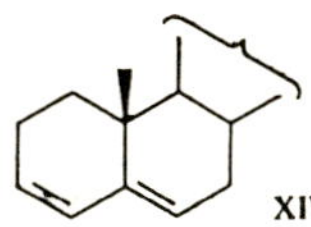

XIV

Regular steradienes

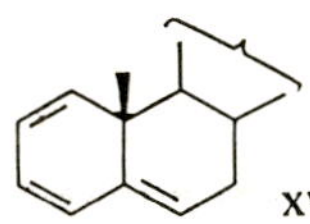

XV

Regular steratrienes

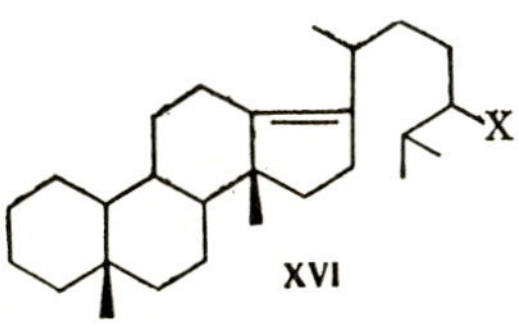

XVI

Rearranged sterenes

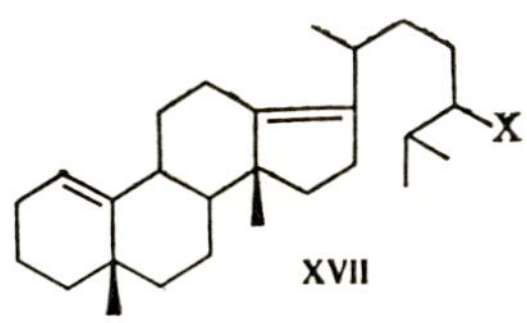

XVII

Rearranged steradienes

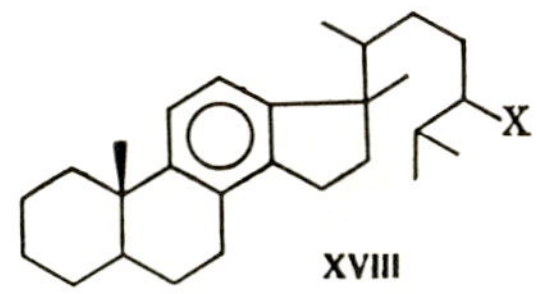

XVIII

Monoaromatic steroids

In summarizing the experimental results in Table 55, we suggest the following scheme of sterol transformations (Scheme 3).

Scheme 3

$\xrightarrow{\text{Dehydration, saturation}}$ sterenes → steranes → isosteranes

$\xrightarrow{\text{Sterols, dehydration}}$ steratrienes → monoaromatic steroids

$\xrightarrow{\text{Rearrangement}}$ rearranged sterenes and steradienes $\xrightarrow{\text{saturation}}$ rearranged steranes

As can be observed in the scheme, sterols undergo a number of transformations in parallel directions, each consisting of several consecutive steps.

Of particular significance, in our opinion, is the intensive redistribution of hydrogen, accompanied by the formation of monoenes and trienes from dienes (primary, unstable products of sterol dehydration). Formation of steranes and monoaromatic steroids is a natural continuation of these reactions.

Initial (at early stages of catagenesis) formation of rearranged hydrocarbons is important for understanding the mechanism of petroleum hydrocarbon formation. Since skeletal rearrangement of sterenes is characteristic only of unsaturated systems, then the latter should be terminated after saturation by hydrogen redistribution. In contrast, the formation of isosteranes is a secondary process proceeding in saturated systems. Hence, the ratio of iso/α-steranes analyzed in Chapter 3, is indeed a characteristic factor in evaluating the depth of catagenetic maturation of crude oils.

Besides the comprehensively reviewed experiments with cholesterol by Pustilnikova et al. (1982), experiments on the transformation of ergosterol and sitosterols under similar conditions were also conducted. Experiments with sitosterol provided results closely compatible with those observed in cholesterol transformations. However, ergosterol was mainly transformed into the respective monoaromatic steroid. This is not surprising, since the initial product is known to contain two double bonds more than cholesterol. Thus, the presence of C_{28} steranes in crude oils evidently depends on campesterol, steroid which is cholesterol's methyl homolog.

In conclusion, it may be noted that the formation of a relatively small quantity of saturated hydrocarbons in experiments was based on the high unsaturation of cholesterol dehydration products (cholestadienes). At the same time, if cholestanol (i.e. a saturated analog) is used during thermocatalysis, rather than cholesterol, the saturated hydrocarbons yield increases manifold. Since unsaturated steroid hydrocarbons are known to be absent in crudes, then prior microbiological reduction (at the stage of diagenesis) of sterols into stanols becomes feasible and desirable (Edmunds et al. 1979). It is noteworthy that besides the above examined hydrocarbons, large quantities of D-homodiacholestane (XIX), easily identifiable by the key ion, *m/z* 287 (Sieskind et al. 1979), were produced in experiments with cholestanol:

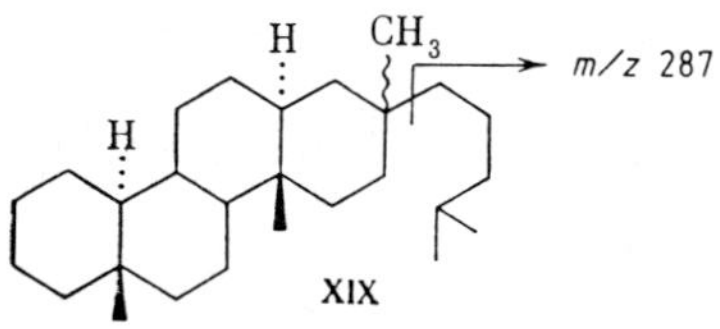

Thus, transformations of sterols (as well as stanols) over aluminosilicates lead to the production of the entire range of appropriate petroleum hydrocarbons: α- and β-steranes, isosteranes, rearranged steranes, D-homodiasteranes and monoaromatic steroids.

The present chapter was largely devoted to reactions of petroleum hydrocarbon formation proceeding in accordance to a carbonium-ion mechanism. However, it is also assumed that biogenic, molecule transformations into petroleum hydrocarbons may also proceed with the participation of free radicals, appearing to a large degree through restructuring of the crystal lattice of the source rocks. The free-radical mechanism of dispersed organic matter transformation allows one to decrease petroleum formation temperatures to 20 °– 50 °C, which is important in various genetic correlations. This subject is comprehensively treated in the monograph of Galimov (1973).

CHAPTER 6

Transformations of Petroleum Hydrocarbons in Nature (Chemical and Biochemical Alteration of Crude Oils)

The present chapter is devoted to the transformation of petroleum hydrocarbons occurring reservoirs, i.e. those that are characteristic of crude oils proper. As will be demonstrated, these transformations may be substantial in causing important modifications in the original hydrocarbon composition of crudes. Ways and mechanisms of these reactions proceeding under a thermal effect (thermolysis), catalytic influence (catalysis), as well as under oxidative impact of microorganisms (biodegradation) have been discussed for years.

The two former reactions are occasionally combined under the general term of "catagenesis", while the oxidative effect of microorganisms is called "hypergenesis". The actual potential of biodegradation, which is most influential in changing the chemical type of crudes in reservoirs, has been much better understood in recent years. However, let us first analyze petroleum transformations under the influence of thermal stress, i.e. thermolysis of thermal evolution (aging) of petroleum.

Alteration of Crude Oil Under Thermal Stress (Thermolysis)

Processes of thermal transformations of primarily high-molecular-weight hydrocarbons attracted the attention of researchers long ago. Extensive material on this problem, as well as thermodynamical analysis of the above transformations were presented in the monograph of Andreev et al. (1958). However, the analytical techniques of that time did not allow direct experimental investigations of thermal transformations of crudes of different chemical types. Moreover, any, albeit most thorough, thermodynamic appraisals may represent only possible and not actual transformations of specific hydrocarbons in crude oils. Thus, experimental research acquires special importance in this area.

In recent years a number of studies have been undertaken devoted to modifications in the composition of hydrocarbons under thermal stress (Safonova 1974; Connan et al. 1975; Tchakhmakhtchev 1983). Their results demonstrated that upon heating certain transformations occur in the compositions of crudes, including an increase in light components, new normal alkane formation, reduction of isoprenoid alkane concentrations, etc. In general, these transformations failed to alter the chemical type of the crude oils analyzed. However, since the influence of thermolysis on petroleum composition can not be excluded entirely, it was imperative to evaluate the actual parameters of such transformations. To this end a series of investigations on thermal transformations of different crude types was undertaken in the laboratory headed by the author. The main task consisted in analyzing compositional modifications in petroleum proper, i.e. its medium- and high-molecular-weight fractions; the compositions of gasolines produced were also tested. Primary attention was paid to experimental feasibility of modifying the type of crudes in accordance to the well-known methanization theory, i.e. evolution in the direction, category B→category A (see Chap 1).

Simulation of natural geological processes under laboratory conditions is always accompanied by difficulties, primarily the impossibility of simulating geological time. Researchers are thus compelled to perform experiments of thermal transformations at higher temperatures, when raising temperature (naturally within reasonable limits) is the only way to compensate the prolonged effect of a lower temperature on the geological scale.

Evidently, such an approach is prone to serious and substantial criticism. Available data on the temperatures of the petroleum reservoir indicate that they do not exceed 150°–170 °C. Unfortunately, no visible changes in the hydrocarbon composition of crudes under laboratory conditions could be registered at these temperatures.

In order to choose an optimal temperature for petroleum thermolysis, a series of experiments was performed at various temperatures: 300°–350°–400 °C. It was established that no appreciable modifications occurred upon heating of petroleum to 300 °C for 20 days. With the temperature level raised to 400 °C, a sharp increase in the rate of petroleum hydrocarbon decomposition was registered, leading to intense gas formation. Moreover, a large amount of unsaturated hydrocarbons was registered in the mixture of reaction products, which is untypical of natural crude oils. With this in mind, and for further investigation of thermolysis, we adopted the temperature range of 350°–360 °C, when the rate of petroleum decomposition increases 30–60-fold as compared to 300 °C, but hardly any formation of unsaturated hydrocarbons occurs.

The extent and direction of petroleum transformation were evaluated on the basis of quantitative and qualitative modifications in the composition of saturated hydrocarbons, established through GC and MS. The length of heating was chosen in such a way as to have approximately 10–15% gasoline in the decomposition products, which corresponds to the average gasoline content in A^1 crudes. Usually an experiment lasted 1–2 days. This allowed one to establish the general direction of the alteration in the crudes' chemical composition under thermolysis, and to simulate certain simple reactions, which occur in oil reservoirs, under laboratory conditions.

Twenty different topped crudes, representing all four chemical types, were used as starting material. Although thermolysis of category B crudes bears utmost importance in ascertaining the theory of methanization (Andreev et al. 1958), category A crudes were also subjected to a thermal treatment, in order to evaluate the dependence of various classes of hydrocarbons on the chemical origin of petroleum from the view point of their transformations.

Experimental results of thermolysis of the most representative crudes of all four types (A^1, A^2, B^2, and B^1) are represented in Tables 56–59. Traditional analysis of the original crudes and their thermolytic products demonstrate that

Table 56. Relative distribution of heptanes in a mixture of petroleum thermolytic products (%)[a]

Hydrocarbons	Reservoir, Crude type					
	Kenkiyak B[1]	Starogroznenskoe B[2]	Karajanbas, A[2]	Salymskoe, A[1]	Salymskoe (catalysis)	Salymskoe (crude)
n.Heptane	36.9	42.9	36.1	55.6	0.7	47.3
2-Methylhexane	19.4	20.0	17.1	20.1	43.6	20.5
3-Methylhexane	29.0	27.2	28.6	20.5	33.3	24.3
3-Ethylpentane	1.5	2.7	1.9	Trace	Trace	Trace
Σ Monosubstituted	49.9	49.9	47.6	40.6	77.2	44.8
2,4-Dimethylpentane	0.6	–	0.5	Trace	9.0	Trace
2,3-Dimethylpentane	10.4	6.8	12.7	2.9	13.0	6.7
Σ Disubstituted	11.0	6.8	13.2	2.9	22.0	6.7
2,2-Dimethylpentane	0.7	Trace	0.9	–	–	0.6
3,3-Dimethylpentane	1.1	0.4	1.9	0.5	–	0.4
2,2,3-Trimethylbutane	0.4	Trace	0.3	0.4	0.4	0.2
Σ Hydrocarbons with quaternary carbon atom	2.2	0.4	3.1	0.9	0.4	1.2
Σ Monosubstituted / Σ Disubstituted	3.9	6.9	3.0	11.9	3.5	5.8

Note: Columns 6 and 7 represent the distribution of isomers in the original crude and in a mixture of catalytic cracking products in the presence of an aluminosilicate catalyst.
[a] In the sum total of isomers.

the formation of gasoline fractions and the decrease in the content of high-boiling hydrocarbons occurs in all experiments. At the same time, the amount of the medium fraction (200°–430 °C) remains basically unaltered.

Table 57. Relative distribution of C_7 cycloalkanes in a mixture of petroleum thermolytic products (%)

Hydrocarbons	Reservoir, Crude type					
	Kenkiyak, B^1	Starogrosn-enskoe, B^2	Karajan-bas, A^2	Salym-skoe, A^1	Salymskoe (catalysis)	Salmyskoe (crude)
Ethylcyclo-pentane	12.2	11.9	11.4	16.9	17.6	3.3
1,1-Dimethyl-cyclopentane	1.2	1.4	1.3	–	–	–
1,3-Dimethyl-cyclopentane, *trans*	9.2	4.7	6.9	10.4	10.6	5.6
1,3-Dimethyl-cyclopentane, *cis*	15.2	9.3	11.0	12.9	11.8	6.4
$\sum$ 1,3-Di-methylcyclo-pentanes (*trans* + *cis*)	24.4	14.0	17.9	23.3	22.4	12.0
1,2-Dimethyl-cyclopen-tanes, *cis*	7.9	2.3	6.9	6.8	2.8	1.7
1,2-Dimethyl-cyclopen-tanes, *trans*	21.3	25.3	20.8	15.3	10.6	11.7
$\sum$ 1,2-Di-methylcyclo-pentanes (*trans* + *cis*)	29.2	27.6	27.7	22.1	13.4	13.4
Methylcyclo-hexane	33.0	45.1	41.7	37.7	46.6	71.3

Compositional Analysis of Light Petroleum Fractions

Tables 56 and 57 are devoted to the composition of light-boiling hydrocarbons, produced in the thermolysis of various types of crudes. We restricted ourselves to C_7 hydrocarbons, which are nonetheless quite representative for the destruction mechanism [data on the general composition of gasolines has been published earlier (Sokolov et al. 1972; Martin et al. 1964)].

As the Tables demonstrate, light hydrocarbons emerging in thermolysis have considerable similarity and depend only slightly of the type on the original crude. In all cases thermolytic gasolines corresponded to natural gasolines of A^1 crudes. Comparative data on such a gasoline are provided in Tables 56 and 57 (see column 7). Thermolytic gasolines were rich in alkanes, while the alkane/cyclane ratio varied within 1.5–2.0. This ratio is largely compatible with that which

Table 58. Characteristics of 200° – 430 °C fractions in crude oils and their thermolytic products

Deposit	Sample type[a]	Crude type	Fractional composition (%)		Pristane / Phytane	C_i (see Table 1)	N_b	$\Sigma(n\text{-}C_{13}-n\text{-}C_{15}) / \Sigma(n\text{-}C_{25}-n\text{-}C_{27})$
			Alkanes + naphthenes	Aromatics				
Keniyak	I	B^1	67	33	–	–	–	–
	II	–	60	40	1.11	0.1	0.4	5.1
Russkoe	I	B^1	58	42	–	–	–	–
	II	–	49	51	1.67	0.6	0.6	2.6
Starogroznenskoe	I	B^2	72	28	1.54	–	–	–
	II	–	61	39	1.67	0.4	2.5	7.7
Balakhanskoe	I	B^2	77	23	1.11	–	–	–
	II	–	68	32	1.43	0.6	1.1	5.6
Karajanbas	I	A^2	61	39	0.91	3.8	0.6	1.5
	II	–	41	59	1.05	0.5	2.3	2.8
Oval-Toval	I	A^2	81	19	1.33	3.4	3.2	1.6
	II	–	50	50	1.6	0.2	5.8	3.5
Duvan-more	I	A^1	67	33	1.43	0.8	10.9	1.0
	II	–	54	46	1.43	0.6	7.8	2.2

[a] Here and in Table 59: I original crude, II thermolytic products.

Table 59. Compositional modifications in saturated hydrocarbons in the 200° – 430 °C fraction under thermolysis

Deposit	Sample type	Crude type	Fractional composition (%)		Alkanes (Rel. %)			Naphthenic fingerprint (%)				
			Alkanes	Naphthenes	Normal	Isopr.	Iso	Monocyclic	Bicyclic	Tricyclic	Tetracyclic	Pentacyclic
Kenkiyak	I	B^1	13.0	87.0	–	–	100	19.1	30.1	23.7	17.0	10.1
	II		16.0	84.0	16.0	5.0	79.0	16.3	29.0	23.6	19.4	11.7
Russkoe	I	B^1	10.0	90.0	–	–	100	17.6	30.0	27.6	17.5	7.3
	II		12.0	88.0	16.7	4.2	79.1	16.9	29.8	28.6	17.7	7.0
Starogroznenskoe	I	B^2	35.6	64.4	–	14.0	86.0	28.9	30.3	19.0	14.8	7.0
	II		38.2	61.8	10.7	2.6	86.7	28.0	29.3	20.3	15.2	7.2
Balakhanskoe	I	B^2	22.6	77.4	–	16.4	83.6	20.6	26.5	21.6	21.2	10.1
	II		25.0	75.0	14.4	3.2	82.4	18.3	25.8	22.8	22.0	11.1
Karajanbas	I	A^2	27.2	72.8	12.9	25.6	61.5	24.4	29.8	22.8	14.0	9.0
	II		35.1	64.9	21.6	8.6	69.8	25.6	30.0	21.5	14.0	8.9
Oval Toval	I	A^2	35.3	64.7	12.8	18.3	68.9	24.9	25.8	17.2	22.6	9.5
	II		40.6	59.4	34.2	7.6	58.2	22.9	25.8	18.4	20.9	12.0
Duvan-More	I	A^1	43.3	56.7	38.9	12.5	48.6	20.2	25.8	20.6	22.1	11.3
	II		43.0	57.0	37.9	6.7	55.4	20.5	24.1	21.2	21.5	12.7

Dobryanskii expected of petroleum methanization under thermolysis (Andreev et al. 1958).

In analyzing the distribution of heptane isomers (see Table 56) it can be seen that all experiments gave the highest yields of n-heptane, which accounted for 37–55% of all isomers. It may also be noted that in gasolines derived from B crudes, i.e. those which either have nor gasolines or contain very small amounts of n-heptane, concentrations of this hydrocarbon in thermolytic products were slightly lower (37–43%). Monomethyl-substituted isomers among the branched alkanes dominate in thermolytic products. The concentration of *gem*-substituted alkanes is insignificant.

All gasolines produced had similarities in the cycloalkane distributions (see Table 57). Thus, methylcyclohexane accounted for 33–45%; 1,3-dimethylcyclopentanes were represented by *cis*- and *trans*-isomers in a 60:40 ratio; the more stable *trans*-isomer usually predominated among 1,2-dimethylcyclopentanes. The *gem*-substituted cyclane content is usually insignificant.

Accordingly, it was demonstrated that the relative distribution of hydrocarbons in thermolytic gasolines does not depend on the original crude's chemical type, and emerging mixtures are compositionally similar to natural gasolines of A^1 crudes. Hence, the origin of gasolines typical of certain naphthenic crudes (such as Anastasievsko-Troitskoe or Gryazevaya Sopka, see Chaps. 2 and 3), with high concentrations of di- and tri-substituted C_7-C_8 alkanes and cyclanes, as well as of *gem*-substituted hydrocarbons, is still a mystery. As expected, by substituting the catalytic decomposition for the thermal one (see Tables 56 and 57) we could only affect the yield of normal alkanes. However, this procedure failed to reproduce complex and peculiar B-type gasolines.

Compositional Changes in Medium Petroleum Fractions

Let us now review those modifications which occur in the 200°–430 °C fraction upon heating. Appropriate GC, MS results are presented in Tables 58 and 59 and Figs. 77–81.

Type B^1 Crudes. Although these crudes undergo fairly extensive transformations upon heating, exemplified by intense gasoline formation (up to 15% of the initial crude), the difference in group compositions of the 200°–430 °C fraction in original and thermolytic petroleums is insignificant. Alkanes increased only by 2–3%, while aromatic hydrocarbons, by 7–9%. In comparing chromatograms of the original crude oil and its thermolytic products, as reproduced in Figs. 77 and 78, we can observe some formation of new, normal and isoprenoid alkanes in the heating of type B^1 crudes. Normal alkanes thus produced are comprised of 12 to 25 carbon atoms. In comparison, isoprenoid hydrocarbons appeared even in smaller quantities not exceeding 0.3–0.5%. Characteristically, all alkanes emerged in the first hours of experimentation and further heating failed to increase their concentrations.

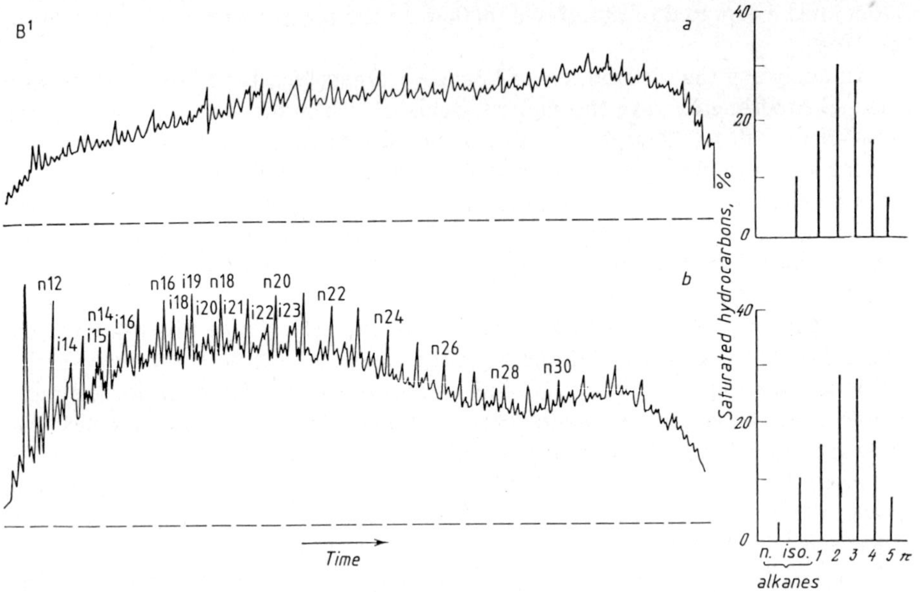

Fig. 77a, b. Compositional alterations in Kenkiyak (Jurassic) crude upon heating. **a** Original (B^1); **b** thermolytic products. Distribution of saturated hydrocarbons according to structure types presented on the *right*. Normal and isoprenoid alkane peaks are identified. Here and in Figs. 78–86 *n* denotes the number of rings in cyclanes. Capillary Column 30 m, Apiezon; linear temperature programming, 100°→320 °C at 3°/min

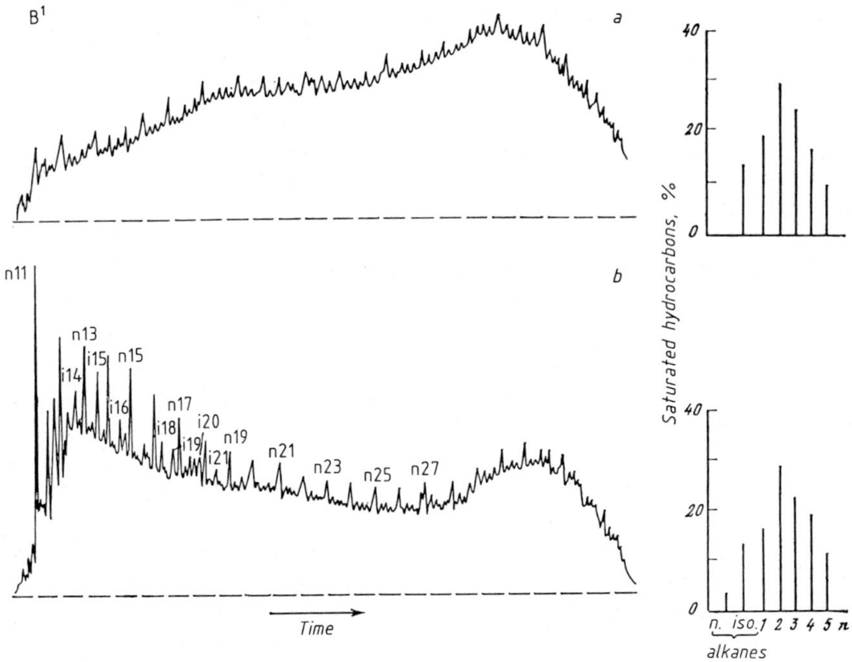

Fig. 78a, b. Compositional alterations in the Russkoe crude upon heating. **a** Original (B^1) crude; **b** thermolytic products

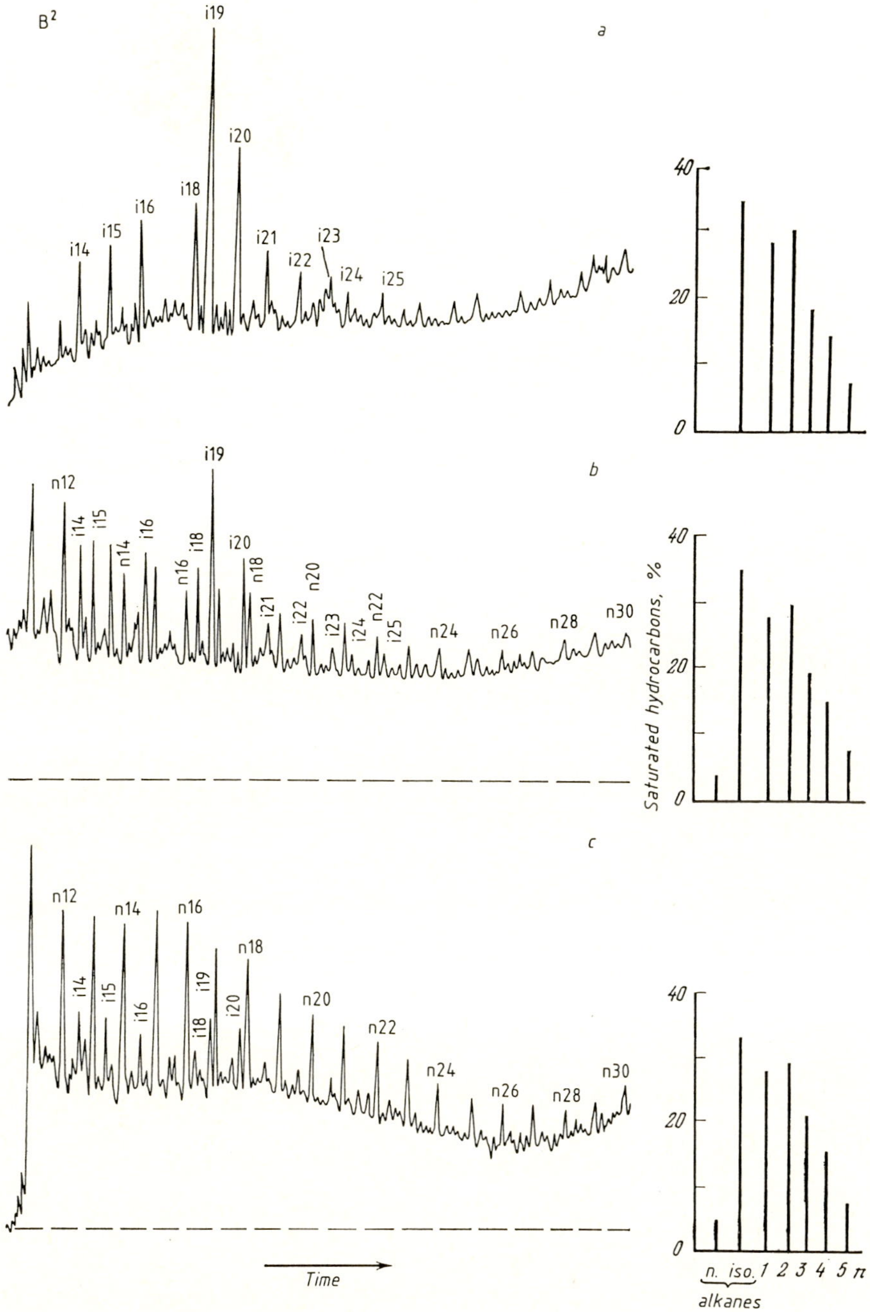

Fig. 79a–c. Compositional alterations in the Starogroznenskoe crude upon heating. **a** Original crude (B^2); **b, c** thermolytic products (20 and 45 h)

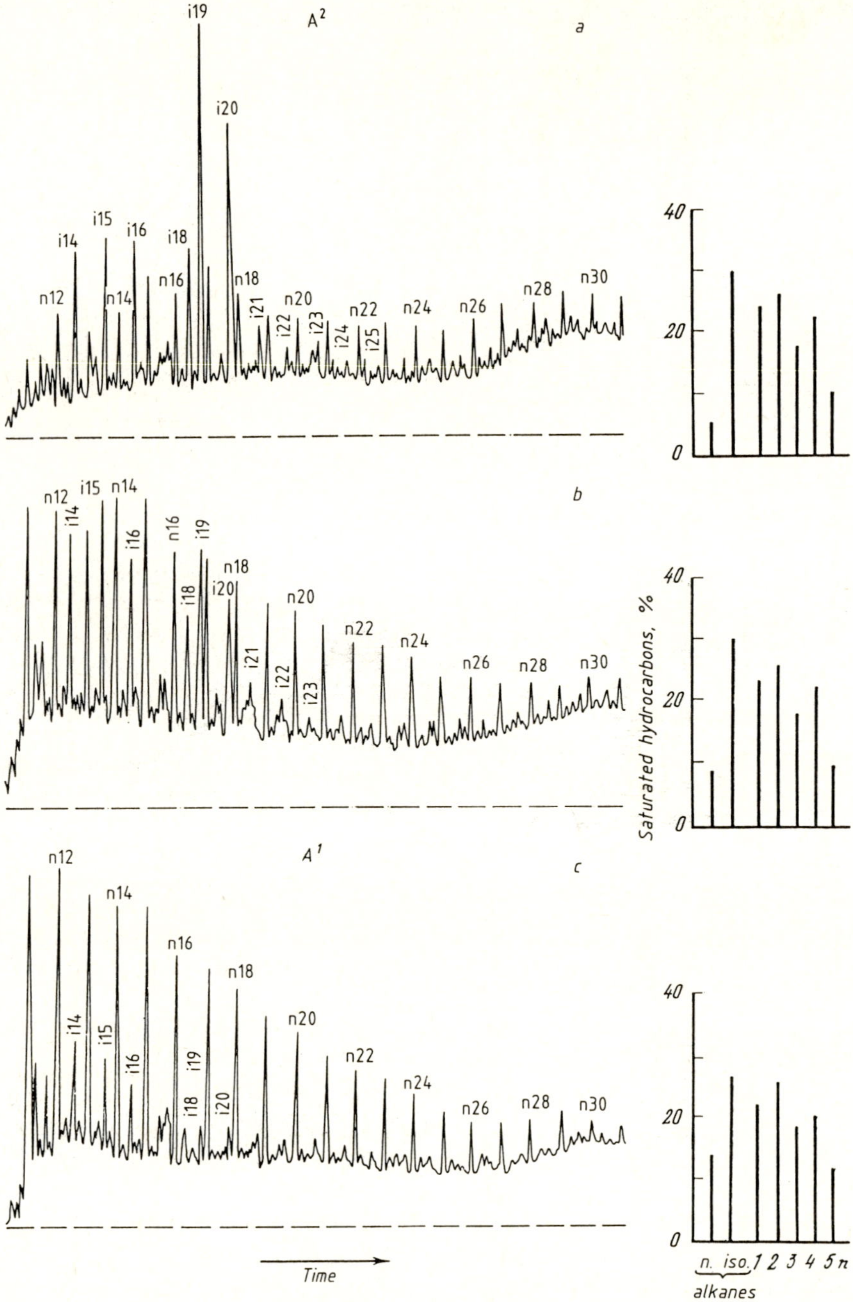

Fig. 80 a–c. Compositional alterations in the Oval-Toval crude upon heating. **a** Original crude (A^2); **b, c** thermolytic products (15 and 35 h)

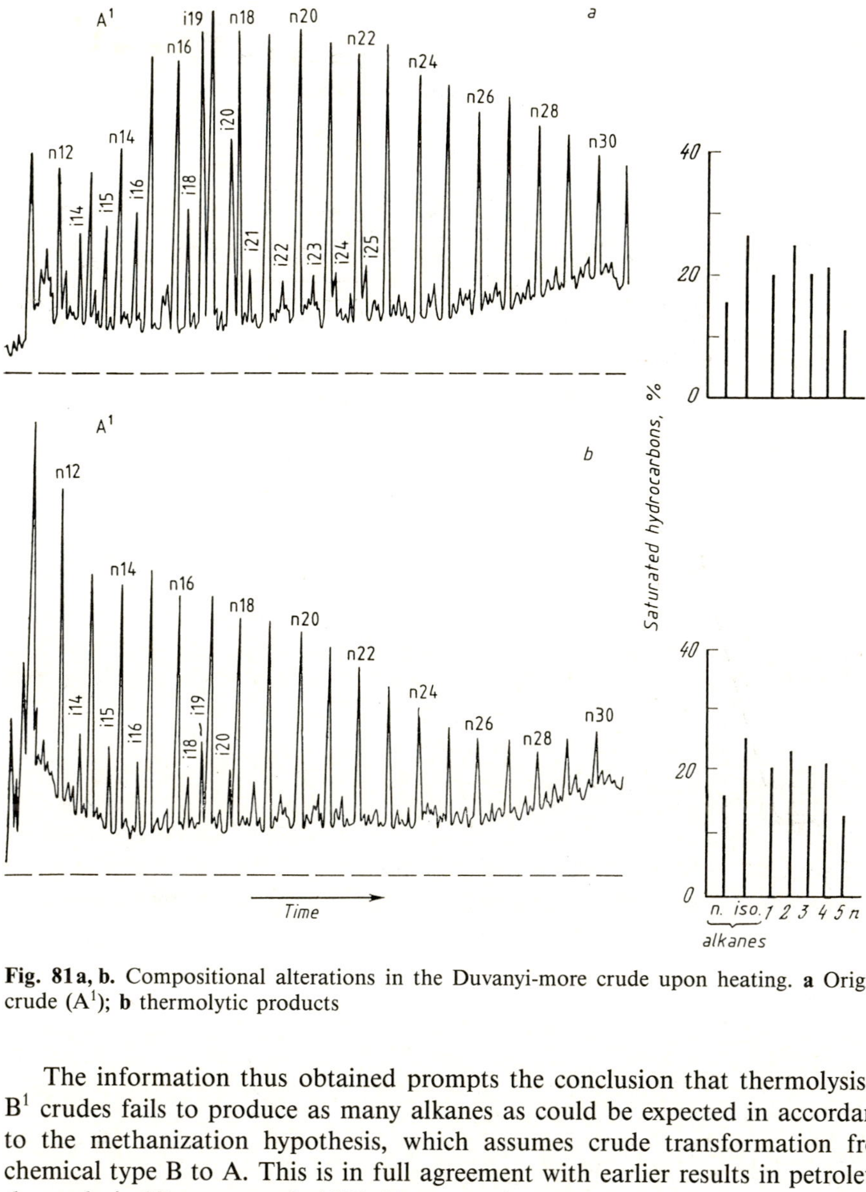

Fig. 81a, b. Compositional alterations in the Duvanyi-more crude upon heating. **a** Original crude (A^1); **b** thermolytic products

The information thus obtained prompts the conclusion that thermolysis of B^1 crudes fails to produce as many alkanes as could be expected in accordance to the methanization hypothesis, which assumes crude transformation from chemical type B to A. This is in full agreement with earlier results in petroleum thermolysis (Connan et al. 1975; Evans et al. 1971).

As compared to original crudes, naphthenic fingerprints of the decomposition products are affected insignificantly. A slight decrease in mono- and bicyclic naphthenes only could be registered. It should be emphasized that no analogs of the destruction products of B^1 crude could be detected among natural oils (Figs. 77 b and 78 b), since in their main parameters these mixtures fail to correspond to any natural crudes identified.

Hence, there may be no transformation of type B^1 crude into type A under thermal stress. An analogous situation is obtained in the heating of B^2 crudes.

Type B^2 Crudes. Results of experiments with B^2 crudes are represented in Fig. 79 and Table 59. Some (limited) quantities of $C_{12}-C_{30}$ normal alkanes were also identified in mixtures of B^2 thermolytic products. However, the isoprenoid content decrases considerably under heating (fivefold). The content of other branched alkanes remains unaffected, which is not true of the relative concentrational distributions of isoprenoids. Heating brings about a reduction in total concentrations of pristane and phytane, while, at the same time, isoprenoids of smaller molecular mass increase in their relative number.

The pristane/phytane ratio tends to increase in the course of an experiment (from 1.1 to 1.6). It was, however, observed that up to a certain point in time, temperature affects this correlation only slightly.

Remarkably, in the first few hours of heating, some increase in isoprenoids in a mixture of reaction products could even be observed. This signifies that thermolysis was accompanied not only by the destruction of isoprenoids, but by their new formation as well, apparently due to losses of alkyl chains in cyclanes and arenes and the cracking of higher isoprenoid alkanes. Characteristically, an entire series of $C_{12}-C_{25}$ isoprenoid alkanes, in quantities reaching 2%, with phytane and pristane comprising the majority, was produced in the heating of the 200°–430 °C fraction of the Oval-Toval crude.

In analyzing the MS results of initial B^2 crudes it was demonstrated that, similar to B^1, naphthenic fingerprints are only slightly altered by heating, which testifies to the relative stability of the cyclic systems proper. Though normal alkane peaks are traced with more ease in chromatograms of the thermolytic products of B^2 crudes, new crudes should not be interpreted as being type A^2. As demonstrated in Chapter 1, A^2 crudes are chromatographically characterized by a relatively low naphthenic background and higher phytane and pristane peaks. However, B^2 thermolytic products have a fairly high naphthenic background and small pristane and phytane peaks (see Fig. 79).

Consequently, even in this case new oil properties fail to correspond to those of A^2 crudes (low values for C_i and N_b). Such oils are not found in nature. Hence, the transformation of B^2 crudes into A ones through heating appears unfeasible.

Type-A^2 Crudes. The study of the thermolysis of A^2 crudes demonstrated that total concentrations of alkanes are increased by 5–7% in reactions, mostly due to normal alkanes. However, the overall isoprenoid content decreases as a result of thermolysis. As in all previous experiments, the content of other isoalkanes remains practically unaltered (see Table 59). Heating of A^2 crudes results in a dramatic decrease of the C_i coefficient (two- to sevenfold). A comparison of the individual contents of normal alkanes in the decomposition mixture (Fig. 80) reveals a shift in concentrational maxima. Thus, if it belongs to normal $C_{15}-C_{17}$ alkanes in the original crude, then upon thermolysis it shifts towards $C_{12}-C_{13}$ normal alkanes while the Σn-C_{13}–n-C_{15}/Σn-C_{35}–n-C_{27} ratio also increases. Isoprenoid hydrocarbons are also characterized by shifts of the concentrational maxima towards light hydrocarbons: from $C_{19}-C_{20}$ to $C_{13}-C_{15}$. The pristane/phytane ratio increases slightly.

However, heating of A^2 crudes fails to produce a new type of crude (A^1), since A^1 crudes with such low concentrations of pristane and phytane are scarce in nature[1]. It should also be recalled that crudes of chemical type A^2 are quite rare in nature. As in the case of other types of crudes, A^2 naphthenic fingerprints remain practically unaltered in thermolysis.

Type-A^1 Crudes. As can be observed in Table 58, group composition changes in the 200°–430 °C fraction after heating of A^1 crudes are insignificant. In effect, there are also no changes in the normal alkane content in the fraction. Isoprenoid content decreased almost twofold. The C_i coefficient was also diminished. High-boiling, normal alkane and isoprenoid peaks are markedly lower in the chromatograms of thermolytic products (Fig. 81). Maxima shifts from n-C_{15} – n-C_{17} and iso-C_{19} – iso-C_{20} to n-C_{11} – n-C_{12} and iso-C_{13} – iso-C_{15} is registered. The naphthenic fingerprint, as in all other experiments, remains unaltered in the process of thermolysis.

Discussion of the Results. In summarizing the experimental data on thermolysis, the conclusion may arise that heating of various type crudes is accompanied by characteristic (albeit insignificant) transformations in their composition. In particular, the formation of new, both normal and (to a lesser degree) isoprenoid alkanes occurs. Concentrations of isoprenoids decrease in those crudes which already possess them. Pristane and phytane contents especially decrease drastically. Diminution of isoprenoids may be explained by a more restricted source of isoprenoid structures occurring in crudes, compared to structures which may serve as potential sources of normal alkanes (Tissot et al. 1971). On the whole, all alkanes of various types participate in two opposite reactions: formation of new and decomposition of old structures.

The mean molecular mass of both normal and isoprenoid alkanes decreases in the process of thermolysis. Alterations in the individual composition of alkanes upon thermolysis may be exemplified by the following indicators: C_i; Σi-C_{14} – i-C_{18}/Σi-C_{19} – i-C_{25}; n-C_{13} – n-C_{15}/n-C_{25} – n-C_{27}; pristane/phytane ratio. It has been established that in the heating of any type of crude, the first indicator tends to decrease and the three remaining increase.

It should also be emphasized that naphthenic fingerprints remain intact or are modified only slightly under thermolysis. As demonstrated in Chapter 1, the same situation may be observed in natural, multilayer reservoirs with crudes of common genetic origin, but different chemical types.

Quantitatively, the process of thermolysis depends on fractions boiling above 430 °C. The higher the concentration of these hydrocarbons in crudes, the mose extensive their compositional transformations. The least changes in alkane compositions were registered in experiments with A^1-crude oils, while the greatest occurred in A^2, with B crudes occupying an intermediate position. Beyond doubt, A^1 crudes are also affected considerably, which is, however, less obvious in analysis, since compositionally, new hydrocarbons are mostly analogous to that occurring in the initial crude.

---

[1] Moreover, these (paraffinic) crudes have maximal concentrations of normal alkanes within the C_{20} – C_{30} range.

Table 60. Relative destruction rate constants of C_{24} hydrocarbons of various structural type at 450 °C

Hydrocarbon	I	II	III	IV	V	VI	VII	VIII	IX	X	XI	XII	XIII	$n-C_{24}$
k	1.6	1.5	1.0	1.0	1.0	1.0	0.9	0.8	0.7	0.7	0.7	0.7	0.7	1.0

Other hydrocarbons													
Hydro-carbon			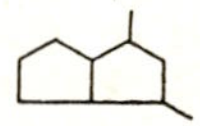	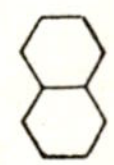	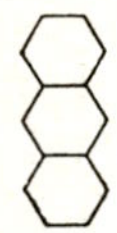	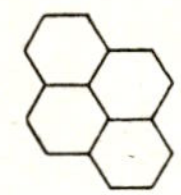	Adian-tane	Hopane	Homo-hopane	Bis-homo-hopane	Tris-homo-hopane	α-Choles-tane	Pris-tane
k	0.01	0.1	0.5	0.2	0.4	0.2	0.3	0.35	0.4	0.6	1.0	1.1	0.5

Normal alkanes										
Hydrocarbon	$n-C_9$	$n-C_{10}$	$n-C_{17}$	$n-C_{18}$	$n-C_{19}$	$n-C_{20}$	$n-C_{21}$	$n-C_{24}$	$n-C_{28}$	$n-C_{30}$
k	0.23	0.25	0.40	0.50	0.70	0.80	0.85	1.0	2.5	3.2

Note: k is the decay rate constant of n- tetracosane taken as 1. The structure of hydrocarbons I – XIII see page 211.

The main reaction involved in thermolysis is the destruction of alkanes and long (normal and isoprenoid) aliphatic chains of cyclanes and alkanes, which systematically leads to the formation of lighter alkanes. The presence of such hydrocarbons in crudes was mentioned in previous chapters. Moreover, numerous data on infrared spectra reconfirm the presence of long normal and isoprenoid aliphatic chains in cyclic hydrocarbons (Kuklinskii and Pushkina 1974).

It is noteworthy that upon heating, crudes retain constant ratios among separate groups of saturated hydrocarbons in the 200°–430 °C fraction, which is compatible with the properties of different natural crudes of common genetic origin. This peculiarity may be explained by stronger C-C bonds in rings than in aliphatic chains (see information on relative decay constants of various hydrocarbons in Table 60). Therefore, naphthenic hydrocarbons are less amenable to ring destruction than C-C bond cleavage in aliphatic substituents, which leads to the formation of naphthenes of smaller molecular mass.

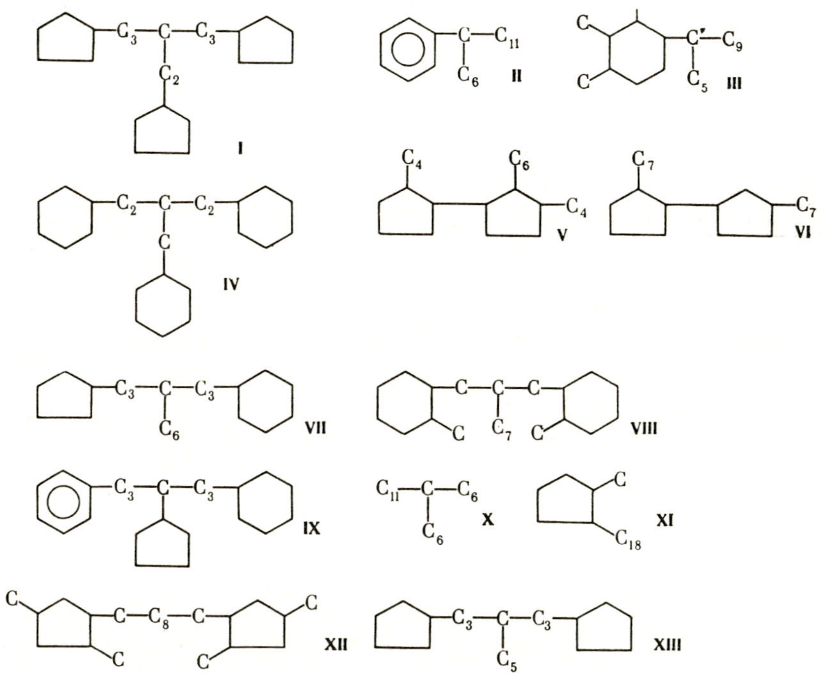

I, II, III, IV, V, VI, VII, VIII, IX, X, XI, XII, XIII

In those cases when the availability of long chains is relatively limited (type B crudes), no extensive formation of new normal alkanes in thermolysis may occur. As opposed to B crudes, the process of microbiological degradation[2] of alkanes (or other hydrocarbons with long, unbranched chains) in A^2 oils is yet incomplete. These crudes contain numerous potential sources of normal alkanes, hence, maximal amounts of alkanes appear upon their heating.

The above data lead to the conclusion that a relatively small amount of alkanes emerges in the cracking of aliphatic chains upon heating of B crudes. Due

[2] For the mechanism of microbiological alteration of crudes, see further.

to biodegradation in an advanced stage, the availability of aliphatic structures is insufficient in this case for the formation of appreciable quantities of normal and isoprenoid $C_{12}-C_{25}$ alkanes. This prompts the conclusion that it is practically impossible (due to the absence of the required quantities of aliphatic hydrocarbons in long chains) to alter the chemical type of oils in reservoirs solely by increasing temperature and in accordance to successive transformations within the scheme: $B^1 \rightarrow B^2 \rightarrow A^2 \rightarrow A^1$, i.e. so as to first produce crudes rich in isoprenoids (especially pristane and phytane), and then those in normal alkanes.

The best results of this sort may be obtained in thermolysis of asphaltenes. However, in this case it is not the processes of chemical evolution of crude oils that are being simulated, but rather their formation from kerogen, which is compositionally similar to petroleum asphaltenes. It is noteworthy that asphaltenes, which may only slightly be affected by biodegradation, regularly comprise sufficient long aliphatic chains of both normal and isoprenoid types (see below).

Undoubtedly, catagenetic transformations in natural crudes to occur, however, they affect only their fractional composition and usually proceed in deeply buried crudes of type A^1. This situation finds ample substantiation in the publication of Arefyev et al. (1980) which is devoted to genetically homogeneous crudes of Eastern Siberia.

Increasing the light fraction of crudes is, in effect, a most indisputable consequence of thermal stress. This is accompanied by a decrease in the total medium ($C_{19}-C_{20}$) isoprenoids and a slight increase of both normal and isoprenoid light alkanes, without, however, affecting the chemical type of the crudes. A good catagenetic indicator for genetically homogeneous crudes is represented by the diminution of the C_i coefficient and the increase in the pristane/phytane ratio, as well as in the proportion of light isoprenoids ($C_{14}-C_{18}$) and medium alkanes ($C_{12}-C_{18}$).

Under natural conditions it is usually difficult to separate crude catagenesis from the formation of various oils from kerogen at different stages of maturation.

Alterations in the Chemical Composition of Crudes Under Microbiological Influence (Petroleum Biodegradation)

For 50 years microbiologists have been familiar with the oxidation of petroleum hydrocarbons by certain types of microorganisms. Extensive scientific literature is devoted to this subject. However, the actual impact of microorganisms on crude oil in reservoirs remained disputable for a long period. There were those who supported (Uspenskii 1970) and opposed (Andreev et al. 1958) this concept. Though microbiological procedures found wide application in the processing of normal alkanes and in producing special lubricating oils, serious experimental information was required in order to substantiate the concept that microorganisms are capable of altering crudes in reservoirs.

Previous studies proved the actual possibility of biological oxidation of crudes both in aerobic and anaerobic environments (Uspenskii 1970). It was established that biological alteration results in the gradual transformation of paraffinic crudes into naphthenic ones due to selective consumption of methane aliphatic

Table 61. Compositional alteration of saturated hydrocarbons in the 200° – 430 °C fraction by biodegradation

Experiment number (time)	Crude type	Loss in weight (%)	Fractional composition (%)		Alkanes (%)			Naphthenic passport (%)					C_i	N_b	Pristane/ phytane
			Alkanes	Naphthenes	Normal	Iso	Isopr.	Monocyclic	Bicyclic	Tricyclic	Tetracyclic	Pentacyclic			
Dunga															
Control	A[1]	–	61.6	38.4	53.5	38.4	8.1	39.4	25.6	15.3	11.7	8.0	0.1	21.1	1.85
Experiment 1 (8 months)	A[1]	13.0	46.2	53.8	26.0	63.2	10.8	38.6	26.3	15.8	11.2	8.1	0.8	4.8	1.85
Kenkiyak[a]															
Control	A[1]	–	41.6	58.4	29.2	64.8	6.0	28.6	26.2	16.8	16.3	12.1	9.6	8.4	1.26
Experiment 2 (3 months)	A[2]	21.0	31.6	68.4	2.5	89.6	7.9	28.0	26.7	17.4	16.0	11.9	12.5	0.6	1.43
Kara-Tyube															
Control	A[1]	–	29.1	70.9	24.8	64.2	11.0	25.1	29.0	21.3	16.2	8.4	0.8	4.1	1.67
Experiment 3 (39 months)	B[1]	31.4	6.1	93.9	–	100	–	10.3	24.6	28.3	24.0	12.8	–	–	–
Kenkiyak[b]															
Control	B[1]	–	13.8	86.2	–	100	–	19.7	25.7	22.2	20.1	12.3	–	–	–
Experiment 4 (5 months)	B[1]	16.7	10.0	90.0	–	100	–	14.4	24.1	22.8	23.7	15.0	–	–	–

[a] Triassic
[b] Jurassic

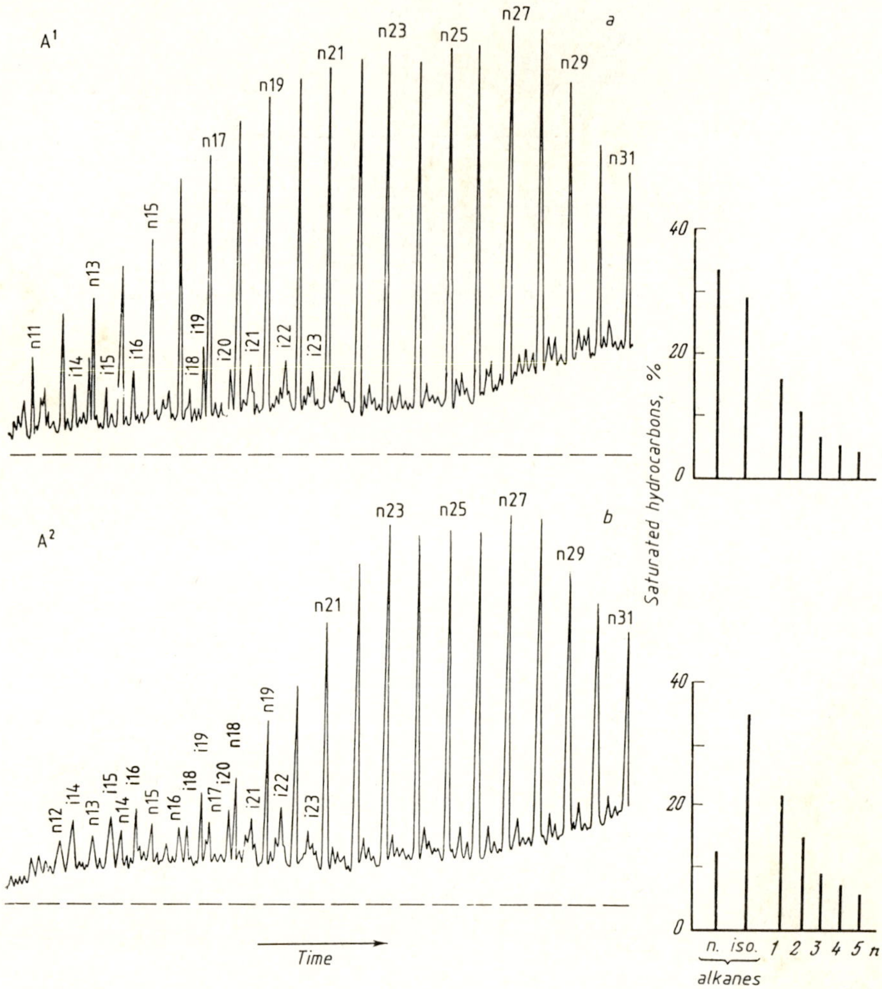

Fig. 82 a, b. Compositional alterations in Dunga crude by biological degradation. **a** Original crude (A¹); **b** initial biodegradation stage

hydrocarbons by microorganisms. Thus, biodegradation leads to an increase in petroleum density and resinous compounds. Residual accumulations increase relative concentrations of naphthenic and aromatic hydrocarbons.

However, detailed chemical analysis of biodegradation processes could only be done on the molecular level, which was originally done be Bailey et al. (1973a). Since then additional studies have been conducted to convincingly prove the potential and limits of microbiological alteration of crude oils in reservoirs (Tissot and Welte 1981; Milner et al. 1977; Reed 1977; Rubinstein et al. 1977; Bailey et al. 1973; Jobson et al. 1979; Philippi 1977; Connan et al. 1979). Recently, an extensive analysis on the biological alteration of crudes in reservoirs was published by Connan (1984). Analogous research was undertaken in relation to

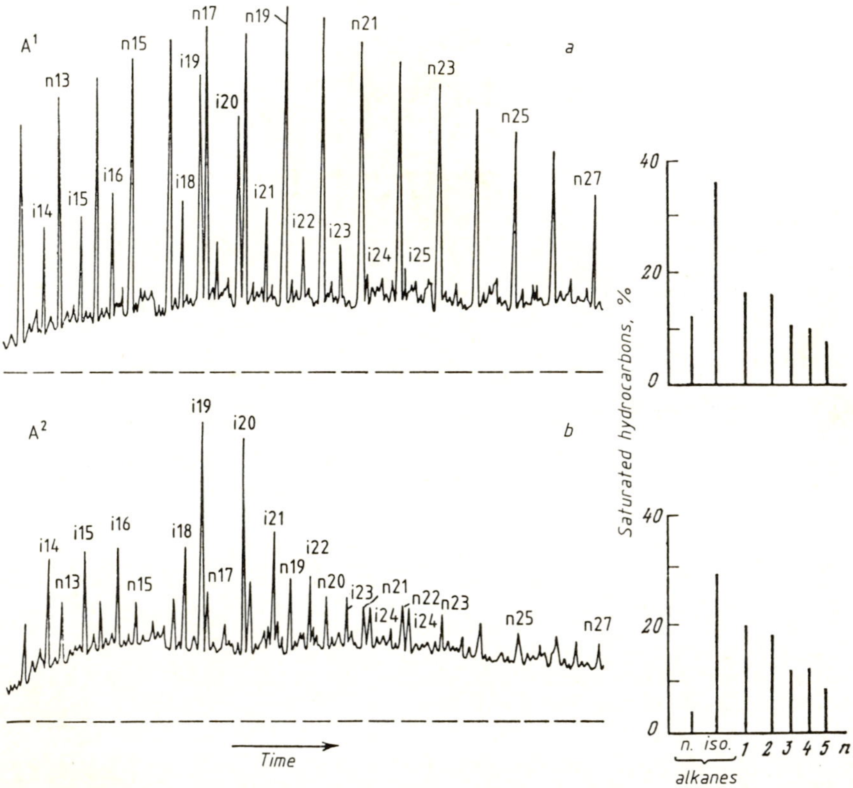

Fig. 83 a, b. Compositional alterations in the Kenkiyak crude (Triassic) by biological degradation. **a** Original crude (A^1); **b** biodegradation product

Soviet crudes in a joint effort by the laboratory of petroleum geochemistry (IGiRGI) and the microbiological laboratory (VNIGRI) (Arefyev et al. 1978; Zabrodina et al. 1981). Results of this research will be reviewed in detail.

Since there were reasons to believe that A^1 crudes belong to those of primary generation (i.e. unaltered by biodegradation), they were mostly used as initial samples. For comparison, biodegradation of a type B^1 crude (Kenkiyak and Kara-Tube reservoirs) was also analyzed. Appropriate results are presented in Table 61 and Figs. 82–84. Compositional analysis was conducted in accordance to the scheme described in Chapter 1.

Results of Experiments on Microbial Alteration of Crude Oils

Research studies have demonstrated that, the microbial alteration of petroleums proceeds fairly rapidly. Depending on the original crude's chemical type and the length of the experiment (from 3 to 39 months), microorganisms destroyed from 13% to 31% of the experimental oil by weight (see Table 31) (Arefyev et al. 1978).

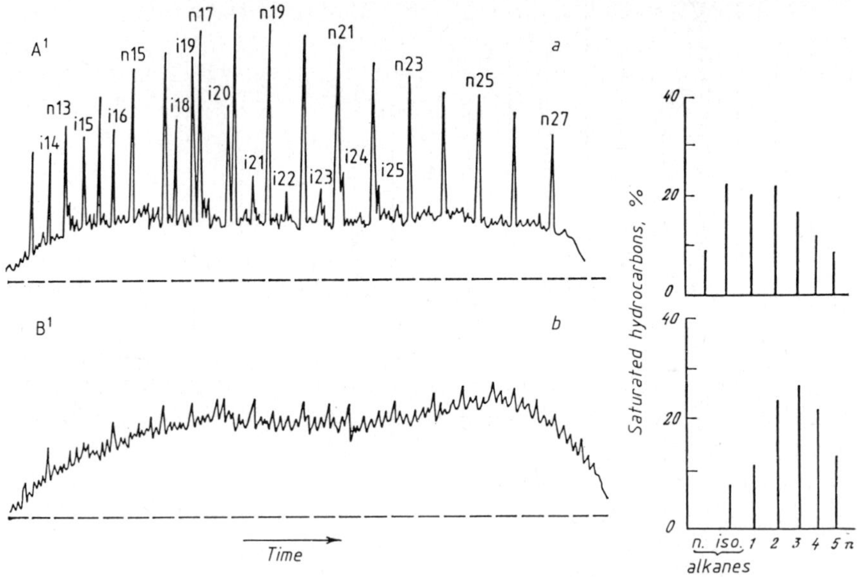

Fig. 84a, b. Compositional alterations in the Kara-Tube crude by biological degradation. **a** Original crude (A^1); **b** biodegradation product

Oxidation of oils was accompanied by changes in their physical and chemical properties and group and individual hydrocarbon composition. High-boiling fractions usually increased, probably due to enrichment of residual resins and the formation of various, new carboxylic acids (Arkhangelskaya et al. 1978).

A better illustration of compositional transformations occurring in petroleum hydrocarbons may be obtained through GC and MS. Thus, during initial stages in the oxidation of the Dunga crude (experiment 1), selective elimination of normal $C_{12}-C_{20}$ alkanes (Fig. 82b) occurred. However, the pristane/phytane ratio, as well as the naphthenic fingerprint and crude type (A^1) remain unaltered. This crude type, with a characteristic "dip" in the normal $C_{12}-C_{18}$ alkane concentration, signifying the initial stages in biological oxidation, is found in nature. Further bacterial oxidation brings about considerable alteration of the crudes' chemical composition, resulting in the alteration of their chemical types.

Thus, the total concentration of normal alkanes decreased from 29.2% to 2.5% during oxidation of the Kenkinyak crude (experiment 2). Comparison of chromatograms of the original crude (control) and oxidation products demonstrated that normal alkane peaks become much smaller, while isoprenoid peaks remain unaltered. The naphthenic fingerprint remained intact. As a result, the A^1-type crude turned into an A^2 crude and corresponded to natural crudes of this chemical type in all parameters.

Even more severe oxidation (of the Kara-Tyube crude) occurred in experiment 3, which lasted 39 months. The alkane content decreased by 23% in this instance. As can be seen in the chromatograms, not only normal, but isoprenoid alkanes as well, were affected by oxidation. Other isoalkanes, eluting in sharp peaks, also disappeared due to oxidation. Accordingly, by the end of the experiment, the crude's chromatogram appeared as a solid background which corresponded already to type B^1 crudes. Certain alterations were also registered among naphthenes. In particular, initial degradation affected the naphthenic fingerprint by decreasing the amount of monocyclic hydrocarbons, which may be explained by the oxidation of long aliphatic chains on these hydrocarbons.

Consequently, prolonged oxidation of type A^1 crudes results in the formation of B^1-crude oils widespread in nature.

B^1-crude proper (experiment 4) resists bacterial oxidation, since structures which may easily be oxidized had already been used up by microorganisms in the reservoirs. To conclude, experimental data demonstrate that biodegradation causes severe transformations in the composition of saturated high-boiling petroleum hydrocarbons. However, in order to graphically illustrate the step-by-step character of the biochemical alteration of crudes in accordance to the scheme $A^1 \rightarrow A^2 \rightarrow B^2 \rightarrow B^1$, it is necessary to learn the order of hydrocarbon oxidation in the process of biodegradation. (Incidentally, in our opinion, the term "biodegradation" fails to reflect the essence of the process. We are dealing here with chemical, or, to be exact, the biochemical alteration of crude oils, which changes their qualities, but not always negatively. For example, this process may lead to the formation of non-paraffinic crudes used in producing excellent lubricating oils. etc.)

Presently, it has been recognized that hydrocarbons of any homologous series may be destroyed/degraded under appropriate conditions, and the order of degradation depends largerly on the culture of microorganisms. In order to clarify the sequence of the involvement of hydrocarbons in the process of biodegradation, we conducted a special, year-long analysis devoted to the dynamic alteration of the chemical composition of crude (Zasrodina et al. 1981). In the course of the experiment, successive analyses of petroleum samples were conducted. Samples taken by the end of the 2nd, 5th, 9th, and 18th month, were studied more thoroughly.

An A^1 Starogroznenskoe crude (Tertiary reservoir) was the starting material in the experiment. It was selected because there are crudes of all four chemical types in that particular oil field, as well as an A^1 crude with traces of early biodegradation (see Fig. 86). Moreover, their geological and geochemical characteristics have been fairly well studied (Vassoyevitch 1958). Experimental data and chromatograms representing the progress in compositional transformations of the Starogroznenskoe crude under bacterial oxidation are given in Table 62 and Fig. 85.

As could be observed in the experiment, the microbial impact on the crude's hydrocarbon composition was of systematic character. As usual, initial (2 months) oxidation involves normal $C_{12}-C_{18}$ alkanes. Further bacterial action continued decreasing the content of these alkanes, which involved an extended carbon number range up to C_{34} in oxidation, which may easily be observed in the chromatogram (Fig. 85b). By the end of the 5th month microorganisms had used up over 90% of normal alkanes of the original crude. The overall concentration of branched alkanes was also slightly diminished at that stage. According to chromatograms, this was caused by the involvement of monomethyl-substituted structures (iso- and anteisoalkanes) into the oxidation process. The relative con-

Table 62. Compositional alteration of saturated hydrocarbons in the 200° – 430 °C fraction of the Starogroznenskoe crude in the process of biodegradation (laboratory simulation)

Experiment Number (Time)	Crude type	Weight loss	Group composition (%)		Alkanes (%)			Naphthenic fingerprint (%)					C_i	N_b	Pristane/ Phytane
			Alkanes	Naphthenes	Normal	Isopr.	Iso.	Monocyclic	Bicyclic	Tricyclic	Tetracyclic	Pentacyclic			
Original crude[a]	A^1	–	56.7	43.3	35.8	7.7	56.5	27.0	30.2	18.4	15.3	9.1	0.55	20.8	1.61
Experiment 1 (2 months)	A^{1-2}	21	46.8	53.2	16.1	11.5	72.4	28.3	29.5	17.2	16.5	8.5	1.7	7.5	1.56
Experiment 2 (5 months)	A^2	39	39.3	60.7	8.3	13.0	78.7	29.5	29.7	16.9	16.5	7.4	8.2	1.3	1.56
Experiment 3 (9 months)	B^2	43	34.6	65.4	None	11.5	88.5	28.3	29.3	18.5	15.7	8.2	–	–	1.56
Experiment 4 (18 months)	B^1	60	20.4	79.6	None	None	100	25.5	29.4	19.5	17.5	8.1	–	–	None

[a] Depth 1320 m.

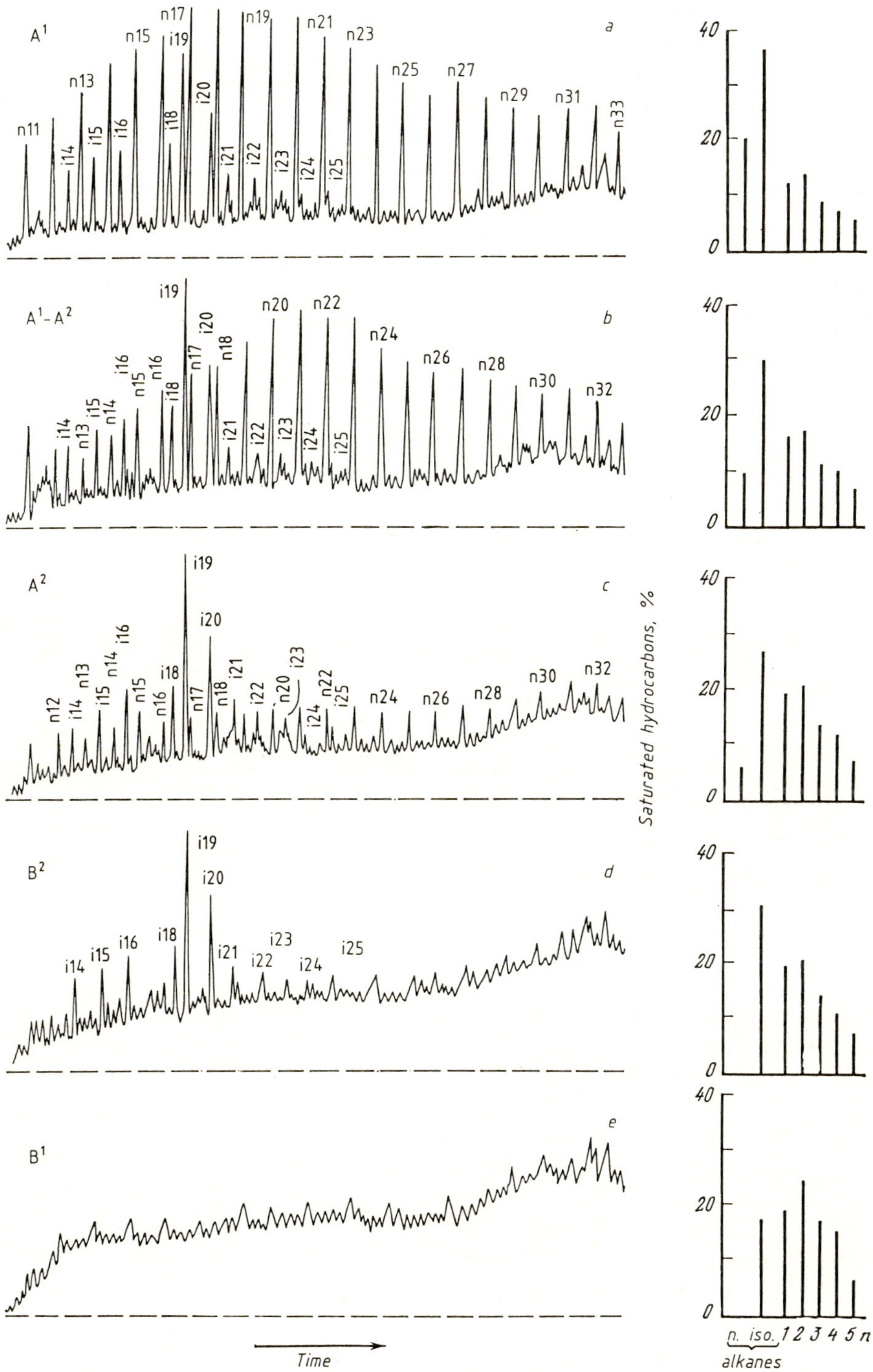

Fig. 85 a – e. Dynamics of compositional alterations in Starogroznenskoe crude by biological degradations. **a** Original curde (A[1]); **b, c, d, e** different stages of biodegradation (2, 5, 9 and 18 months)

tent of isoprenoids was constantly increasing due to accumulation in the residual oil. Since isoprenoids had not yet been subjected to metabolism, neither their concentrational distribution, nor pristane/phytane ratio were altered. However, the value of C_i increased considerably, thus an type A^2 crude emerged.

Further oxidation of the crude under analysis brought about total destruction of normal alkanes. In 9 months these hydrocarbons were completely oxidied. By this time micoorganisms had used up some of the isoprenoids, although the pristane/phytane ratio remained unchanged. The naphthenic fingerprint was also unaltered. This was type B^2 petroleum. Finally, by the end of the experiment (18 months) normal and isoprenoid alkanes as well as a major part (75% of the original content) of isoalkanes belonging to other types were subjected to complete destruction.

It should be emphasized that in this case, as well as in previous tests, naphthenic hydrocarbons were the least affected by biological alterations. The naphthenic fingerprint (see Fig. 85 and Table 62), in the first three stages of biological degradation, remained practically unchanged. It was only in the last stage, when almost all alkanes of various structures had been degraded by microorganisms, that the oxidation of monocyclic hydrocarbons began.

It may be observed that hydrocarbons, eluting in sharp peaks over the background, are the first to be destroyed by biodegradation, followed by background hydrocarbons.

Depending on the ease of oxidation, saturated hydrocarbons my be arranged in the following sequence: normal alkanes; iso- and anteisoalkanes; isoprenoid alkanes; monocyclic naphthenes; other hydrocarbons. Due to the selective nature of biodegradation, consisting in the gradual consumption of aliphatic hydrocarbons by microorganisms, petroleum was enriched in cyclic hydrocarbons, and due to the intensity and duration of this process, there was a stepwise alteration of the crude's chemical type according to the scheme:

$$A^1 \rightarrow A^2 \rightarrow B^2 \rightarrow B^1.$$

Formation of all chemical types of crudes (A^2, B^2, B^1) from the A^1-type under microbial influence represents convincing experimental proof of a direct genetic link between oils of different chemical types. Similar stepwise alterations in a paraffinic crude oil under experimental conditions were convincingly demonstrated by Connan et al. (1979).

Naturally enough, there are less favorable conditions for microorganism activities in natural oil reservoirs. However, the reduced intensity of biological processes may be compensated by a time factor. An especially important role may be played in this case by subsurface waters carrying oxygen.

Another no less important factor determining the degree, if not the very possibility of biological transformations in petroleum, is represented by paleotemperatures (and, possibly, existing temperatures) in oil reservoirs. As mentioned in Chapter 1, maximal occurrence of paraffinic crudes (A^1-type) is registered at 2000-m depths which corresponds to an average formation/subsurface temperature of 90 °C. At the same time, maximal concentrations of A^2, B^2 and B^1 crudes (i.e. those altered by biodegradation) occur at much shallower levels (approximately 1200 m), which correspond to an average temperature of

40 °C. These observations are in agreement with the data of Philippi (1977), who found an obious link between crude properties and their reservoir temperatures using the sample crudes from different oil- and gas-bearing basins. In all instances, the border line between paraffinic and biologically transformed crudes lies within 60 °–70 °C, which is evidently predicated by the limits of intense microbial activity estimated to be within 50 °–60 °C.

Compositional Alteration of Crudes in Nature (Comparison of Experimental Data)

The assumption of the influence of biodegradation on crude oil composition and properties, but mostly on the formation of various type crudes, may be substantiated by factual data on different oils. This becomes especially obvious in comparing crudes of the same age located in different geochemical zones. Examples of such multilayer oil fields are numerous: Starogroznenskoe, Kotur-Tepe, Dagadjik, Samotlor, Novoportvoskoe, Djyer, Western Tebuk, Duvanyi-more and others. Appropriate data on these crudes are tabulated in Table 2a (see Chap. 1). Noteworthy examples of biodegradation from foreign sources are presented in the works of Deroo, Seifert and many other analysts (see Tissot and Welte 1981; Bailey et al. 1973b; Seifert and Moldowan 1979).

These publications demonstrate that at shallower depth, a consistent alteration of the crudes' chemical type in the A^1-B^1 direction occurs.

Convincing data on petroleum biodegradation were gathered in the study of the multilayer Starogroznenskoe deposit (Zabrodina et al. 1981), where geological information had long proved the hypergenetic transformation of crude oil properties (Vassoyevitch 1958). Five crudes of various chemical types were selected for comparing these properties. One of these crudes (A^1) was used as the starting material in laboratory investigations of biodegradation (see above). Results of the analysis of these crudes are presented in Table 63 and Fig. 86. Appropriate information demonstrates that this deposit contains crude types (A^2, B^2, B^1) which may be produced through biodegradation, including even its initial stages (mixed A^1-A^2 type). Hence, such chemical types as A^2, B^2, and B^1 are but separate stages in the biodegradation of A^1-type crudes, which are primary crudes.

An especially informative relationship between changes in chemical properties within the A^1-B^1 sequence and gradual reduction of alkane concentrations exists: from 56.7% to 15.5%. In contrast, as in laboratory simulation, the naphthenic fingerprint of crudes remains practically unaltered under natural conditions. At especially severe stages of oxidation (B^1), a certain alteration in the concentrations of monocyclic naphthenes occurs.

The general outline of alterations in crude properties in nature and laboratory experiments becomes quite evident in comparing the data in Table 63 and Fig. 86 with those in Table 62 and Fig. 85.

Such a peculiar character of the compositional alterations of high-boiling hydrocarbons can only be explained through bacterial influence. In particular, no thermal or thermocatalytic influence can result in the formation of large quantities of isoprenoids, pristane and phytane especially in a shift from B^1 crudes to

Table 63. Relative distribution of saturated hydrocarbons in the 200° – 430 °C fraction in Starogroznenskoe crudes (natural conditions)

Depth (m)	Crude type	Fractional composition (%)		Alkanes (%)			Naphthenic fingerprint (%)					C_i	N_b	Pristane/ Phytane
		Alkanes	Naphthenes	Norm.	Isopr.	Iso.	Monocyclic	Bicyclic	Tricyclic	Tetracyclic	Pentacyclic			
1320 m	A^1	56.7	43.3	35.8	7.7	56.5	27.0	30.2	18.4	15.3	9.1	0.55	20.8	1.61
609 – 622 m	A^{1-2}	48.6	51.4	12.2	16.4	71.4	25.1	27.6	19.3	18.3	9.7	2.6	3.2	1.42
	A^2	37.5	62.5	12.5	18.0	69.5	28.5	28.2	20.0	15.7	7.6	5.6	2.2	1.51
720 – 756 m	B^2	35.6	64.4	–	13.5	86.5	28.9	30.3	19.0	14.8	7.0	–	–	1.64
330 m	B^1	15.5	84.5	–	None	100	22.7	28.3	19.9	19.1	10.0	–	–	None

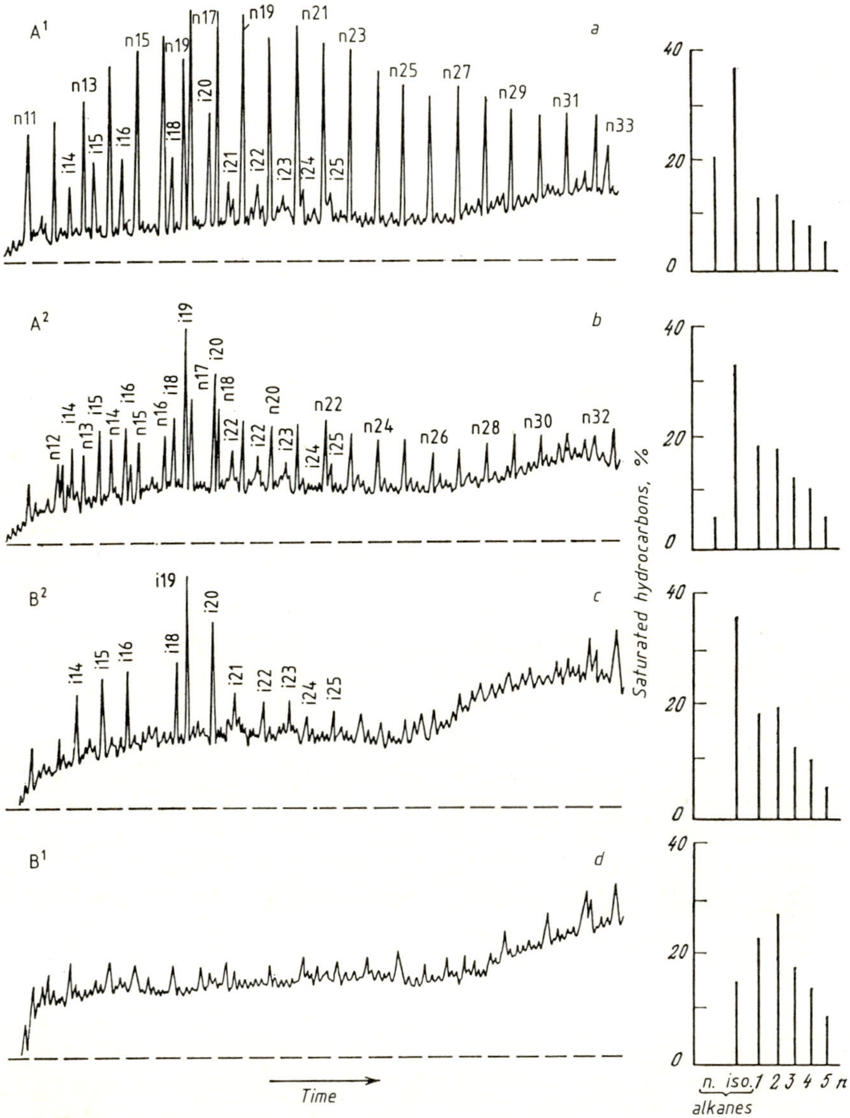

Fig. 86a–d. Chromatograms of different types of crudes in the Starogroznenskoe deposit. **a** type A^1; **b** type A^2; **c** type B^2; **d** type B^1

those of B^2 and A^2 types. And finally, nothing, except the bacterial impact, can explain the existence of crudes oils practically devoid of normal alkanes, especially since the entire column of nearby sediments is usually literally saturated with normal alkanes in the dispersed organic material.

With respect to the aforementioned differences in hydrocarbon compositions of various Starogroznenskoe crudes, important similar features may also be traced. All crudes (see Table 63) are characterized by constant pristane/phytane ratios (1.43–1.61), and similar isoprenoid distribution (naturally at the A^1-B^2

stages). Maximal concentrations of normal alkanes (in A^1 and A^2 crudes) are also identical. As mentioned, similarities also extend to their naphthenic fingerprints.

Compositional resemblences between crudes of the Starogroznenskoe deposit prompt the conclusion of their genetic identity, i.e. that they were all produced from one another, namely from A^1-type oil accumulated in lower horizons. (Additional data on genetic uniformity of Starogroznenskoe crudes are found in a later section devoted to the pyrolysis of asphaltenes). Comparison of gradational transformations in the hydrocarbon composition of these crudes (from deep to shallow), with similar modifications in the laboratory simulation of biodegradation gives every reason to accept the irreversible nature of all the alterations.

As a result of the investigations undertaken, it becomes perfectly clear that biodegradation can be regarded as the main geochemical process leading to alteration of the chemical types of crude oils, thus providing their considerable chemical variety in nature.

In summarizing these results, the conclusion may be drawn that, on the whole, the composition of crudes is influenced by three main factors: compositional characteristics of the initial organic matter and sedimentation conditions (genetic factor), catagenetic transformations and biodegradation. These factors vary in significance even within a single petroleum basin.

The original organic matter has an important influence on the composition of the initial crude and stipulates characteristic genetic features of crudes in a given sedimentary basin. These features become especially apparent in the quantitative distribution of biological markers (problems of genetic classification are treated later). Undoubtedly, thermal (thermocatalytic) processes affect the composition of natural crude oils. However, as research has demonstrated, this factor alone is unable to alter the chemical type of a crude, although it does cause appreciable variations in the content of light hydrocarbons, increase of normal alkane concentrations and decrease of isoprenoids, especially pristane and phytane.

Basic alterations in the chemical type of crudes occur in the zone of hypergenesis due to biodegradation [see also the publication of Gabrielyan (1981)].

Since biodegradation is selective in nature, successive elimination of normal, then branched and isoprenoid alkanes occurs, which is clearly discernable in chromatograms. This peculiarity helped establish the separate chemical types of crudes, corresponding to different stages in their biodegradation, as mentioned before.

On the "Age" and "Maturity" of Crudes. Severe transformations in the chemical types of crudes under the influence of biodegradation significantly alter existing perceptions as to "young" or "old" oils. In fact, even in earlier times these concepts had no clear definitions. In effect, it is next to impossible to establish the true age of crudes, since this often differs from the geological age of the reservoir rocks. Furthermore there are still no reliable procedures of attributing crudes of specific composition to particular geological epochs.

There are two basic ways of establishing the age of crude. Chemically, we can identify the specific geological period to which original biogenic compounds belong and relate this to the time when source rocks (kerogen) were deposited. From the geological standpoint, the time of petroleum generation, i.e. the catagenesis of kerogen, expulsion of crudes and further migration into traps, is of greater importance. There may be a considerable time lag between the first and the second stage.

Therefore, while speaking of "young" and "old" crudes, we usually mean catagenetically unmodified or, as the case may be, catagenetically modified oils (according to A. Dobryanski).

Since in the process of biodegradation naphthenic crudes (category B) are produced from paraffinic crudes (category A), the entire situation is reversed. Hence, in our opinion, the term "young" is unfortunate. Evidently, a concept should be introduced to help differentiate between oils of various stages of generation from a given source. The composition of these crudes may apparently alter since at early stages of generation, crudes rich in high-molecular-weight biomarkers emerge. These crudes may be called young (as opposed to those which appear from the same source at later stages of kerogen transformations). These processes occur similarly in source rocks belonging to different geological epochs.

A few observations are warranted on "immature" (i.e. catagenetically immature) crudes. This therm may be found in scientific literature. And again it is rather unfortunate. So far no catagenetically immature crudes have been discovered. Strictly speaking, these should not exist at all, since hydrocarbons, isolated in a separate liquid phase, must undergo a cycle (phase) of petroleum formation, i.e. mature catagenetically. The latter process is obligatory in order to have hydrocarbons separated from kerogen, i.e. it is the very essence of petroleum formation. Judging by their chemical composition, immature crude oils are most probably those at early stages of generation, however, they lack factors characteristic of catagenetically immature organic matter. Although catagenetic transformations do occur in oil fields, they invariably affect catagenetically mature crudes. Hence, we may only speak of catagenetic alterations (transformations) of crudes, and not their catagenetic maturation.

Microbial Oxidation of Steranes and Hopanes. Since it is mostly biological markers that disappear in biodegradation, the invariability of sterane and hopane compositions under these conditions deserve special treatment. So far there is no unanimity on the subject in the literature, although it is clear that these hydrocarbons are rather resistant to biological influences. Thus, it was established (Rubinstein et al. 1977; Connan et al. 1979; Zabrodina et al. 1981) that steranes and hopanes remain unaltered under biological impact, and their molecular mass distribution in crudes of various chemical type may server as an additional criterion of the genetic uniformity of particular crudes. Seifert insisted (Seifert and Moldowan 1979) that the geochemical analysis of natural oils reveals the susceptibility of steranes and hopanes to biological influences. It was proved in one of the reports at the Tenth International Meeting on Organic Geochemistry in 1981 that highly active microorganisms exist that also destroy steranes and hopanes (Goodwin et al. 1981).

However, in particularly harsh environments it is evidently possible that any petroleum hydrocarbon can be completely oxidized. In any case, the disappearance of steranes and hopanes is rare and may occur only at extreme stages of biodegradation (B^1 stages). Nevertheless, sterane and hopane hydrocarbons are present in the majority of category B crudes. Moreover, their absence in certain cases may be explained not by severe biodegradation, but by the absence (or small concentrations) of these hydrocarbons in the original crude of the area. However, there are obvious changes in hopane and sterane structures typical of heavily oxidized crudes, which apparently appeared at later stages of biodegradation. These changes may include the formation of 10(25)- and 18-normethyl hopanes, as well as relatively high concentrations of rearranged steranes and $C_{19}-C_{26}$ tricyclic hydrocarbons of the perhydropenanthrene series (Bailey et al. 1973b; Connan et al. 1979; Kayukova et al. 1981). The following order of hydrocarbon oxidation by microorganisms has been proposed (McKirdy et al. 1981): normal alkanes > isoprenoid alkanes > regular steranes (20*R* > 20*S*) > rearranged steranes > hopanes.

However, such advanced stages of oxidation are uncommon for most crudes, therefore the distribution of steranes and hopanes is a reliable genetic indicator, which is identical in any type of crude in a given region (Hufnagel et al. 1979). Additional information on the distribution of steranes and hopanes in biodegraded crude oils can be acquired in the thermolysis of asphaltenes (see below).

General Genetic Scheme of the Formation of Different Chemical Type Crudes. On the basis of the results of thermolytic and biodegradation research, the following

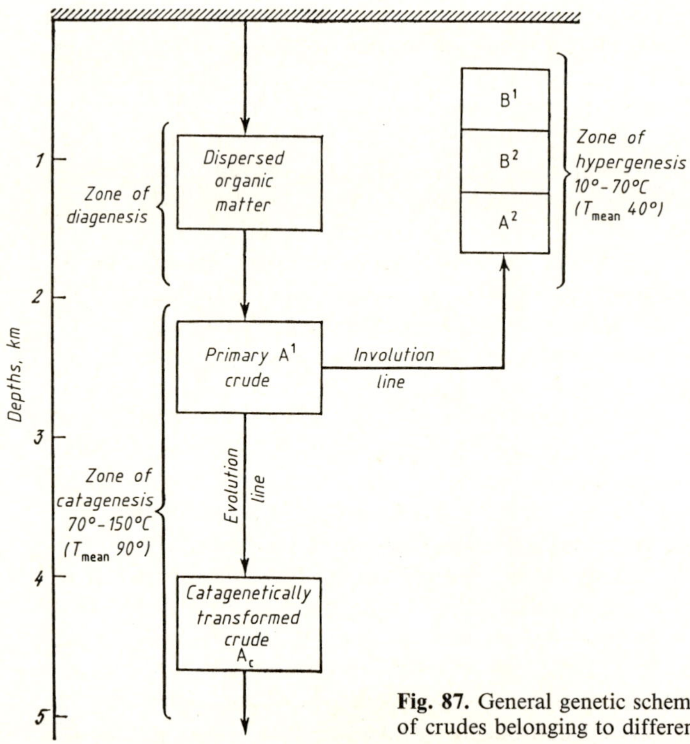

Fig. 87. General genetic scheme of the natural formation of crudes belonging to different chemical types in nature

general, genetic scheme of the formation of crudes of different chemical types in nature may be suggested (Fig. 87).

Type A^1 crudes are the primary ones, generated in the zone of kerogen catagenesis, which yields large amounts of normal and isoprenoid alkanes. Their group and fractional composition varies to a certain degree, depending on the composition of the initial biomass and geochemical conditions of its transformation. Further alteration of primary crudes proceeds along two lines.

1. The evolution line which depends on the residence of crudes in a zone of intense catagenesis, accompanied by their fractional alteration, i.e. increase of light and decrease of heavy components (through thermal transformations). The chemical composition and modification remain within the limits characteristic of A^1 crudes. The evolutionary process is successfully described by the well-known methanization scheme of Dobryanskii.

2. The involution line based on the biodegradation of A^1 crudes after their entry into a zone of hypergenesis (as a result of migration or tectonic uplift). It results in successive alterations of chemical types in accordance to the $A^1 \rightarrow A^2 \rightarrow B^2 \rightarrow B^1$ scheme, as well as in the crudes becoming heavier due to enrichment of resins and asphaltenes.

Methane and graphite are the end products in the evolution line, while carbon dioxide and water are the end products in the involution line.

It should be reemphasized that crudes of different chemical types may be genetically related.

Asphaltenes and the Geochemical History of Crudes

The above-mentioned material on microbiological oxidation of crudes required additional confirmation in that B crudes used to be A^1 crudes at one time, i.e. that they contained n-alkanes, but lost their "chemical identity" due to biodegradation. Such information was obtained in the study of the products of pyrolysis of asphaltenes (Arefyev et al. 1980b; Rubinstein et al. 1979; Kayukova et al. 1982). It was established that asphaltenes (remnants of kerogen, which failed to turn into petroleum) contain information on all structure types characteristic of a given crude appearing in its genesis. This was especially valuable when it was proven that the hydrocarbon part of the asphaltenes is resistant to microbiological oxidation (Rubinstein et al. 1979; Kayukova et al. 1982). Upon heating (300 °C) for a few hours, asphaltenes generate hydrocarbons (~ 20%), gas and pyrobitumen, insoluble in regular solvents. Emerging hydrocarbons may be analyzed by usual means (GC and MS).

In analyzing hydrocarbons belonging to asphaltenes of B-type crude, the latter's initial chemical composition may be established, including such important geochemical indicators as distribution of normal alkanes and isoprenoids, pristane/phytane ratio, C_i and relative distribution of steranes and hopanes (Kayukova et al. 1982; Arefyev et al. 1980b).

Figure 88 describes the results of pyrolysis of asphaltenes, isolated in Starogroznenskoe crude. The analysis is centered on asphaltenes of non-paraffinic crudes (B^2- and B^1-types). Data on pyrolysis of asphaltenes belonging to an

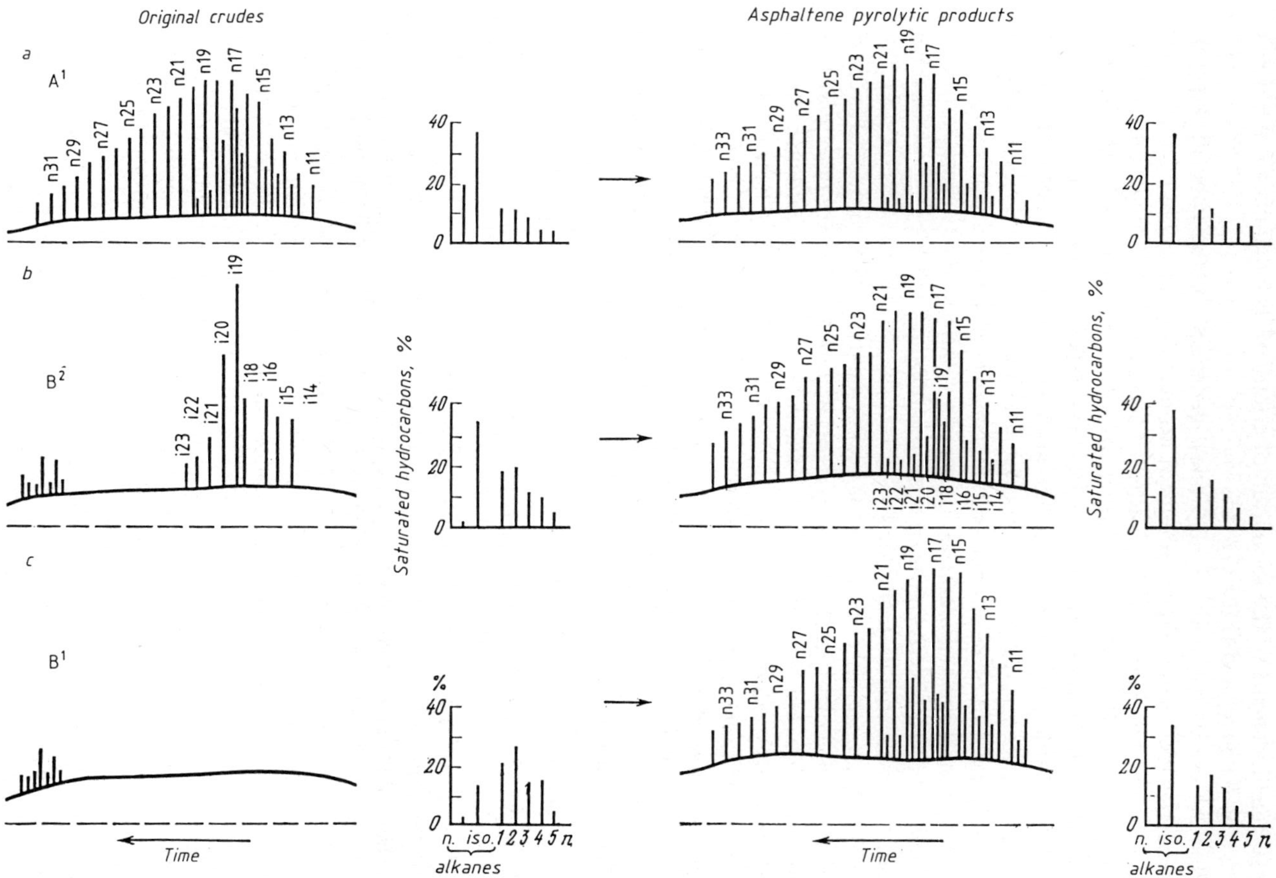

Fig. 88 a–c. Normalized chromatograms of various crudes from the Starogroznenskoe deposit and the asphaltene pyrolytic products isolated from them. Distribution of saturated hydrocarbons by structure types are given to the *right*. Normal and isoprenoid alkane peaks are identified. Here and in Fig. 89 n denotes the number of carbon atoms in alkanes

A^1 natural crude, and those isolated in B^2 crude through laboratory biodegradation (of A^1 crude), are presented for comparison. It should be emphasized that research results demonstrated the complete compositional identity of B^2 asphaltenes of different origin (natural or laboratory).

Data in Fig. 88 demonstrate that pyrolysis of B-crude asphaltenes allows one to realistically reconstruct the "chemical identity" of original petroleum unaffected by biodegradation.[3] This reconstruction becomes even more clear in comparing normalized chromatograms of B^1 and B^2 pyrolytic products with that of the original natural crude oil of A^1-type (upper left). The similarity of the samples' naphthenic fingerprints should also be underlined.

It is noteworthy that despite pyrolysis, the concentration of high-molecular-weight normal alkanes in crudes obtained from asphaltenes is not lower than that in natural A^1 crudes. It may be suggested that the relative value of the content of paraffinic chains in asphaltenes of various crudes (as components least affected by biodegradation) may be used as an additional criterion in establishing a facial (genetic) type of crude oils. For example, asphaltenes isolated from ancient crudes of Eastern Siberia of early marine origin comprised no paraffinic chains longer than C_{25}. At the same time, asphaltenes of Mesozoic crudes from Western Siberia included paraffinic chains of up to C_{40}, which indicated the presence of higher plant residues in the original organic matter.

Generally, all results demonstrate that asphaltenes represent that part of petroleum which is least amenable to bacterial oxidation. Hence, the preserved information on the composition of saturated hydrocarbons is fairly extensive, allowing one to make assured determination of the chemical type of the original crude altered by biodegradation.

Figure 89 exemplifies the compositional reconstructions of two crudes (the Russkoe and Kursai deposits), characterized by intensive biodegradation. As opposed to Starogroznenskoe crudes, they are comprised of appreciable quantities of regular steranes and hopanes.

As proved the pyrolysis of asphaltenes, these are, in effect, strongly biodegraded paraffinic crudes (A^1-type). Their initial composition is given in the chromatograms on the left side of Fig. 89. It is noteworthy that the pristane/phytane ratio in the reconstructed Russkoe crude corresponds to the commonly observed ratio of these hydrocarbons in other Mesozoic crudes of Western Siberia. Heavy biodegradation of two latter crudes should be reemphasized. The amount of alkanes (normal and iso-) generated by asphaltene pyrolysis is slightly lower than that from the asphaltenes of Starogroznenskoe crudes (40% and 60%, respectively), which may apparently be explained by partial oxidation of paraffinic chains, even in asphaltenes. It is also noteworthy that the Kursai crude belongs to rare B^1 oils of deep occurrence (over 4 km) which may be regarded as a paleobiodegraded one.

Thus, the results presented demonstrate that even asphaltenes of heavily biodegradated crudes retain important information on the original petroleum

---

[3] With regards to graphic comparison, GC results of original crudes and their asphaltene pyrolytic products are given as so-called normal distributions, reproducing exclusively the normal and isoprenoid alkane peak heights over the background.

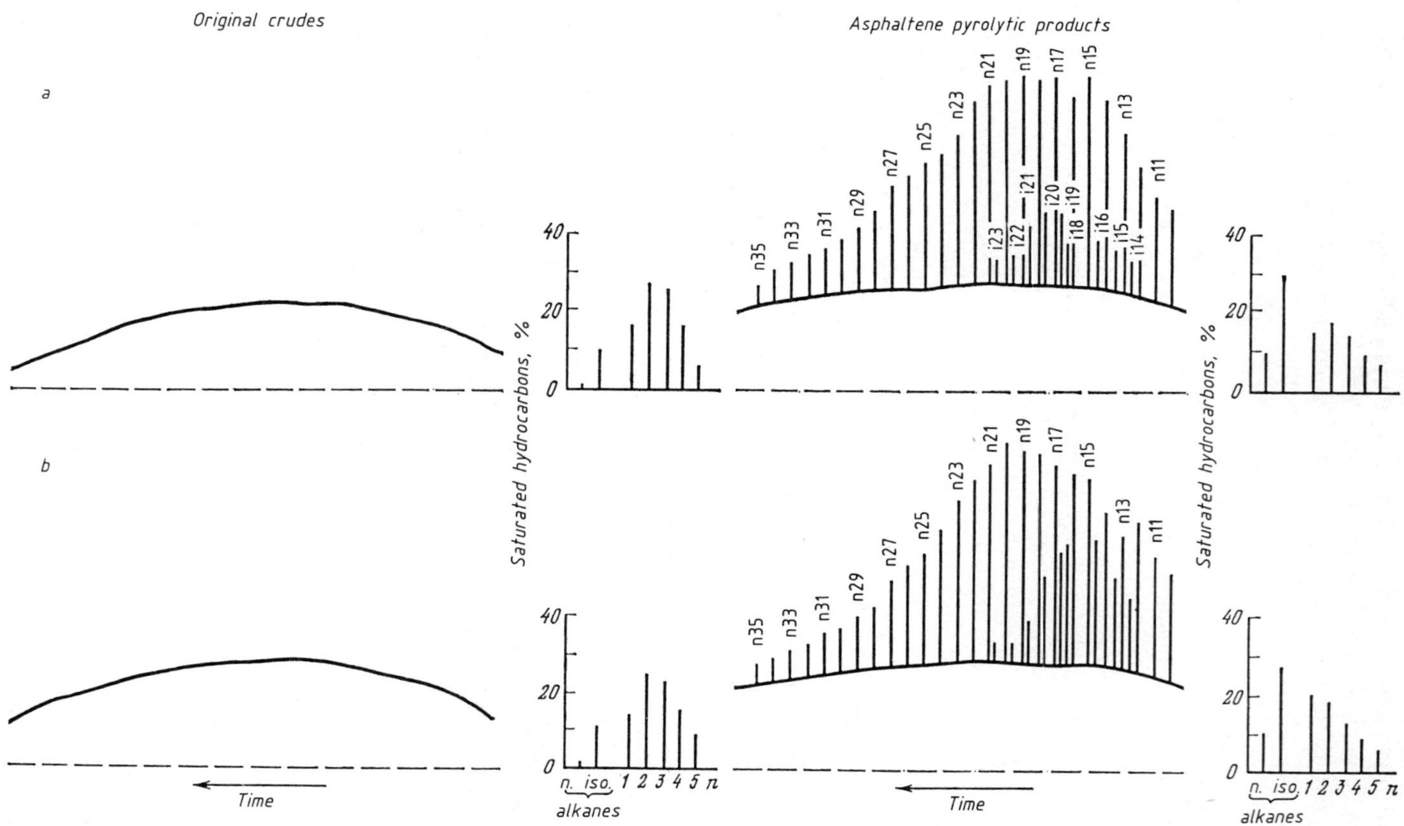

Fig. 89 a, b. Normalized chromatograms of crudes and the asphaltene pyrolytic products isolated from them. **a** Russkoye deposit (B[1]); **b** Kursai deposit (B[1])

Table 64. Relative distribution of 17 αH-hopanes in a mixture of asphaltene pyrolytic products from Baylinskoe and Mordovo-Karmalskoe crudes (%)

Object of analysis	Hopanes						Adiantane Hopane
	C_{27}	C_{29}	C_{30}	C_{31} [a]	C_{32} [a]	C_{33} [a]	
Bavlinskoe crude (A[1])	7.5	16.7	16.8	21.4	22.1	15.6	0.99
Asphaltene pyrolytic products	9.8	19.8	14.0	22.2	21.4	12.8	1.41
Mordovo-Karmalskoe crude (B[2])	7.4	22.3	16.6	25.4	16.6	9.7	1.34
Asphaltene pyrolytic products	12.8	21.5	30.1	16.1	13.9	7.6	0.71

[a] Sum of 22*R* and 22*S* epimers.

type. Besides the establishment of group-type composition, this information helps specify the distribution of individual biological markers typical of original crudes. It is of interest that, as a rule, asphaltenes retain all of the genetic features characteristic of crude hydrocarbon composition. Of note in this connection is the presence of 12- and 13-methylalkanes in various crudes and in asphaltene pyrolytic products belonging to East Siberian crudes. A typical example may also be found in the relative distribution of 17αH-hopanes in Tataria crudes and their asphaltene pyrolytic products, as represented in Table 64.

A thorough compositional analysis of asphaltene pyrolytic hydrocarbons has been reported (Rubinstein et al. 1979; Kayukova et al. 1982). Besides 17αH-hopanes, a considerable amount of $C_{19}-C_{21}$ and $C_{26}-C_{30}$ hydrocarbons, belonging to 5αH-, 5βH- and 14βH,17βH-steranes, were identified among pyrolytic products from biodegradated crudes). It is peculiar that besides 17αH-hopanes, a small amount of 17βH-hopanes, which are usually not found in crudes, was identified. This fact indicates a slightly lower degree of asphaltene catagenetic transformation (maturity), as compared to petroleum hydrocarbons. Certain peculiarities occurred in the composition of aromatic hydrocarbons as well. Thus, monoaromatic steroids were not found in pyrolytic products. However, on the whole, hydrocarbon compositions of asphaltene pyrolytic products and of natural crudes were largely compatible.

Accordingly, even in the case of heavily biodegradated crudes, there is always a possibility of reconstructing the chemical type of the original crude and of establishing its genetic precursor, composition and relative distribution of the most important biological markers, which disappeared in biodegradation. Mild pyrolysis of asphaltenes is the most reliable method in such a situation.

Biological Markers and the Genetic Classification of Crudes

In conclusion, let us again consider the important role of biological markers in petroleum chemistry and geochemistry.

The geochemical significance of these hydrocarbons was thoroughly analyzed in the monograph of Tissot and Welte (1981).

As mentioned, the composition of petroleum hydrocarbons depends on three main factors: catagenesis (of crudes and kerogen), biodegradation, as well as compositional and structural peculiarities of the original organic matter. The first factor was comprehensively analyzed in previous chapters. At this point we remphasize the important role of stereochemical transformations in establishing the degree of maturity (catagenesis) of the initial biological molecules on the way to petroleum. The last factor, i.e. the original biomolecule composition and structure, may be called a genetic one, and deserves special consideration.

Beyond any doubt, the composition of the original biomass and geochemical conditions of its transformation cannot fail to reflect the composition of petroleum hydrocarbons. Moreover, each sedimentary basin, as a potential source of oil fields, is characterized by a particular set of certain initial compounds, hence, certain resulting hydrocarbons. Compositional characteristics of the original organic matter in a given area provide highly important information used in oil exploration, i.e. the correlation of crudes in oil fields and the establishment of petroleum sources.

However, the destinies of organic molecules in the geological history of the Earth are quite intricate. They do not always turn into petroleum hydrocarbons. On the contrary, as a rule, these molecules undergo considerable structural and stereochemical transformations, hence, their source is sometimes difficult to determine.

Still the problem of genetic classification of crudes must be resolved, or at least clarified in its main concepts.

What is meant by the "genetic type" or "genotype" of a particular crude? Contrary to the chemical type, the term "genotype" is inconclusive, since hydrocarbons used for these purposes sometimes comprise only a small portion in petroleum. In this respect oil cannot be compared to coal, since the coal "genotype" clearly depends on the peculiarities of coal compositions.

In our opinion, rather than specifying the concept of petroleum "genotype", it is worthwhile to determine the genetic classification of crudes on the basis of certain characteristic features in the structure and relative destribution of biological markers inherited from source rocks in a given sedimentary basin. Crude oils should be considered genetically homogeneous is they contain similar (qualitatively and quantitatively) sets of biological markers, as well as other genetic indicators, such as carbon and sulfur isotopes typical of the organic material producing these crudes.

It has already been mentioned that concepts pertaining to the "chemical" and "genetic types" of crudes are different in nature, since crudes exist, which belong to the same source, but are of absolutely different chemical nature ($A^1 \rightarrow A^2 \rightarrow B^2 \rightarrow B^1$).

However, even after heavy biological degradation crudes will retain, if only infinitesimal, concentrations of molecules structurally linked to the genesis of a given crude oil.

The chemical classification of crudes is based on regularities in the relative distribution of different classes of hydrocarbons: alkanes, cyclanes, arenes. The chemical composition of crudes is seriously affected by various secondary processes, such as biodegradation, catagenesis, migration, dissolution in compressed gas, etc.

In contrast, the *genetic classification of crudes,* in our opinion, should be based on compositional regularities and molecular-mass distribution of chemofossils in crudes. Biological markers can be considered most promising in this respect.

A most important property of chemofossils used in genetic classification is their stability towards catagenesis and biodegradation. Although the overall content of chemofossils in crudes belonging to a given oil area is small, they are characteristic "fingerprints".

The importance of genetic classification of crude oils and the identification of separate genotypes has been emphasized on many occasions in the works of Botneva and Shulova (1981) and Maksimov (1981). The genetic classification in these publications was based mostly on group-type methods of evaluating the petroleum hydrocarbon structure. Hopefully, in the near future, the genetic classification of crudes will be raised to the molecular level.

Which petroleum hydrocarbons can be used in genetic classification? In principle, hydrocarbons of various molecular masses may fit this role. However, the larger the molecule, the more valuable the geochemical information it contains. Here are some examples.

Genetic information in light (gasoline) components is borne by the ratio of isomeric cyclopentanes and cyclohexanes, easily established in the case of C_7 and C_8 molecules.

Generally, cyclopentanes are of special genetic value since, firstly, they are thermodynamically unstable and, secondly, have a limite occurrence in biological compounds. The high concentration of cyclopentane hydrocarbons in gasolines is rooted in a peculiar mechanism of cyclic hydrocarbon formation in crudes. From the genetic point of view relative concentrations of *gem*-substituted C_7 and C_8 alkanes and cyclanes acquire special interest. High concentrations of these hydrocarbons are found both in alkanes and cycloalkanes.

Genetic information in medium fractions (200°–430°C) is borne by naphthenic and aromatic fingerprints, isoprenoid distribution, pristane/phytane ratios (for samples of moderate maturity) and the distribution of $C_{19}-C_{26}$ tricyclic terpanes. Important in the case of high-boiling fractions are sterane/hopane ratios, the relative distributions of steranes and hopanes, adiantane/hopane ratios, the ratios of rearranged and regular steranes, relative concentrations of 4-methylsteranes, etc. n-Alkanes distributions of whole oils is clearly a reliable genetic indicator for category A crudes.

An important role is played especially by reconstructed mass chromatograms of key fragment ions from computerized GC/MS analysis: *m/z* 217 for sterane hydrocarbons and *m/z* 191 for hopanes and tricyclic terpanes (polymethylalkylperhydrophenanthrenes), *m/z* 113 for isoprenoid alkanes, etc.

The reconstructions help avoid tedious procedures of isolating and concentrating polycyclic petroleum alkanes. Figure 90 exemplifies such GC-MS reconstructions of the saturated hydrocarbon fraction > 400°C in two Starogroznenskoe crude oils. Regardless of the strongly different of these crudes (A^1 on the hand; B^1 on the other) hopane distributions in both instances are similar, which suggests their common source.

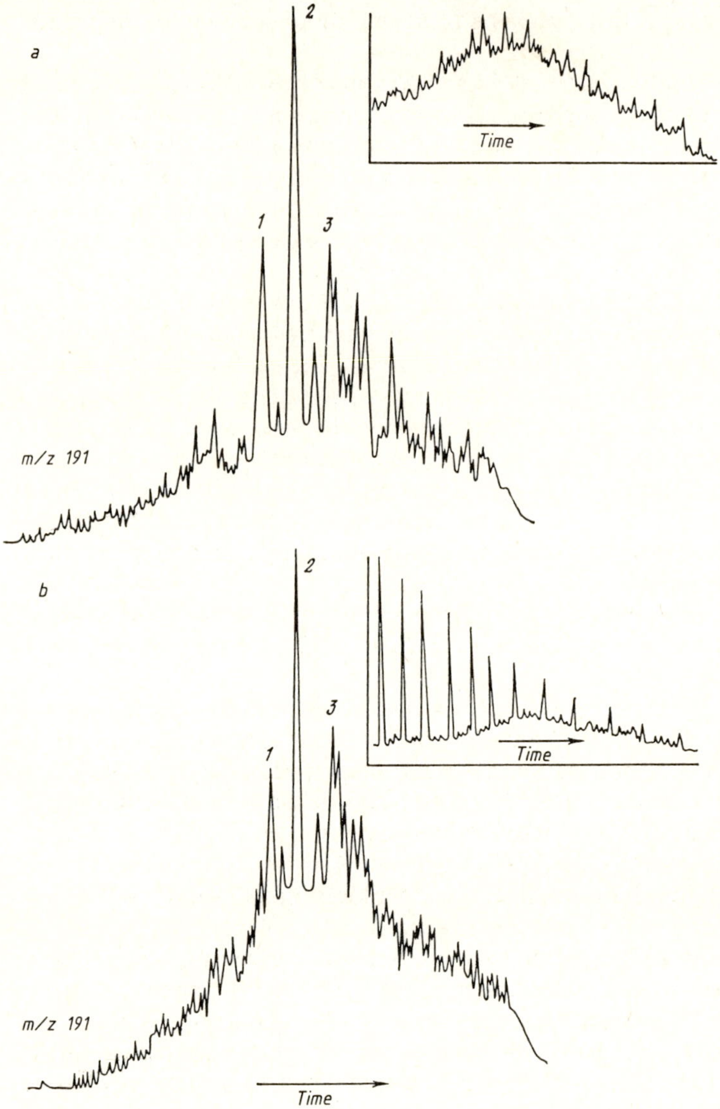

Fig. 90 a, b. Mass fragmentograms of two different chemical crude types from the Starogroznenskoe deposit, reconstructed m/z 191 mass chromatogram (fraction >400 °C). **a** type B^1 crude; **b** type A^1 crude. *1* Adiantane; *2* hopane; *3* C_{31} hopanes. Gas chromatograms of the same crudes presented to the *right*

The identification of unique biological markers, absent in other crudes, is of special genetic importance. Such hydrocarbons indicate sources of petroleum formation. This group of hydrocarbons may include 12- and 13-methylalkanes in East Siberian crudes, botryococcane in Idonesian crude, C_{28} hopane in Sivinski crude, bisnormethyladiantanes in North Sea crudes, irregular cyclic isoprenoids and trimethylaklycyclohexanes with isoprenoid chains in Buzachi peninsular crude, hexacyclic monoaromatic hopanoid hydrocarbons in hypersaline sediments of the Pripyat region and many other compounds.

Obviously, the time has come to establish, within the next few years, comprehensive genetic characteristics (genetic classification) of crude oils belonging to the most important regions of petroleum production in the USSR. This endeavor would certainly contribute both to successful oil prospecting and quality evaluations of new crudes. The determination of the basic type of organic molecules – marine or continental – is a simpler matter.

Crudes derived from marine sediments are characterized by low relative concentrations of normal alkanes, small N_β, i.e. a high chromatographic ratio of the naphthenic background to normal alkane peaks (no predominance of $C_{25}-C_{31}$ normal alkanes and presence of steranes); pristane/phytane ratios that equal or are below 1.

In contrast, crudes derived from an organic matter of continental origin have typically high concentrations of n-alkanes, low isoprenoid concentrations, large N_β values, $C_{25}-C_{31}$ normal alkane predominance, lack of steranes, pristane/phytane ratios >1.

Taxonomic indications, i.e. the establishment of the original biological molecules' geological age, is more complex than the genetic classification of crudes. However, this issue is of special importance since it pertains to peculiarities of the occurrence of various organic compounds in living organisms belonging to contemporary and ancient epochs. These problems are extensively analyzed by the special science of organic geochemistry. Complications in this area of research arise for a number of reasons. First, the type of organic molecules produced in photosynthesis remains the same, though a certain evolution (in a molecular aspect) takes place (Guseva et al. 1976; Leifman and Guseva 1978). Stability of photosynthesis may be proved by the fact that in the billions of years of their existence, living cells continue synthesizing a specific phytol enantiomer, from eight available possibilities, and just one of the 256 possible enantiomers of cholesterol (!).

The identification of ancient biogenic compounds, which served as a source for a particular petroleum hydrocarbon, is complicated by microbiological reworking of the original biomass at early stages of diagenesis. Most often, this processing leads not only to the disappearance or transformation of initial molecules, but also to the emergence of entirely new compounds, as in the synthesis of bacteriohopane, and a $C_{27}-C_{35}$ hopane series on its basis (Tissot and Welte 1981).

Nevertheless, the study of chemofossil fuels deservedly attracts much attention. Although their determination in crudes is accompanied by certain difficulties, resulting scientific information is invaluable. Noteworthy research in this respect is being done in order to establish ecological conditions of sedimentation (see for example Wen-Yen Kuang and Meinschein 1979), as well as in petroleum exploration (Mc Kirdy et al. 1983).

Evidently, besides the already mentioned petroleum-petroleum and petroleum-source rock correlations, the study of chemofossil compositions will help solve taxonomic problems of establishing the geological age of crudes in the near future. Considerable research must be done in analyzing and comparing organic molecule compositions in nature and fossil sediments. Today, as never before, we realize the correctness of V. I. Vernadskii's concept that "petroleum properties are conceived in organisms."

Chemofossils – biological markers – acquired considerable importance in chemical classification, hence, chemical processing of crudes.

Thus, biological markers find application in various spheres of petroleum production: oil exploration and production (in particular in the scientifically oriented collection of crudes of different quality and composition). These hydrocarbons are indispensible in chemical processing and petrochemistry.

Relict petroleum hydrocarbons are compounds of great potential.

Conclusion
Venues of Further Research on Petroleum Hydrocarbon Composition and Structure

Obvious progress has been registered in recent years in establishing the composition and structure of petroleum hydrocarbons. Evidence is provided by the impressive number of hydrocarbons identified in crudes (over 700). However, the complexity of emerging problems has also increased manifold. It suffices to mention the determination of the sterical structure of isosteranes originally discovered in petroleums. However, this is not the sole example of how a new type of hydrocarbon is identified in petroleum. A similar occurrence 50 years ago was the discovery of adamantane, which served as the basis for a new and original scientific discipline – the chemistry of polyhedranes.

However, it would be a mistake to assume that there are no further problems in the chemistry of petroleum hydrocarbons. The situation is much more complicated, and as usual in science, each new achievement is not the end, but poses new tasks in research.

Thus, the following issues await clarification in the near future.

1. The determination of the structure of branched alkanes which are usually present in large concentrations in the medium fractions of A crudes.
2. The study of the structure of polycyclic saturated hydrocarbons in type-B crudes (as opposed to A crudes, this type is not as well described at the molecular level). In particular, we have no information at all on the structure of C_{15} and higher tetracyclanes of non-steroid type, which obviously belong to bridged hydrocarbons with no more than 13–14 carbon atoms in the polycyclic nucleus. They are widely represented in a number of category B crudes (naphthene-based).
3. Detailed research devoted to the composition of naphtheno-aromatic hydrocarbons, of indane and tetralin series in particular, in medium petroleum fractions. And, naturally, further research is needed in order to identify new biomarkers and rearranged hydrocarbons of saturated and aromatic types.

Petroleum geochemistry is also confronted with difficult tasks, such as the already mentioned cardinal problem of establishing precise criteria for setting limits in the genetic homogeneity of crudes and dispersed organic matter, as well as petroleum taxonomy.

In other words, researchers are faced with important and challenging tasks, the fulfilment of which would contribute to the progress in such areas of practice as petroleum exploration and prospecting and the chemical processing of crudes.

Pioneer work in the synthesis and study of different physical and chemical properties of standard (reference) hydrocarbons of the petroleum type is an important prerequisite for progress in petroleum chemistry and geochemistry.

References

Chapter 1

Andreev PF, Bogomolov AI, Dobryanskii AF, Kartsev AA (1958) Petroleum transformations in nature. Gostoptekhizdat, Leningrad, 416 pp (in Russian)

Dobryanskii AF (1961) Petroleum chemistry. Gostoptekhizdat, Leningrad, 223 pp (in Russian)

Guseva AN, Leyfman IE, Vassoevitch NB (1976) On biogeochemistry of lipids, lipoids and related polymeric compounds. In: Analysis of organic matter in recent and fossil sediments. Nauka, Moscow, pp 25–56 (in Russian)

Kartsev AA (1978) Principles of petroleum and gas geochemistry. Nedra, Moscow, 279 pp (in Russian)

Kontorovitch AE, Stasova OF (1978) Crude types in the sedimentary care of the earth. Geol Geofiz 8:3–13 (in Russian)

Mackenzie A (1984) Application of biological markers in petroleum geochemistry. Adv Petrol Geochem, vol 1. Academic Press, London, pp 115–124

Maksimov SP, Botneva TA (eds) (1981) Parameter catalog of regional characteristics of the chemical and molecular compositions of crude oils in the Soviet Union. Nedra, Moscow, 294 pp

Martin R, Winters J, Williams J (1964) Occurrence of hydrocarbons in petroleum and its genesis. In: New research in the area of petroleum and gas genesis. CNIITE Neftegas, Moscow, pp 38–78 (in Russian)

Nametkin SS (1955) Petroleum chemistry, 3rd edn. The USSR Acad Sci Publ, Moscow, 800 pp (in Russian)

Petrov AA (1974) Chemistry of alkanes. Nauka, Moscow, 243 pp

Philippi GT (1977) On the depth, time and mechanism of origin of the heavy to medium-gravity naphthenic crude oils. Geochim Cosmochim Acta 41:33–52

Polyakova AA (1973) Molecular mass-spectral analysis of crudes. Nedra, Moscow, 181 pp (in Russian)

Proskuryakov VA, Drabkina AE (eds) (1981) Petroleum and gas chemistry. Khimiya, Leningrad, 359 pp (in Russian)

Safonova GI (1980) Relict structures in petroleum hydrocarbons of various stratigraphic subdivisions. Nedra, Moscow, 260 pp (in Russian)

Sokolov VA (1965) Processes of petroleum and gas formation and migration. Nedra, Moscow, 276 pp (in Russian)

Solodkov VK, Galkina GA, Dragunskaya VS, Kamyanov VF (1977) On petroleum alkane geochemistry. Isv TSSR Acad Sci Ser Phys Chem Geol 4:81–84

Tissot B, Welte D (1984) Petroleum formation and occurrence. Springer, Berlin Heidelberg New York

Vassoevitch NB, Berger MG (1968) On identification of crudes and their fractions according to hydrocarbon compositions. Geol Nefti Gaza 12:38–41 (in Russian)

Weisberger A (ed) (1967) Physical and chemical methods of establishing the structure of organic compounds, vol 1. Khimiya, Moscow, 531pp (in Russian, transl. from English)

Zabrodina MN, Arefyev OA, Makushina VM, Petrov AA (1978) Chemical types of crudes and petroleum transformations in nature. Neftekhimiya 18:280–289 (in Russian)

Chapter 2

Albaiges J (1979) Identification and geochemical significance of long chain acyclic isoprenoid hydrocarbons in crude oils. In: Advances in organic geochemistry: physics and chemistry of the earth, vol 12. Pergamon, Newcastle, pp 19–28

Albaiges J, Borbon G, Salagre P (1978) Identification of series of $C_{25}-C_{40}$ acyclic isoprenoid hydrocarbons in crude oils. Tetrahedron Lett 6:595–598

Blumer M, Snyder W (1965) Isoprenoid hydrocarbons in recent sediments. Presence of pristane and probably absence of phytane. Science 150:1588–1590

Brooks P, Maxwell J, Cronforth J et al. (1975) The stereochemistry of farnesane from crude oil. In: Advances in organic geochemistry. ENADIMSA, Madrid, pp 81–97

Calvin M (1971) Chemical evolution. Moscow, 240 pp

Chappe B, Michaelis W, Albrecht P (1979) Molecular fossils of Archaebacteria as selective degradation products of kerogen. In: Advances in organic geochemistry: physics and chemistry of the earth, vol 12. Pergamon, Newcastle, pp 265–274

Cox R, Maxwell J, Ackman R, Hoopper S (1972) Stereochemical studies of acyclic isoprenoid compounds III. The stereochemistry of naturally occurring (marine) 2,6,10,14-tetramethylpentadecane. Can J Biochem 50:1238–1241

Eglinton G, Murphy M (1969) Organic geochemistry. Methods and results. Springer, Berlin Heidelberg New York

Granwell A (1973) Branched-chain and cyclopropanoid acids in a recent sediment. Chem Geol 11:307–313

Habgood H, Harris W (1964) Retention indices in programmed temperature gas chromatography. Anal Chem 36/3:663–665

Han G, Calvin M (1969) Occurrence of $C_{22}-C_{25}$ isoprenoids in bell creek crude oil. Geochim Cosmochim Acta 33:733–742

Hoeven W, Haug van P, Burlingame A, Calvin M (1966) Hydrocarbons from an Australien oil, 200 million years old. Nature 211:1361–1366

Kartsev AA (1978) Principles of petroleum and gas geochemistry. Nedra, Moscow, 279 pp (in Russian)

Krasavtchenko MI, Zemskova ZK, Mikhnovskaya AA, Pustilnikova SD, Petrov AA (1971) Mono-substituted alkanes of $C_{10}-C_{16}$ Composition in Paraffinic Oils. Neftekhimiya 11:803–809 (in Russian)

Kreps EM (1981) Lipids of cell membranes. Nauka, Leningrad, 242 pp (in Russian)

Mair B, Ronen Z, Eisenbraun E, Horodysky A (1966) Terpenoid precursors of hydrocarbons from the gasoline range of petroleum. Science 154:1339–1341

Makushina VM, Arefyev OA, Zabrodina MN, Petrov AA (1978) New relict petroleum alkanes. Neftekhimiya 18:847–854 (in Russian)

Martin R, Winters J, Williams J (1964) Occurrence of hydrocarbons in petroleum and its genesis. In: New research in the area of petroleum and gas genesis. CNIITE Neftegas', pp 38–78 (in Russian)

Maxwell J, Cox R, Ackman R, Hoopper S (1971) The diagenesis and maturation of phytol. The stereochemistry of 2,6,10,14-tetramethylpentadecane from ancient sediment. In: Advances in organic geochemistry, vol 33. Pergamon, Hannover, pp 277–291

Methods of analysing organic petroleum compounds, their mixtures and derivatives (1969) Col 2. Nauka, Moscow, 211 pp (in Russian)

Michaelis W, Albrecht P (1979) Molecular fossils of Archaebacteria in kerogen. Naturwissenschaften 66:420–421

Moldowan J, Seifert W (1979) Head-to-head linked isoprenoid hydrocarbons in petroleum. Science 204:169–171

Moldowan J, Seifert W (1980) First discovery of botryococcane in petroleum. J Chem Soc Chem Commun, pp 912–914

Patience R, Rowland S, Maxwell J (1978) The effect of maturation of the configuration of pristane in sediments and petroleum. Geochim Cosmochim Acta 42:1871–1875

Patience R, Yon D, Ryback G, Maxwell J (1979) Acyclic isoprenoid alkanes and geochemical maturation. In: Advances in organic geochemistry: physics and chemistry of the earth, vol 12. Pergamon, Newcastle, pp 287–293

Petrov AA (1960) Catalytic isomerization of hydrocarbons. The USSR Acad Sci Publ, Moscow, 215 pp (in Russian)

Petrov AA (1974) Chemistry of alkanes. Nauka, Moscow, 243 pp (in Russian)

Petrov AA (1981) Stereochemistry of saturated hydrocarbons. Nauka, Moscow, 254 pp (in Russian)

Petrov AA, Tsedilina AL, Pustilnikova SD, Krasavchenko MI, Abrutina NN, Yakubson ZV (1973) Isoprenoid petroleum hydrocarbons. Neftekhimiya 12:779–785 (in Russian)

Proskuryakov VA, Drabkin AE (ed) (1981) Petroleum and gas chemistry. Khimiya, Leningrad, 359 pp (in Russian)

Rashid M (1979) Pristane-phytane ratios in relation to source and diagenesis of ancient sediments from the Labrador shelf. Chem Geol 25:109–122

Rowland S, Yon D, Lewis C, Maxwell J (1985) Occurrence of 2,6,10-trimethyl-7-(3-methyl-butyl) dodecane and related hydrocarbons in the green Alga and sediments. Org Geochem 3/3:207–213

Safonova GI (1980) Relict structures in petroleum hydrocarbons of various stratographic subdivisions. Nedra, Moscow, 260 pp (in Russian)

Sanin PI (1976) Petroleum hydrocarbons. Uspekhi Khimii 45/8:1361–1394 (in Russian)

Seifert W (1975) Carboxylic acids in petroleum and sediments. In: Herz W et al (eds) Progress in the chemistry of organic natural products. Springer, Berlin Heidelberg New York, 49 p

Smith H, Rall H (1953) Relationship of hydrocarbons with six to nine carbon atoms. Indust Eng Chem 45:1491–1497

Sokolov VA, Bestuzhev MA, Tikhomolova TV (1972) Chemical composition of crudes and natural gases depending on origin. Nedra, Moscow, 276 pp (in Russian)

Stransky K, Streibl M, Sorm F (1966) Über einen neuen Typ verzweigter Paraffine aus dem Wachs der Honigbiene. Coll Czech Chem Commun 31:4694–4702

Ten Fu Yen, Chilingarian G (eds) (1976) Oil shale. Elsevier, Amsterdam

Tissot B, Welte D (1981) Petroleum formation and occurrence. Mir, Moscow, 501 pp (in Russian)

Tocanne G (1972) Application à la détermination de la configuration absolue de l'acide lectobacillique. Tetrahedron 28:363–371

Vorobyova NS, Zemskova ZK, Petrov AA (1982) Petroleum isoprenoid alkanes of irregular and pseudo-regular structure types. Neftekhimiya/22:587–591 (in Russian)

Vorobyova NS, Zemskova ZK, Poladov K, Ernepesov KN, Petrov AA (1985) Lycopane and isolycopane in petroleum. Neftekhimiya, 25/6

Yon D, Maxwell J, Ryback G (1982) Tetrahedron Lett 23/2:2143–2146

Chapter 3

Anders D, Robinson W (1971) Cycloalkane constituents of the bitumen from Green River shale. Geochim Cosmochim Acta 35:661–678

Aquino Neto F, Trendel J, Restle A, Connan J, Albrecht P (1983) Occurrence and formation of tricyclic and tetracyclic terpanes in sediments and petroleum. Adv Org Geochem. Wiley, pp 659–667

Arefyev OA, Zabrodina MN, Makhushina VM, Petrov AA (1980) Relict tetra- and pentacyclic hydrocarbons in ancient oils of the Siberian platform. Isv USSR Acad Sci Ser Geol 3:135–140 (in Russian)

Bagrii EI, Sanin PI, Vorobyova NS, Petrov AA (1967) Hydrocarbon composition of extracts isolated in petroleum fractions through extractional crystallization with thiocarbamide. Neftekhimiya 7:515–518 (in Russian)

Bendoraitis J (1973) Hydrocarbons of biogenic origin in petroleum, aromatic triterpenes and bicyclic sesquiterpenes. In: Advances in organic geochemistry. Technip Paris, pp 209–224

Berman SS, Yakubson ZB, Petrov AA (1973) Relative thermodynamic stability of trimethylbicyclo(4.4.0)decanes at 298 and 573 K. Neftekhimiya 13:473–477 (in Russian)

Chappe B, Michaelis W, Albrecht P (1979) Molecular fossils of Archaebacteria as selective degradation products of kerogen. In: Advances in organic geochemistry: physics and chemistry of the earth, vol 12. Pergamon, Oxford, pp 205–274

Connan G, Restle A, Albrecht P (1979) Biodegradation of crude oil in the Aquitaine basin. In: Advances in organic geochemistry. Physics and chemistry of the earth, vol 12. Pergamon, Oxford, pp 1–17

Dididze AV, Arefyev OA, Shakarashvili TS, Bekaury NG (1977) Analysis of $C_{11}-C_{13}$ composition saturated hydrocarbons in Georgian deposits' crudes, vol 85. Rep Acad Sci Georgian SSR, pp 633–636 (in Russian)

Dididze AV, Pustilnikova SD, Arefyev OA, Petrov AA (1979) Analysis of individual compositions of high-molecular saturated hydrocarbons in Georgian crudes. Neftekhimiya 19:336–343 (in Russian)

Dimler A, Cyr T, Strausz O (1984) Identification of bicyclic terpenoid hydrocarbons in the saturated fraction of Athabasca oil sand bitumen VI Org Geochem 7, 3/4:231–238

Eglinton G, Murphy M (eds) (1974) Organic geochemistry. Nedra, Leningrad, 488 pp (in Russian; transl. from English)

Ekweozor G, Strausz O (1981) Tricyclic terpanes in the Athabasca oil sands. Adv Org Geochem 746–766

Ekweozor C, Okogun J, Ekong D, Maxwell J (1979) Preliminary organic geochemical studies of samples from the Niger Delta (Nigeria) I. Analyses of crude oils for triterpanes. Chem Geol 27:11–28

Ensminger A, van Dorsselaer A, Spyckerelle C et al. (1973) Pentacyclic triterpenes of the hopane type as ubiquitous geochemical markers: origin and significance. In: Advances in organic geochemistry. Technip, Paris, pp 245–260

Ensminger A, Joly G, Albrecht P (1978) Rearranged steranes in sediments and crude oils. Tetrahedron Lett 18:1575–1578

Grantham P, Posthuma J, de Groot K (1979) Variation and significance of the C_{27} and C_{28} triterpane content of a North Sea core and various North Sea crude oils. In: Advances in organic geochemistry: physics and chemistry of the earth, vol 13. Pergamon, Newcastle, pp 29–38

Kagramanova GR, Pustilnikova SD, Pekhk TI, Denisov UV, Petrov AA (1976) Sesquiterpane petroleum hydrocarbons. Neftekhimiya 16:18–22

Kaneda M, Takahashi R, Itaka I, Shibata S (1972) Retigeranic acid, a novel sesterterpene. Tetrahedron Lett 4609–4611

Kayukova GP, Pustilnikova SD, Abryutina NN, Golovkina LS, Petrov AA (1980a) Equilibrium composition and properties of epimers of rearranged cholestane. Neftekhimiya 20:183–193 (in Russian)

Kayukova GP, Pustilnikova SD, Abryutina NN, Petrov AA (1980b) Rearranged steranes in crudes. Neftekhimiya 21:643–650 (in Russian)

Kayukova GP, Pustilnikova SD, Abryutina NN, Golovkina LS, Petrov AA (1981a) Stereochemistry and equilibrium ratios of epimeric 4-methylcholestanes, rearranged 4-methylcholestanes and 4,4-dimethylcholestanes. Neftekhimiya 21:483–495 (in Russian)

Kayukova GP, Pustilnikova SD, Abryutina NN, Golovkina LS, Mekhteeva VL, Petrov AA (1981b) Structurally modified hydrocarbons of hopane type. Neftekhimiya 21:803–811 (in Russian)

Kimble B, Maxwell J, Philp R, Eglinton G (1974) Identification of steranes and triterpanes in geolipid extracts by highresolution chromatography and mass-spectrometry. Chem Geol 14:173–198

Kuliev AM, Petrov AA, Levshina AM, Shepeleva TV, Pustilnikova SD, Abryutina NN (1984) Steranes of the Naftalanski crude. Azerbaidzhanski Khimitcheskii Zhurn 2:48–53

Lindeman L, Tourneau R (1963) New information on the composition of petroleum. In: Proc 6th World Petrol Congr (Frankfurt a. M.) sect 5, pap 14

Mackenzie A, Patience R, Maxwell J et al. (1980) Molecular parameters of maturation in the Toarcian shales. Paris basin, France I. Changes in the configurations of acyclic isoprenoid alkanes, steranes and triterpanes. Geochim Cosmochim Acta 44:1709–1721

Matveeva IA, Sokolova IM, Pekhk TI, Petrov AA (1975) Synthesis and isomeric transformations of dimethyl-bicyclo(3.2.1)octane. Neftekhimiya 15:17–23 (in Russian)

Moldowan J (1984) C_{30}-Steranes, novel markers for marine petroleum and sedimentary rocks. Geochim Cosmochim Acta 48:2767–2768

Moldowan J, Seifert W, Haley M, Djerassi C (1980) Proof of structure by synthesis of 5a, 14β, 17β(H)-cholestan (20R) a major petroleum sterane. Correction of previous assignment. Geochim Cosmochim Acta 44:1613

Moldowan J, Seifert W, Gallegos E (1983) Identification of an extended series of tricyclic terpanes in petroleum. Geochim Cosmochim Acta 47:1531–1534

Moldowan J, Seifert W, Edward A, Clazdy J (1984) Structure proof and significance of stereoisomeric 28,30-bisnorhopane in petroleum. Geochim Cosmochim Acta 48 8:1651–1661

Mulheirn L, Ryback G (1975a) Isolation and structure analysis of steranes from geological sources. In: Advances in organic geochemistry. ENADIMSA, Madrid, pp 173–192

Mulheirn L, Ryback G (1975b) Stereochemistry of some steranes from geological sources. Nature 256:301–302

Noble R, Know J, Alexander R, Kagi R (1985) Identification of tetracyclic diterpane hydrocarbons in Australian crude oils and sediments. J Chem Soc Chem Commun 1:32–33

Oelert H, Severin D, Windhager H (1973) Saturated hydrocarbon types from a Tia Juana residue. Erdöl und Kohle-Erdgas-Petrochem Brennst Chem 26:397–401

Olenina ZK, Petrov AA (1969) Compositional analysis of saturated hydrocarbons in fraction 125–150° of direct discillation gasolines. Neftekhimiya 9:129–136 (in Russian)

Ourisson G, Albrecht P, Rohmer M (1979) The hopanoids. Paleochemistry and biochemistry of a group of natural products. Pure Appl Chem 51:709–729

Pekhk TI, Pustilnikova SD, Abryutina NN, Kayukova GP, Petrov AA (1982) NMR ^{13}C spectra of epimeric steranes. Neftekhimiya 22:21–29 (in Russian)

Petrov AA (1971) Chemistry of naphthenes. Nauka, Moscow, 388 pp (in Russian)

Petrov AA (1981) Stereochemistry of saturated hydrocarbons. Nauka, Moscow, 254 pp

Petrov AA, Pustilnikova SD, Abryutina NN, Kagramanova GR (1976) Petroleum steranes and triterpanes. Neftekhimiya 16:411–427 (in Russian)

Pryce R (1971) The occurrence of bound water-soluble squalene, 4,4-dimethyl sterols, 4-methyl sterols and sterols in leaves of Kalanchoe blossfeldiana. Phytochemistry 10:1303–1307

Pum J, Ray J, Smith G, Whitehead E (1975) Petroleum triterpane fingerprinting of crude oils. Anal Chem 47 9:1617–1622

Pustilnikova SD, Abryutina NN, Kagramanova GR, Petrov AA (1976) Hopane series hydrocarbons in curdes. Geochimiya 3:460–468 (in Russian)

Pustilnikova SD, Abryutina NN, Kayukova GP, Petrov AA (1980) Equilibrium composition and properties of epimeric cholestanes. Neftekhimiya 20:20–33 (in Russian)

Pustilnikova SD, Abryutina NN, Kayukova GP, Petrov AA (1981) Equilibrium composition and properties of androstane and pregnane epimers. Neftekhimiya 21:182–185 (in Russian)

Reed W (1977b) Molecular compositions of weathered petroleum and comparison with its possible source. Geochim Cosmochim Acta 41:237–247

Rossini FD (1967) Petroleum hydrocarbons. Neftekhimiya 7:906–916 (in Russian)

Rubinstein J, Albrecht P (1975) The occurrence of nuclear methylated steranes in a shale. J Chem Soc Chem Commun 24:957–958

Rullkötter J, Philp P (1981) Extended hopanes is to C_{40} in thoronton bitumen. Nature 292:616–618

Rulkötter J, Wendisch D (1982) Microbial alteration of 17α(H)-hopanes in Madagascar asfhalts: removal of C-10 methyl group and ring opening. Geochim Cosmochim Acta 46:1545–1553

Ryback G (1976) Chromatography of saturated steroid hydrocarbons (steranes) on alumina. J Chromatogr 116:207–210

Sanin PI, Bagrii EI, Tsitsugina NN, Sutchkova AA, Musayev IA, Kurashova EK (1974) On petroleum alkyladamantanes. Neftekhimiya 14:333–340 (in Russian)

Seifert W (1975) Source rock/oil correlations by $C_{27}-C_{30}$ biological marker hydrocarbons. In: Advances in organic geochemistry. ENADIMSA, Madrid, pp 21–44

Seifert W (1978) Steranes and terpanes in kerogen pyrolysis for correlation of oils and source rocks. Geochim Cosmochim Acta 42:473–484

Seifert W, Moldowan J (1978) Applications of steranes, terpanes and monoaromatics to the maturation, migration and source of crude oils. Geochim Cosmochim Acta 42:77–95

Seifert W, Moldowan J (1979a) The effect of biodegradation on steranes and terpenes in crude oils. Geochim Cosmochim Acta 43:111–126

Seifert W, Moldowan M (1979b) The effect of thermal stress on source-rock quality as measured by hopane stereochemistry. In: Advances in organic geochemistry: physic and chemistry of the earth, vol 12. Pergamon, Oxford, pp 229–237

Seifert W, Moldowan J (1981) Paleoreconstruction by biological markers. Geochim Cosmochim Acta 45:783–794

Seifert W, Moldowan J, Smith G, Whitehead E (1978) First proof of structure of a C_{28}-pentacyclic triterpane in petroleum. Nature 271:436–437

Seifert W, Moldowan J, Jones R (1979) Application of biological marker chemistry to petroleum exploration. In: Spec Pap 8, Proc 10th World Petrol Congr Bucharest, pp 425–440

Sevostyanova GV, Sanin PI, Musayev IA, Ivtchenko EG, Vaisberg KM (1967) Analysis of medium fractions in the Arlanski Crude. Neftekhimiya 17:695–702 (in Russian)

Sieskind O, Joly G, Albrecht P (1979) Simulation of the geochemical transformations of sterols: superacid effect of clay minerals. Geochim Cosmochim Acta 43:1675–1679

Sokolova IM, Berman SS, Gervits ES, Matveeva IA, Petrov AA (1981) On the composition of saturated hydrocarbons in petroleum fractions 150–175°. Neftekhimiya 21:20–27 (in Russian)

Tchakhmatchev VA, Vinogradova TL, Babenyshev AP (1978) The influence of the geochemical modifications in petroleum on the composition of light hydrocarbons. Geol Nefti Gasa 5:44–52 (in Russian)

Ten Fu Yene, Chilingarian G (eds) (1976) Oil shal Elsevier, Amsterdam

Tissot B, Welte D (1981) Petroleum formation and occurrence. Springer, Berlin Heidelberg New York

Ushakova IB, Zaikin VG, Genekh IS, Smirnov BA, Sanin PI (1975) Triterpanes of the Baku Crude. Neftekhimiya 15:635–640 (in Russian)

Van Dorsselaer A, Ensminger A, Spyckerelle C et al. (1974) Degraded and extended hopane derivatives ($C_{27}-C_{35}$) as ubiquitous geochemical markers. Tetrahedron Lett 14:1349–1352

Van Dorsselaer A, Albrecht P, Connan J (1975) Changes in composition of polycyclic alkanes by thermal maturation (Yallourn Lignite, Australia). In: Advances in organic geochemistry. ENADIMSA, Madrid, pp 53–59

Vorobyova NS, Zemskova ZK, Petrov AA (1973) $C_{14}-C_{26}$ Composition of polycyclic naphthenes in the Siva Crude. Neftekhimiya 13:345–351 (in Russian)

Vorobyova NS, Arefyev OA, Pekhk TI, Denisov UV, Petrov AA (1975) Mechanism and kinetics of tricycloundecanes' isomerization. Neftekhimiya 15:659–666

Vorobyova NS, Zemskova ZK, Pekhk TI, Petrov AA (1977) Tricyclododecane isomerization. Neftekhimiya 17:22–30 (in Russian)

Vorobyova NS, Zemskova ZK, Petrov AA (1979) Tri- and tetracyclic saturated petroleum $C_{11}-C_{13}$ hydrocarbons composition. Neftekhimiya 11:3–6

Vorobyova NS, Zemskova ZK, Petrov AA (1980) Monocyclic alkanes of isoprenoid construction in Karajanbas Crude. Neftekhimiya 20:483–489 (in Russian)

Weidenhoffer Z, Hala S (1971) Progress in the chemistry of adamantane (review of literature 1964–1965). In: Sci Pap Inst Chem Technol. Inst Chem Technol, Prague, pp 5–84

Whitehead E (1973) The structure of petroleum pentacyclanes. In: Advances in organic geochemistry. Technip, Prague, pp 225–243

Yakubson ZB, Arefyev OA, Petrov AA (1973) Tetracyclic saturated petroleum hydrocarbons of $C_{11}-C_{13}$ composition. Neftekhimiya 13:345–351 (in Russian)

Chapter 4

Aizenshtat A (1973) Perylene and its geochemical significance. Geochim Cosmochim Acta 37:559–567

Anders DE, Doolittle FG, Robinson WE (1973) Analysis of some aromatic hydrocarbons in a benzene soluble bitumen from Green River shale. Geochim Cosmochim Acta 37:1213–1228

Bendoraitis J (1974) Hydrocarbons of biogenic origin in petroleum-aromatic triterpenes and bicyclic sesquiterpenes. In: Advances in organic geochemistry. Technip, Ed. Paris, pp 209–224

Carruthers W (1957) The constituents of high-boiling petroleum distillates IV. Some polycyclic aromatic hydrocarbons in a Kuwait oil. J Chem Soc 278–281

Carruthers W, Cook J (1954) The constituents of high-boiling petroleum distillates I. Preliminary studies. J Chem Soc 2047–2052

Carruthers W, Walkins D (1963) Identification of 1,2,3,4-tetrahydro-2,2,9-trimethylpicene in an American crude oil, vol 34. Chem Indust, Leningrad, pp 1433

Dooley J, Hirsch D, Thompson C, Ward C (1974) Comparisons of heavy distillates from different crude oils. Hydrocarbon Proc 53/11:187–194

Gala S, Kurash M, Kubelka V (1977) Establishment of hydrocarbons' structure in groups of isoalkanes and aromatics in the kerosene fraction of the Romashkin crude oil. In: Collection of the Prague Chemical Technological Institute Fuels' Technology. State Pedagogical Publishers, Prague, pp 151–180 (in Russian)

Giger W, Schaffner C (1978) Determination of polycyclic aromatic hydrocarbons in the environment by glass capillary gas chromatography. Anal Chem 50:243–249

Greiner A, Spyckerelle C, Albrecht P (1976) Aromatic hydrocarbons from geological sources I. New naturally occurring phenanthrene and chrysene derivatives. Tetrahedron 32:257–260

Greiner A, Spyckerelle C, Albrecht P, Ourisson G (1977a) Aromatic hydrocarbons from geological sources V. Mono- and diaromatic hopane derivatives. J Chem Res 334

Greiner A, Spyckerelle C, Albrecht P, Ourisson G (1977b) Hydrocarbons aromatiques d'origine géologique. J Chem Res 3829–3871

Kayukova GP, Pustilnikova SD, Abryutina NN et al. (1981) Structurally modified hydrocarbons of hopane type. Neftekhimiya 21:803–811 (in Russian)

Kolesnikova LP (1972) Gas chromatography in the study of natural gases, petroleum and condensates. Nedra, Moscow, 135 pp (in Russian)

Laflamme R, Hites R (1978) The global distribution of polycyclic aromatic hydrocarbons in recent sediments. Geochim Cosmochim Acta 42:289–303

Laflamme R, Hites R (1979) Tetra- and pentacyclic, naturally occurring aromatic hydrocarbons in recent sediments. Geochim Cosmochim Acta 43:1687–1691

Lee M, Vassilaros D, White C, Novotny M (1979) Retention indices for programmed-temperature capillary-column gas chromatography of polycyclic aromatic hydrocarbons. Anal Chem 51:768–773

Lekveishvili EG, Melikadze LD, Tevdorashvili MN, Kartvelishvili EV (1979) On the study of petroleum phenanthrenic hydrocarbons. Neftekhimiya 19:689–695 (in Russian)

Mackenzie A (1984) Application of biological markers in petroleum geochemistry. Advances in petroleum geochemistry, vol 1. Academic Press, London, p 190

Mackenzie A, Hoffman C, Maxwell J (1981) Molecular parameters of maturation in the Toarcian shales, Paris basin, France III. Changes in aromatic steroid hydrocarbons. Geochim Cosmochim Acta 45:1345–1355

Mair B (1964) Terpenoids fatty acids and alcohols as source materials for petroleum hydrocarbons. Geochim Cosmochim Acta 28:1303–1321

Mair B, Barnewall J (1964) Composition of the mononuclear aromatic material in the light gas oil range, low refractive index portion, 230 °C to 305 °C. J Chem Eng Data 9:282–292

Mair B, Martinez-Pico J (1962) Composition of the trinuclear aromatic portion of the heavy gas oil and light lubricating distillate. Proc Am Petrol Inst 42, 3:172–185

Maksimov SP (ed) (1981) Critera in crudes' and gases' qualitative composition prognosis, issue 223. Nedra, Moscow, 164 pp (in Russian)

Martin R, Winters J, Williams J (1964) Occurrence of hydrocarbons in petroleum and its genesis. In: New research in the area of petroleum and gas genesis. CNIITE nefte Gas, Moscow, pp 38–78 (in Russian)

Melikadze LD, Lekveshvili EG (1977) Photochemical condensation of maleic anhydride with hydrocarbons of phenanthrene series. Metsniereba, Tbilisi, 103 pp (in Russian)

Meney J, Joung-Ho Kim, Stevrenson R, Margulis TA (1973) A new steroid aromatization rearrangement involving inversion of side-chain configuration. Tetrahedron 29:21–30

Nekrasov AS, Ptitsina NV (1957) Aromatic hydrocarbons of kerosene fractions in Crimean Crude Oils. Tr Inst Nefti USSR Acad Sci 10:74–95 (in Russian)

Ostroukhov SB, Arefyev OA, Makushina VM, Zabrodina MN, Petrov AA (1982) Monocyclic aromatic hydrocarbons with isoprenoid chains. Neftekhimiya 22:723–728 (in Russian)

Ostroukhov SB, Arefyev OA, Pustilnikova SD, Zabrodina MN, Petrov AA (1983a) High-molecular alkylbenzenes with normal alkyl chains. Neftekhimiya 23:20–30 (in Russian)

Ostroukhov SB, Arefyev OA, Petrov AA (1983b) Hexacyclic monoaromatic petroleum hydrocarbons. Neftekhimiya 23:152–159 (in Russian)

Petrov AA (1981) Stereochemistry of saturated hydrocarbons. Nauka, Moscow, 255 pp (in Russian)

Polyakova AA, Tevdorashvili MN, Lekveishvili EG, Melikadze LD, Semanyuk RN, Kogan LO (1982) The study of aromatic hydrocarbons in the Norio curde through mass- and gas chromatography-mass-spectrometry methods. Rep Acad Sci Georgian SSR 105/N 1: pp 65–68 (in Russian)

Radke M, Wilsch H, Leuthaeuser D, Teichmüller M (1982) Aromatic components of coal: relation of distribution pattern to rank. Geochim Cosmochim Acta 40:1831–1848

Schaefle J, Ludwig B, Albrecht P, Ourisson G (1978) Aromatic hydrocarbons from geological sources VI. New-aromatic steroid derivatives in sediments and crude oil. Tetrahedron Lett 43:4163–4166

Seifert W. Moldowan J (1978) Applications of sterane, terpanes and monoaromatics to the maturation, migration and source of crude oils. Geochim Cosmochim Acta 42:77–95

Seifert W, Carlson R, Moldowan J (1981) Geomimetic synthesis, structure assignment and geochemical correlation. Application of monoaromatized petroleum steranes. In: 10th Int Meet Organ Geochem, programme and Abstract, Bergen, p 78

Sevastyanova GV, Sanin PI, Musaev IA, Ivtchenko EG, Vaisberg KM (1967) Analysis of medium fractions of the Arlanski crude. Neftekhimiya 7:685–702 (in Russian)

Simoneit B, Burlingame A (1973) Study of the organic matter in the DSDP (JOIDES) cores, legs 10–15. In: Advances in organic geochemistry. Technip, Prague, pp 629–648

Sokolov VA, Bestuzhev MA, Tikhomolova TV (1972) Chemical composition of crudes and natural gases in connection with their origin. Nedra, Moscow, 276 pp (in Russian)

Solli H, Larter S, Douglas A (1979) Analysis of kerogens by pyrolysis-gas chromatography-mass spectrometry using selective ion monitoring III. Long-chain alkylbenzenes. In: Advances in organic geochemistry: physics and chemistry of the earth, vol 12. Pergamon, Newcastle, pp 591–597

Spyckerelle C, Greiner A, Albrecht P, Ourisson G (1977a) Aromatic hydrocarbons from geological sources III. A tetrahydrochrysene derived from triterpenes in recent and old sediments: 3,3,7-trimethyl-1,2,3,4-tetrahydrochrysene. J Chem Res 330–331

Spyckerelle C, Greiner A, Albrecht P, Ourisson G (1977b) Aromatic hydrocarbons from geological sources IV. An octahydrochrysene derived from triterpenes in oil shale: 3,3,7,12a-tetramethyl-1,2,3,4,4a,11,12,12a-octahydrochrysene. J Chem Res 332–333

Stall D, Westrum E, Sinke G (1964) The chemical thermodynamics of organic compounds. Wiley, London

Ten Fu Yen, Chilingarian G (eds) (1976) Oil Shal Elsevier, Amsterdam

Tissot B, Welte D (1981) Petroleum formation and occurrence. Mir, Moscow (in Russian)

Wakeham S, Schaffner C, Giger W (1980a) Polycyclic aromatic hydrocarbons in recent Lake sediments I. Compounds having anthropogenic origins. Geochim Cosmochim Acta 44:403–413

Wakeham S, Schaffner G, Giger W (1980b) Polycyclic aromatic hydrocarbons in recent Lake sediments II. Compounds derived from biogenic precursors during early diagenesis. Geochim Cosmochim Acta 44:415–429

Yew F, Mair B (1966) Isolation and identification of C_{13} to C_{17} alkylnaphthalenes, alkylbiphenyls and alkyldibenzofurans from the 275 °C to 305 °C dinuclear aromatic fraction of petroleum. Anal Chem 38:231–237

Zubenko VG, Gordadze GN, Petrov AA (1979) High-boiling aromatic hydrocarbons in the Anastasiyevskoe crude. Neftekhimiya 19:833–838 (in Russian)

Zubenko VG, Pustilnikova SD, Abryutina NN, Petrov AA (1980) Petroleum monoaromatic hydrocarbons of steroid type. Neftekhimiya 20:490–497 (in Russian)

Zubenko VG, Vorobyova NS, Zemskova ZK, Pekhk TI, Petrov AA (1981) On the equilibrium of cis- and trans-isomers in octahydrophenanthrenes – structural fragments of monoaromatic steranes. Neftekhimiya 21:323–328 (in Russian)

Chapter 5

Andreev PF, Bogomolov AI, Dobryanskii AF, Kaktsev AA (1958) Petroleum transformations in nature. Gostoptekhisdat, Leningrad, 416 pp (in Russian)

Bogomolov AI, Shimanskii VK (1966) Genesis of light methane petroleum hydrocarbons in view of their compositional regularities. Geokhimiya 1:115–121 (in Russian)

Brooks P, Maxwell J (1973) Early stage fate of phytol in a recently-deposited Lacustrine sediment. In: Advances in organic geochemistry. Technip, Paris, pp 977–991

Connan J (1973) Diagenèse naturelle et diagenèse artificielle de la matière organique à éléments végétaux prédominants. In: Advances in organic geochemistry. Technip, Prague, pp 73–95

De Leeuw J, Correia V, Schenck P (1973) On the decomposition of phytol under simulated geological conditions and in the toplayer of natural sediments. In: Advances in organic geochemistry. Technip, Prague, pp 993–1004

Edmunds K, Brassell S, Eglinton G (1979) The short-term diagenetic fate of 5-cholesten-3-ol: in situ radiolabelled incubations in algal mats. In: Advances in organic geochemistry: physics and chemistry of the earth, vol 12. Pergamon, Newcastle, pp 427–434

Eglinton G (1971) Laboratory simulation of organic geochemical processes. In: Advances in organic geochemistry, vol 33. Pergamon, Hannover, pp 29–48

Elington E, Murphy M (eds) (1974) Organic Geochemistry. Nedra, Leningrad, 488 pp (in Russian; transl. from English)

Frost AV (1945) The role of clays in petroleum formation in earth crust. Uspekhi Khimii 14, issue 6:501–509 (in Russian)

Frost AV (1946) The role of catalysis in petroleum formation in the earth crust. Utchenye Zapiski MGU issue 86:4–12 (in Russian)

Gagosian R, Farrington J (1978) Sterenes in surface sediments from the southwest African shelf and slope. Geochim Cosmochim Acta 42:1091–1101

Gagosian R, Heinzer F (1979) Stenols and stanols in the oxic and anoxic waters of the Black Sea. Geochim Cosmochim Acta 43:471–486

Galimov EM (1973) Hydrocarbon isotopes in petroleum and gas geology. Nedra, Leningrad, 384 pp

Gallegos E (1975) Terpane-sterane release from kerogen by pyrolysis gas chromatography-mass spectrometry. Anal Chem 47:1524–1528

Gulyaeva ND, Arefyev UA (1978) Regularities of normal and isoprenoid alkanes' distribution in humus coals. Khimiya Tverdogo Topliva 1:45–52 (in Russian)

Gulyaeva ND, Arefyev OA, Petrov AA (1976) Regulation in normal and isoprenoid alkanes' distribution in coals at various stages of metamorphism. Khimiya Tverdogo Topliva 1:106–110 (in Russian)

Gulyaeva ND, Arefyev OA, Petrov AA (1978) $C_{27}-C_{31}$ pentacyclic hydrocarbons in brown coals. In: Accumulation and transformation of organic matter in contemporary and fossil sediments. Nauka, Moscow, pp 158–162 (in Russian)

Hoering T (1977) Olefinic hydrocarbons from Bradford, Pennsylvania crude oil. Chem Geol 20:1–8

Kontorovitch AE, Danilova VP (1973) Petroleum formation in coal-bearing sedimentary layers (on Example of mesozoic and paleozoic sedimentation of south-western and mid-Siberia). Tr Sibirskogo Nautchno-Issledovatelskogo Inst Geol Geogr Mineral Syrya, issue 167:73–82 (in Russian)

Kvenvolden K, Weiser D (1967) Mathematical model of a geochemical process. Geochim Cosmochim Acta 31:1281–1310

Leythaeuser D, Welte D (1968) Relation between distribution of heavy n-paraffins and coalification in carboniferous coals from the Soar District, Germany. In: Advances in organic geochemistry. Pergamon, Amsterdam, pp 429–442

Mackenzie A, Lewis C, Maxwell J (1981) Molecular parameters of maturation in the Toarcian shales, Paris basin, France IV. Laboratory thermal alteration studies. Geochim Cosmochim Acta 45:2369–2376

Maxwell J. Cox R, Ackman R, Hooper S (1971) The diagenesis and maturation of phytol. The stereochemistry of 2,6,10,14-tetramethylpentadecane from an ancient sediment. In: Advances in organic geochemistry, vol 33. Pergamon, Hannover, pp 277–291

Monin J, Durand B, Vandenbroucke M, Huc A (1979) Experimental simulation of the natural transformation of kerogen. In: Advances in organic geochemistry: physics and chemistry of the earth, vol 12. Pergamon, Newcastle, pp 517–530

Nishimura M (1978) Geochemical characteristics of the high reduction zone of stenols in Suwa sediments and the environmental factors controlling the conversion of stenols into stanols. Geochim Cosmochim Acta 42:349–357

Ourisson G, Albrecht P, Rohmer M (1984) The microbial origin of fossil fuels. Sci Am 251 (2):34–41

Perregaard G, Schiener E (1979) Thermal alteration of sedimentary organic matter by a basalt intrusive (Kimmeridgian shales, Milne Land, East Greenland). Chem Geol 26:331–343

Petrov AA (1960) Catalytic isomerization of hydrocarbons. The USSR Acad Sci Publ, Moscow, 215 pp (in Russian)

Petrov AA (1971) The chemistry of naphthenes. Nauka, Moscow, 388 pp (in Russian)

Petrov AA (1974) The chemistry of alkanes. Nauka, Moscow, 243 pp (in Russian)

Petrov AD, Nikishin GI, Ogibin YN (1960) Free-radical connection of monocarbonic acids and their methyl ethers to α-olefins. Dokl USSR Acad Sci 131/3:580–583 (in Russian)

Philp R, Calvin M, Brown S, Yang E (1978) Organic geochemical studies on kerogen precursors in recently deposited algal mats and oozes. Chem Geol 22:207–231

Pustilnikova SD, Tsedilina AL, Krasavtchenko MI, Petrov AA (1973) Isoprenoid petroleum hydrocarbons' generation out of phytol. Geol Nefti Gaza 12:56–59 (in Russian)

Pustilnikova SD, Abryutina NN, Kayukova GP, Zubenko VG, Petrov AA (1982) Sterols' transformations under aluminosilicates' impact. Geokhimiya 1:144–149 (in Russian)

Rodionova KF, Maksimov SP, Telkova MS, Ilyinskaya VV (1973) On understanding the composition of organic matter in coals at various stages of metamorphism. Tr Vsesoyuznogo Nautchnogo-Issledovatelskogo Geol Neftyanogo Inst, issue 138:175–183 (in Russian)

Rubinstein J, Strausz C (1979) Geochemistry of the thiourea adduct fraction an Alberta petroleum. Geochim Cosmochim Acta 43:1392

Rubinstein J. Sieskind O, Albrecht P (1975) Rearranged sterenes in a shale. Occurrence and simulated formation. J Chem Soc Perkin Trans I, 19/20:1833–1835

Seifert W (1978) Steranes and terpanes in kerogen pyrolysis for correlation of oils and source rocks. Geochim Cosmochim Acta 42:473–484

Seifert W, Moldowan J (1979) The effect of thermal stress on source-rock quality as measured by hopane stereochemistry. In: Advances in organic geochemistry: physics and chemistry of the earth, vol 12. Pergamon, Newcastle, pp 229–237

Sieskind O, Joly G, Albrecht P (1979) Simulation of the geochemical transformations of sterols: superacid effect of clay minerals. Geochim Cosmochim Acta 43:1675–1679

Sokolov VA, Bestuzhey MA, Tikhomolova TV (1972) Chemical composition of crudes and natural gases depending on origin. Nedra, Moscow, 276 pp (in Russian)

Tissot B, Welte D (1981) Petroleum formation and occurrence. Mir, Moscow, 501 pp (in Russian)

Vassoevitch NB, Kortchagina YI, Lopatin NV, Tchernishov VV (1969) Main phase in petroleum formation. Vestnik MGU Ser Geol 6:3–27 (in Russian)

Vassoevitch NB, Burlin IK, Konukhov AI, Karnyushina EE (1975) The role of clays in petroleum formation. Sov Geol 3:15–39

Vassoevitch NB, Guseva AN, Leifman IE (1976) Petroleum biogeochemistry. Geokhimiya 7:1075–1083 (in Russian)

Vleet E, Quinn J (1979) Early diagenesis of fatty acids and isoprenoid alcohols in estuarine and coastal sediments. Geochim Cosmochim Acta 43:289–303

Welte D, Waples D (1973) Über die Bevorzugung geradzahliger n-Alkane in Sedimentgesteinen. Naturwissenschaften 60:516–517

Yen T, Chilingarian H (eds) (1980) Oil shale. Nedra, Leningrad, 262 pp (in Russian)

Chapter 6

Andreev PF, Bogomolov AI, Dobryanskii AF, Kartsev AA (1958) Petroleum transformations in nature. Gostoptekhizdat, Leningrad, 416 pp (in Russian)

Arefyev OA, Zabrodina MN, Norenkova IK, Karpenko MN, Makushina VM, Petrov AA (1978) Biological degradation of crudes. Isv USSR Acad Sci Ser Geol 9:134–139 (in Russian)

Arefyev DO, Zabrodina MN, Makushina VM, Petrov AA (1980a) Relict tetra- and pentacyclic hydrocarbons in ancient crudes of the Siberian platform. Isv USSR Acad Sci Ser Geol 3:135–140 (in Russian)

Arefyev OA, Makushina VM, Petrov AA (1980b) Asphaltenes markers in petroleum geochemical history. Isv USSR Acad Sci Ser Geol 4:124–130 (in Russian)

Arkhangelskaya RA, Norenkova IK, Tarasova TG (1978) Impact of oxidation processes on the chemical composition of resinous petroleum fractions. Geol Nefti Gaza 10:49–54 (in Russian)

Bailey N, Jobson A, Rogers M (1973a) Bacterial degradation of crude oil-comparison of field and experimental data. Chem Geol 11:203–221

Bailey N, Krouse H, Evans C, Rogers M (1973b) Alteration of crude oil by water and bacteria. Evidence from geochemical and isotope studies. Am Assoc Petrol Geol Bull 57:1276

Botneva TA, Shulova NS (1981) Genetic types of the near Caspian depression crudes and their compositional prognosis. In: Organic geochemistry of pre-cambrian oils, gases and organic matter. Nauka, Moscow, pp 41–47 (in Russian)

Connen J (1984) Biodegradation of crude oils in reservoirs. Advances in petroleum geochemistry, vol 1. Academic Press. London, pp 300–335

Connan J, Le Tran K, van der Weide B (1975) Alteration of petroleum in reservoirs. In: Proc 9th World Petrol Congr, Tokyo, vol 2. Appl Sci Publ, Leningrad, pp 171–178

Connan J, Restle A, Albrecht P (1979) Biodegradation of crude oils in the Aquitaine basin. In: Advances in organic geochemistry: physics and chemistry of the earth, vol 12. Pergamon, Newcastle, pp 1–17

Evans C, Rogers M, Batley N (1971) Evolution and alteration of petroleum in western Canada. Chem Geol 8:147–170

Gabrielyan AG (1981) On the problem of petroleum evolution in natural environment. In: Organic geochemistry of pre-cambrian crude oils, gases and organic matter. Nauka, Moscow, pp 119–125

Goodwin N, Park P, Rawlinson T (1981) Crude oil biodegradation. 10th Int Meet Organ Geochem, programme and abstracts. Bergen, p 69

Guseva AN, Leifman IE, Vassoyevitch NB (1976) On biogeochemistry of lipids, lipoids and related polymeric compounds. In: Analysis of organic material of contemporary and fossil sediments. Nauka, Moscow, pp 25–56 (in Russian)

Hufnagel H, Teschner M, Wehner H (1979) Correlation studies on crude oils and source-rocks in the German Molasse basin. In: Advances in organic geochemistry: physics and chemistry of the earth, vol 12. Pergamon, Newcastle, pp 51–66

Jobson A, Cook F, Westlake D (1979) Interaction of aerobic and anaerobic bacteria in petroleum biodegradation. Chem Geol 24:355–365

Kayukova GP, Pustilnikova SD, Abryutina NN, Golovkina LS, Mekhtieva VL, Petrov AA (1981) Structurally modified hydrocarbons of hopane type. Neftekhimiya 21:803–811 (in Russian)

Kayukova GP, Pustilnikova SD, Arefyev OA, Kurbskii GP, Petrov AA (1982) Peculiarities of polycyclic relicts' distribution in the crude of Tatariya. Neftekhimiya 22:579–586 (in Russian)

Kuklinskii AI, Pushkina RA (1974) Isoprenoid links in chains of saturated petroleum hydrocarbons. Neftekhimiya 14:520–525 (in Russian)

Leifman IE, Guseva AN (1978) On compositional alteration of organic matter, original for the formation of combustible fossils, in the course of the vegetable world's evolution. In: Accumulation and transformation of organic matter in contemporary and fossil sediments. Nauka, Moscow, pp 9–17 (in Russian)

Maksimov SP (ed) (1981) Criteria in crudes' and gases' qualitative composition prognosis, issue 22. Nedra, Moscow, 164 pp (in Russian)

Martin R, Winters J, Williams I (1964) Occurrence of hydrocarbons in petroleum and its genesis. In: New research in the area of petroleum and gas genesis. CNIITEneftegas, Moscow, pp 38–78 (in Russian)

McKirdy D, Aldridge A, Ypma P (1981) A geochemical comparison of crude oils from pre-ordovician, carbonate rocks. 10th Int Meet Organ Geochem, programme and abstracts. Bergen, pp 144–145

McKirdy D, Aldridge A, Ypma P (1983) Geochemical comparison of some crude oils from pre-ordovician carbonate rocks. Advances in organic geochemistry, 1981. Wiley, New York, pp 99–107

Milner C, Rogers M, Evans C (1977) Petroleum transformation in reservoirs. J Geochem Explor 7:101–153

Philippi G (1977) On the depth, time and mechanism of origin of the heavy to medium-gravity naphthenic oils. Geochim Cosmochim Acta 41:33–52

Reed W (1977) Molecular compositions of weathered petroleum and comparison with its possible source. Geochim Cosmochim Acta 41:237–247

Rubinstein J, Strausz O, Spyckerelle C et al. (1977) The origin of the oil sand formations of Alberta: a chemical and a microbiological simulation study. Geochim Cosmochim Acta 41:1341–1353

Rubinstein J, Spyckerelle C, Strausz O (1979) Pyrolysis of asphaltenes: a source of geochemical information. Geochim Cosmochim Acta 43:1–6

Safonova GI (1974) Catagenetic alterations of crudes in deposits. Nedra, Moscow 151 pp (Tr Vsesoyuznogo Nauchno-Issledovatelskogo Geol Neftyanogo Inst, issue 145, in Russian)

Seifert W, Moldowan J (1979) The effect of biodegradation on steranes and terpanes in crude oils. Geochim Cosmochim Acta 43:111–126

Sokolov VA, Bestuzhev MA, Tikhomolova TV (1972) Chemical composition of crude oils and natural gases in connection with their origins. Nedra, Moscow, 276 pp (in Russian)

Tchakhmakhtchev VA (1983) Geochemistry of the process of migration of hydrocarbon systems. Nedra, Moscow, 228 pp (in Russian)

Tissot B, Califet-Debyser J, Deroo G, Oudin J (1971) Origin and evolution of hydrocarbons in early toarcian shales. Paris basin, France. Am Assoc Petrol Geol Bull 55/12:2177–2193

Tissot B, Welte D (1981) Petroleum formation and occurrence. Mir, Moscow, 501 p (in Russian)

Uspenskii VA (1970) Introduction to petroleum geochemistry. Nedra, Leningrad, 309 pp (in Russian)

Vassoyevitch NB (1958) Petroleum formation in Terrigenic depositions. Tr Vsesoyuznogo Nautchno-Issledovatelskogo Geol-Razvedotchnogo Inst, issue 128:9–205 (in Russian)

Wen-Yen Kuang, Meinschein W (1979) Sterols as ecological indicators. Geochim Cosmochim Acta 43:739–745

Zabrodina MN, Arefyev OA, Norenkova IK, Makushina VM, Arkhangelskaya RA, Petrov AA (1981) Biodegradation of crudes in the Starogroznenskoe deposit. Isv USSR Acad Sci Ser Geol 9:126–132 (in Russian)

Subject Index